HUMAN

EVOLUTION

HUMAN EVOLUTION

GEORGE D. BROWN, JR.

Boston College

WCB **Wm. C. Brown Publishers**

Dubuque, IA Bogota Boston Buenos Aires Caracas Chicago
Guilford, CT London Madrid Mexico City Sydney Toronto

Book Team

Developmental Editor *Mary Hill*
Production Editor *Audrey A. Reiter/Marla K. Irion*
Art Editor *Brenda A. Ernzen*
Photo Editor *John C. Leland*
Permissions Coordinator *Gail I. Wheatley*
Publishing Services Coordinator *Barbara Hodgson*

Wm. C. Brown Publishers

A Division of Wm. C. Brown Communications, Inc.

Vice President and General Manager *Beverly Kolz*
Vice President, Publisher *Jeffrey L. Hahn*
Vice President, Publisher *Kevin Kane*
Vice President, Director of Sales and Marketing *Virginia S. Moffat*
Vice President, Director of Production *Colleen A. Yonda*
National Sales Manager *Douglas J. DiNardo*
Marketing Manager *Amy Halloran*
Advertising Manager *Janelle Keeffer*
Production Editorial Manager *Renée Menne*
Publishing Services Manager *Karen J. Slaght*
Permission/Records Manager *Connie Allendorf*

Wm. C. Brown Communications, Inc.

President and Chief Executive Officer *G. Franklin Lewis*
Senior Vice President, Operations *James H. Higby*
Corporate Senior Vice President, President of WCB Manufacturing *Roger Meyer*
Corporate Senior Vice President and Chief Financial Officer *Robert Chesterman*

Copyedited by Cathy Di Pasquale

Cover photo © H. Levart/Superstock

Cover design by Terri Ellerbach

The credits section for this book begins on page 255 and
is considered an extension of the copyright page

Library of Congress Catalog Card Number: 94–71592

ISBN 0–697–24307–9

Printed in the United States of America by Wm. C. Brown Communications, Inc.,
2460 Kerper Boulevard, Dubuque, IA 52001

10   9   8   7   6   5   4   3   2   1

CONTENTS

ILLUSTRATIONS vii

PREFACE ix

1 Why? 1
INTRODUCTION 1
INTEREST IN ANCESTRY 1
THE STUDY OF HUMAN ORIGINS 2
THE CONCEPT OF HUMANS 3
HOW TO STUDY HUMAN ORIGIN 5
THE ORIGIN—LINES OF EVIDENCE ... 6
WHEN DID HUMANS APPEAR ON EARTH? ... 9
GOALS AND METHODS OF SCIENCE ... 10
SUMMARY 12

2 Darwin's Idea 14
INTRODUCTION 14
CHARLES DARWIN 14
THE GALÁPAGOS ISLANDS 17
THE DEVELOPMENT OF THE CONCEPT ... 18
THE SYNTHESIS OF IDEAS 22
PUNCTUATED EQUILIBRIA 22
SUMMARY 23

3 The Primates 25
INTRODUCTION 25
PRIMATE CLASSIFICATION 27
PRIMATE LIFESTYLES 36
PRIMATE BEHAVIOR 41
TERRITORIALITY 43
DOMINANCE AND AGGRESSION 44
BABOON SOCIETY 45
CHIMPANZEE SOCIETY 47
GORILLA SOCIETY 48
ORANGUTAN SOCIETY 49
SUMMARY 50

4 Humans and Apes Compared 52
INTRODUCTION 52
PHYSICAL COMPARISONS BETWEEN APES AND
   HUMANS 52
THE BRAIN 60
BIOCHEMICAL RELATIONSHIPS 63
THE NONPHYSICAL COMPARISONS ... 68
THE ORIGIN OF BEHAVIORS 73
SUMMARY 76

5 The Primate Pathway 78
INTRODUCTION 78
THE ANCESTRAL PRIMATES 78
FOSSILS OF NONHUMAN PRIMATES ... 78
THE DRYOPITHECINES 87
SUMMARY 95

6 The Rise of Hominids 96
INTRODUCTION 96
THE EVOLUTIONARY PATH FROM DRYOPITH-
   ECINES TO HUMANS 97
THE HOMINIDS APPEAR 98
DISCOVERY AND LEARNING 104
COOPERATIVE BEHAVIOR 106
SUMMARY 108

7 The Australopithecines 110
INTRODUCTION 110
THE SOUTHERN APE-MEN 111
THE GREAT MAKAPANSGAT BONE
   CONTROVERSY 116
EAST AFRICAN AUSTRALOPITHECINES ... 119
THE IMPORTANCE OF OLDUVAI FOSSILS ... 122
OTHER EAST AFRICAN SITES 123
THE AUSTRALOPITHECINE TYPES ... 129
THE AUSTRALOPITHECINE WAY OF LIFE ... 132
SUMMARY 136

8 The Humans Appear: Homo habilis 138
INTRODUCTION 138
THE FIRST HUMANS 140
ARCHEOLOGICAL EVIDENCE FOR *HOMO
   HABILIS* 142
AUSTRALOPITHECINES AND *HOMO HABILIS*
   COMPARED 147
SUMMARY 151

9 The Great Migration: Homo erectus 153
INTRODUCTION 153
HAECKEL AND THE MISSING LINK ... 153
EUGENE DUBOIS AND JAVA MAN ... 154
PEKING MAN 158
THE RELATIONSHIPS BETWEEN JAVA MAN AND
   PEKING MAN 161
ADDITIONAL FOSSIL EVIDENCE ... 162
EUROPE AND ARCHAIC SAPIENS ... 164

BEHAVIOR 167
COOPERATIVE BEHAVIOR 172
INTELLIGENCE 173
TOOLS 173
SUMMARY 175

10 The Neandertals Arrive 177
INTRODUCTION 177
THE DISCOVERY AND THE DEBATE 179
RECOGNITION OF THE NEANDERTALS AS FOSSIL
   HUMANS 181
OLD WORLD DISTRIBUTIONS 186
THE NEANDERTAL PROBLEM 192
THE ARCHAIC SAPIENS 194
NEANDERTAL TOOLMAKING 196
NEANDERTAL BEHAVIORS AND PRACTICES 196
THE EVOLUTIONARY PATH TO *HOMO SAPIENS* 199
SUMMARY 200

11 The Cro-Magnons 202
INTRODUCTION 202
THE DORDOGNE 203
STONE AGE TOOL CULTURES 204

NATURE OF THE CRO-MAGNONS 205
THE MEANS FOR CULTURAL CHANGE 207
THE GREAT VARIETY OF TOOLS 211
CRO-MAGNON LIFESTYLES 213
CRO-MAGNON DISTRIBUTIONS 214
CAVE ART 215
ORIGIN OF THE CRO-MAGNONS 221
SUMMARY 222

12 The Rise of Modern Humans 224
INTRODUCTION 224
THE ANCESTRAL SPECIES—*HOMO ERECTUS* 225
ARCHAIC SAPIENS IN EUROPE 228
BIOCHEMICAL EVIDENCE FOR HUMAN
   ORIGINS 231
THE TRANSITION TO MODERN MAN 232
CLIMATIC INFLUENCES 238
SUMMARY 240

REFERENCES 243

INDEX 257

Illustrations

Fig. 1.1. Recognizably different humans.

Fig. 1.2. Different human races.

Fig. 1.3. Clarence Darrow addressing the Scopes trial jury in the front two rows. Note the microphone from radio station WGN in the foreground.

Fig. 1.4. Generalized distribution of life through time.

Fig. 2.1. Charles Darwin as a young man.

Fig. 2.2. Map of the HMS *Beagle* voyage.

Fig. 2.3. The Galápagos Islands.

Fig. 2.4. Finches of the Galápagos Islands.

Fig. 2.5. Tool-using Galápagos finch.

Fig. 2.6. James Hutton.

Fig. 2.7. Gregor Mendel.

Fig. 3.1. Distribution of major primate groups through time.

Fig. 3.2. The tree shrew, *Tupaia*.

Table 3.1. Classification of the primates.

Fig. 3.3. Ring-tailed lemur, *Lemur catta*.

Fig. 3.4. Male baboon's threat-display "yawn."

Fig. 3.5. *Oreopithecus* skeleton.

Fig. 3.6. Gibbon in brachiating mode.

Fig. 3.7. Dian Fossey with a gorilla.

Fig. 3.8. Female orangutan with young.

Fig. 3.9. Young baboon showing opposable thumbs and toes.

Fig. 3.10. Prosimian (potto) in a VCL posture.

Fig. 3.11. Baboon mother cuddling offspring.

Fig. 3.12. Grooming activity in baboons.

Fig. 3.13. Male baboon eating a young gazelle after the kill.

Fig. 3.14. Aggressive behavior in a baboon troop.

Fig. 4.1. Gorilla and modern human skeletons in upright posture.

Fig. 4.2. Gorilla and modern human skulls.

Fig. 4.3. Gorilla and modern human jaws with teeth.

Fig. 4.4. The 5-Y dental cusp pattern.

Fig. 4.5. Gorilla and modern human spines.

Fig. 4.6. Pelvic bones of gorilla, *Australopithecus africanus* and modern human.

Fig. 4.7. Legs of modern human, *Australopithecus africanus* and ape.

Fig. 4.8. Feet of gorilla and modern human.

Fig. 4.9. Vertebrate animal brains.

Fig. 4.10. Brain size vs. body weight in advanced primates.

Fig. 4.11. Fossil hominoid and hominid brain casts.

Table 4.1. Primate relationships based on immunological comparisons.

Table 4.2. Immunological distances in some higher primates.

Table 4.3. Albumin reactivity in some higher primates.

Fig. 4.12. Primate lineages based on immunological distance units.

Fig. 4.13. Living primate divergence graph based on genetic distance data.

Fig. 4.14. Hybridization of human and chimpanzee DNA.

Table 4.4. Partial amino acid sequences in primate hemoglobin chains.

Fig. 4.15. Baboon troop foraging.

Fig. 4.16. Koko signing to Dr. Francine Patterson.

Fig. 4.17. Jane Goodall observing grooming activity of chimpanzees.

Fig. 5.1. Reconstruction of *Dryopithecus africanus* skeleton.

Fig. 5.2. Hominoid and hominid lineages.

Fig. 5.3 Skulls of (*a*) *Pronycticebus gaudryi* (X1.25), (*b*) Ring-tailed lemur, *Lemur catta* (X1), (*c*) Tarsier, *Tarsius spectrum* (X2), and (*d*) Loris, *Galago garnetti* (X1.3).

Fig. 5.4. Skeleton of *Plesiadapis*.

Fig. 5.5. Skull of *Adapis parisiensis*.

Fig. 5.6. Skeleton of *Smilodectes*.

Fig. 5.7. Skull of *Aegyptopithecus zeuxis*.

Fig. 5.8. Reconstructed jaw of *Dryopithecus fontani*.

Fig. 5.9. Skull of *Sivapithecus*.

Fig. 5.10. Mary Leakey examining australopithecine footprints at Laetoli.

Fig. 6.1. Gorilla knuckle walking.

Fig. 6.2. Chimpanzee using bipedalism.

Fig. 6.3. Chimpanzee going termite "fishing."

Fig. 6.4. Chimpanzee using a tool, throwing a branch.

Fig. 6.5. Japanese macaque washing food.

Fig. 6.6. Chimpanzee using a tool (stick) to open a box.

Fig. 6.7. Cooperative behavior in early humans.

Fig. 7.1. Neandertal skullcap.

Fig. 7.2. Gibraltar skull.

Fig. 7.3. Raymond Dart with Taung skull.

Fig. 7.4. Skull and brain cast of Taung child.

Fig. 7.5. Skull of *Australopithecus robustus*.

Fig. 7.6. Reconstructions of *Australopithecus robustus* and *A. africanus*.

Fig. 7.7. Skull of *Australopithecus (Plesianthropus transvaalensis) africanus*, Sterkfontein skull V.

Fig. 7.8. Olduvai Gorge where *Australopithecus boisei* was originally found.

Fig. 7.9. Oldowan stone tools.

Fig. 7.10. *Australopithecus boisei* skull.

Fig. 7.11. Specimens of *Homo habilis*.

Fig. 7.12. Stratigraphic section at Olduvai Gorge

Fig. 7.13. *Homo habilis* skull, KNM-ER 1470.

Fig. 7.14. Knees of modern human, *Australopithecus afarensis* and ape.

Fig. 7.15. The skeleton known as Lucy.

Fig. 7.16. Reconstructed skull of *Australopithecus afarensis*.

Fig. 7.17. Skeletons of the gracile and robust types of australopithecines.

Fig. 7.18. Skulls of the gracile and robust types of australopithecines from South Africa.

Fig. 7.19. Australopithecine jaws.

Fig. 8.1. The evolution of stone-tool cultures in Europe.

Fig. 8.2. Oldowan pebble tools from East Africa.

Fig. 8.3. Fossil site at Lake Turkana.

Fig. 8.4. Possible windbreak of stones at Olduvai Gorge.

Fig. 8.5. Mary Leakey with J. Desmond Clark at butchering site in Olduvai Gorge.

Fig. 8.6. Jaw and teeth of *Homo habilis*.

Fig. 9.1. Haeckel's family tree of man.

Fig. 9.2. Skullcap of Java man.

Fig. 9.3. Cast of Java man femur (left) with left femur of modern human.

Fig. 9.4. The Mauer or Heidelberg jaw.

Fig. 9.5. Peking man skull.

Fig. 9.6. Skull from Sangiran.

Fig. 9.7. *Homo erectus* skeleton, KNM-WT 15000.

Fig. 9.8. *Homo erectus* skull, KNM-ER 3733.

Fig. 9.9. Acheulian tools.

Fig. 9.10. The kill site of Olorgesailie.

Fig. 9.11. Reconstructed hut from Terra Amata.

Fig. 9.12. Acheulian tools from Terra Amata

Table 10.1. Pleistocene glacial and interglacial stages with dates and tool cultures. (in text)

Table 10.2. Pleistocene paleomagnetic timescale. (in text)

Fig. 10.1. Reconstruction of Neandertals using Boule's concept.

Fig. 10.2. Modern interpretation of Neandertals.

Fig. 10.3. Neandertal remains from La Chapelle-aux-Saints.

Fig. 10.4. Rhodesian man (Broken Hill) skull.

Fig. 10.5. Solo man skull, Ngandong calvarium No. 6.

Fig. 10.6. Typical Mousterian tools from Mount Carmel.

Fig. 10.7. Skhül V skull.

Fig. 10.8. Reconstructed skull of Piltdown man.

Fig. 10.9. Saccopastore I skull.

Fig. 10.10. Steinheim skull.

Fig. 10.11. Petralona skull.

Fig. 11.1. The Dordogne River valley.

Fig. 11.2. Cro-Magnon scene.

Table 11.1. Paleolithic cultures of Europe. (in text)

Fig. 11.3. Skull of Cro-Magnon male.

Table 11.2. Brain sizes in hominids. (in text)

Fig. 11.4. Cro-Magnon stone tools. (Scale bar = 5cm.)

Fig. 11.5. Cro-Magnon laurel-leaf blade.

Fig. 11.6. Engraved mammoth tusk.

Fig. 11.7. Hunter using an atlatl, a spear-throwing stick.

Fig. 11.8. Drawing of a small horse in Lascaux Cave.

Fig. 11.9. Ceiling of Lascaux Cave showing various animals.

Fig. 11.10. Painting of bison in Altamira Cave.

Fig. 11.11. Figure in the Hall of Bulls, Lascaux Cave.

Fig. 11.12. Willendorf Venus.

Fig. 11.13. Reindeer antler with counting marks.

Fig. 11.14. Human head carved from ivory.

Fig. 12.1. The likely path of human evolution.

Fig. 12.2. The spread of *Homo erectus* from Africa.

Fig. 12.3. Saint-Césaire Neandertal skull.

Fig. 12.4. Omo I skull.

Fig. 12.5. Border Cave I cranium.

Preface

The study of science has, in recent years, become so complex that courses have tended to become independent studies of specific areas of this major portion of human culture. As students majoring in science quickly learn, this independent nature of their particular discipline does not exist in reality; it is an artificially contrived separation for convenience; science simply does not exist in isolation. Majors in the geological sciences must complete supporting courses in mathematics through calculus, chemistry, physics and biology to fully understand their own discipline. The same is true of the other sciences; students must "take" more than their own discipline to understand how the laws of nature and science operate. In many cases, such a demanding approach has led to increases in numbers of students who *do not* major in science. This will have serious consequences in the future for us as a people.

Many colleges and universities have acknowledged the need to produce graduates who have at least been introduced to a variety of human endeavors. This has resulted in "core" courses all must take, including some in science. Typically these are specific courses in specific sciences; but there is a growing belief that such limited requirements and offerings will not produce enough individuals capable of understanding the scientific problems facing our societies today. Important scientific decisions, however, must be made daily. But decision-making leaders are themselves too busy to "learn" all the scientific and social ramifications of their decisions. They must rely on others for reasoned advice. All too often, however, their decisions are made with "votes" in mind rather than the general welfare. It is essential, therefore, that our population become as well educated as possible in scientific matters for our mutual benefit rather than the benefit of the vocal few. The

> ...political consequences of the failure to create a scientifically informed citizenry should be clear. Such failure would be a decisive step in the decline of this nation.[1]

A major scientific topic that can be used to blend a variety of disciplines is the origin of mankind, a subject that has intrigued people everywhere for hundreds, perhaps thousands, of years. It is a perennial question that has attracted every religion and philosophy and, more recently, science. We want to know where we came from and how we came to be on Earth. The obvious answer according to most people is a religious one: God created us! This creation explanation has been an important belief in Christianity for about two thousand years and for much longer in Judaism. It is also typical of many of the non-Judeo-Christian religions. But such an explanation really is not satisfactory because it introduces more questions than it answers. God, as a supernatural being, is beyond the rational understanding of humans who as natural beings, have attempted to arrive at some answers through faith and reason. And to fathom God's objectives is impossible. Accordingly, our explanation for God's handiwork is based on human beliefs and interpretations.

The religious explanations were originally intended for a relatively simple people, farmers and herders, largely uneducated by our standards as was customary at that time. Such people did not require details of comparative anatomy, DNA replication, protein shapes and sizes, the evolutionary pathway, population shifts under geological and climatic influences, and other aspects that we routinely provide today with evidence and examples to our scientifically literate audiences. They needed simple answers to account for many troublesome aspects of their lives that they could not understand. Today, our lives are much more complicated and we are faced with serious problems of overpopulation in some countries, environmental pollution, recycling, the implications of diet and food additives, wars of aggression, racial struggles, national debts, genetic engineering and more, all beyond the routine questions of how and where we can earn our livings. We need solutions to most of the complexities of our everyday lives, and beyond.

Many people have sought a nonreligious approach to the question of human origins and tried to explain it using evidence derived only from the material world that surrounds us. Over the past 150 to 200 years, various naturalists and scientists have offered such explanations for our origin based on their studies of the physical and biological world; they sought a scientific explanation. The scientific

---

[1] Prof. Dudley Shapere, Wake Forest University, "Demystify the lessons of science," *Middlesex News*, p. 9A, Monday, April 12, 1993, Framingham, MA 01701.

explanation for human origins does not admit a supreme being as the creator of life and humans. This is not a rejection or denial of God, merely an acceptance of the limitations of science. By its very nature, science is unable to deal with a supernatural being. Science deals only with nature, with the materials, processes and life forms of the natural world. God is, by definition, beyond the investigative powers of science.

How we approach the question of human origins depends largely on how we view the world and its life forms. It is quite clear that man is only one of millions of different species that have inhabited Earth and shares existence with other organisms, not as a master but as a fellow traveler on a planet spinning through space. Ours is an integrated existence in which all species are dependent upon others for their very survival. The origin of humans is, therefore, a topic well within the investigative powers of science. In fact, science is precisely the discipline to seek answers to human origins. Humans are life forms on Earth, a part of nature. Science investigates nature; therefore, science must seek and provide answers to human origins. Science and scientists have already provided many such answers, even though these explanations are not yet complete.

The pathway for human evolution was presented more than one hundred years ago by Charles Darwin. Many eminent scientists immediately took up the scientific search for human origins around the Old World, primarily by seeking fossil evidence. Since then, the likely pathway has been more than adequately identified. We now have established details of an evolutionary descent from apes to humans over the past 5 to 10 million years using direct anatomical evidence from the appropriate fossils and living primates. The fertile fields of biochemistry and genetics have supported this descent by demonstrating the closeness of our relationships with chimpanzees, gorillas, and other living advanced primates. Great additional support has been provided during the past four decades by direct field studies of the behaviors of these same living, advanced primates. Because these evidences have supported an evolutionary descent of humans from apes, they have allowed a much more complete and better understanding of the nature of modern man.

This work is intended for use with supporting material in an introductory level, multidisciplinary college course. Its audience are students who are not majoring in one of the sciences directly pertaining to human origins, e.g., biol-

ogy, geology, and anthropology. There are many excellent texts on human evolution available for use in such a course. Each tends to emphasize the particular discipline of its author; each imparts its own flavor to the subject. Clearly, there is no single pathway to the study of human origins.

This text is not intended to provide complete coverage of every aspect of human origins, nor does it utilize all the scientific disciplines that defend and offer supporting evidence for human origins. In fact, this text covers only the material for the second half of a one-semester core course in science. Much pertinent supporting scientific material has been bypassed in this text. For example, this work provides no coverage of the origin of the universe, solar system and Earth, yet these pertain directly to the origin of various minerals (elements and chemicals) that had to have been present on Earth when life originated and which are complexly intertwined with this life. There is also no coverage of the various hypotheses that have been postulated for the origin of life, an essential topic but too far removed for direct bearing. Genetics, too, is given virtually no coverage herein, though it plays a fundamental role in the evolution of species. So, too, are the various mechanisms by which species have evolved. As a paleontologist, I could find many reasons for much greater coverage of life of the past, but believe that restricting coverage to the pertinent primate fossils is more productive. It would be very useful to include in this text the history of the vertebrates, especially the evolutionary pathway from fish to amphibians to reptiles and culminating in the mammals. But, these are simply beyond the scope of this work. A good instructor should be able to provide much of this supporting scientific material for a course without comprising the central theme; I cover all of these topics.

This text should provide a better understanding of the fundamental aspects of human evolution by placing them in an integrated context with all of nature, as outlined in the preceding paragraph. I hope that learning more about our origins—about how, why, where and when humans appeared—will have a beneficial effect on the intended audience. Knowing such information may not be directly useful to the audience in terms of career or salary, but if such knowledge improves their understanding of the human condition and provides a greater appreciation of our lives and life around us, then I will have reached my goals.

The search for human origins is an exciting topic. For the researcher it combines the adventure of detective work with the rigors of field work in remote places, and the thrill of being the first to find a "new" fossil. For others it provides the satisfaction of sharing in the hunt with its many challenges. Many individuals in the past devoted years of their lives to seeking the answers to questions of human origins. Others in the future will join in this laudatory search. Some of these explorers will receive their spark from courses on human evolution or related subjects. Others will enter the search from other directions. How they get involved is not important, only that they get involved. Today, despite the coverage presented in this work, only the broadest picture of human origins has been established; many of the finer details have yet to be determined. Someday, researchers will find more of the crucial fossils that will provide us with a complete and accurate picture of the pathway from apes to modern humans. How many of us wish we could be there as active participants?

My thanks go to the very able production staff at Wm. C. Brown Publishers for their efforts in making this textbook a reality, especially Audrey Reiter and Marla Irion, who were overall production editors; Brenda Ernzen, the art editor; John Leland, the photo editor; and Cathy Di Pasquale, who did an excellent job as free-lance copy editor. Thanks also go to Jeffery Hahn who initiated the project and provided the encouragement to see it through.

The time necessary to initiate and complete this textbook was substantial, as any author knows quite well. The variety of topics and disciplines involved required many hours of research at libraries and solitude in my den where it was all put together. Never once did my wife, Barbara, express resentment for my neglect or tardiness of household responsibilities, not to mention the hours I was unable to spend with her. To her go my special gratitude for her patience. Perhaps the royalties, if any, will compensate.

George D. Brown, Jr.
Framingham, Massachusetts

1 Why?

INTRODUCTION

Most higher animals, especially the mammals, are very inquisitive creatures. Humans are the most inquisitive of all. We constantly pry into the mysteries of the world around us seeking the answers to every imaginable question, even beyond our Earth. Young children are particularly known for their propensity and ability to examine everything within reach, to touch things, to taste and to throw them. Adults, however, seek more sophisticated answers to questions—the temperature at which water boils, how mountains form, why rainbows are curved, what kinds of rocks comprise the Moon, whether there is life on other planets, and much more. As professionals and amateurs we study insects, roses, corn, minerals, air, water and innumerable other objects of the material world to learn their natures. We also collect stamps, coins, pictures, machines and a great variety of human-made objects. We collect and examine these for various reasons, as the goals of academic disciplines in schools, as practical studies for our economic benefit and as pursuits for pure pleasure. We do not restrict our curiosity to the natural and material world but are also concerned with abstract concepts like honor, justice and fairness that do not exist except by our own creation. We devote considerable time and energy as well as much money to find the answers to questions of abstract concepts and the material world. We seek knowledge about these because it allows us to understand them and use them for our own needs and ends, perhaps also because knowledge about these objects gives us power over them. Certainly we use this knowledge for our benefit and to help control our destiny. In learning we try to satisfy our innate curiosities and by so doing also hope to find answers to the most fundamental questions about ourselves and our presence on Earth.

Perhaps the most intriguing questions we can ask concern our own ancestry, our origins on Earth, the very origin of humans. We presume that humanity had a beginning on Earth. This is a natural assumption. After all, we are here, and everything we make had a beginning. Is it not true also for ourselves? Perhaps, but is such a presumption adequate for humans? Perhaps we humans have always been present on Earth, from the beginning of time? The answer that most of us have learned, that presented in the Bible's book of Genesis, does not seem satisfactory. There is too much contradictory evidence of recent years to accept this assumption, and we should examine much of this evidence. But if humans have not always been present on Earth, where did we come from, when did we appear or arrive, how did this happen, and, above all, why? These are old questions that must have been asked for centuries—long before the Bible was written, long before Christ walked the fields and towns of Galilee nearly 2,000 years ago and long before people began to leave written records of their questions.

INTEREST IN ANCESTRY

Today we avidly seek answers to questions of our ancestry because knowledge of our family origins carries with it a sense of permanence, stability and history. The searches for our ancestry do not, however, always seek distant origins, more often they are for more recent relationships. Daily newspapers, periodicals, popular magazines and other media carry stories of interest to genealogists, advertisements to help trace your family tree, scientific discoveries of important fossils, and reports of biological advances. The popularity of these reflects our innate curiosity about ourselves and our origins, especially our grandparents, their parents, and as far back as we can trace our lineages. Some of us proudly wear ties and scarves woven in the plaids of Scottish or Irish clans from which we have descended. This interest in our ancestors is not unique to us but is fundamental to most societies. Even among many primitive tribes, important ancestors are among their most prized assets.

Today many social organizations in the United States and elsewhere have ethnic or religious bases as their principal binding force. The Daughters of the American Revolution, the Knights of Columbus, the Croatian Fraternal Union and numerous others emphasize the ancestry of their memberships. Basically, these groups seek to keep their heritages alive within an adopted culture; they are retaining their roots. Native American Indians have tried for many years to retain their identity in a foreign culture even as they have tried to move into the modern world. Unfortunately, they have had only limited success. Over the years, many millions of immigrants arriving in the United States have faced similar problems attempting to assimilate into this American foreign society while retaining traditional values. A recent popular book and movie, *Roots* by Alex Haley, examined the history of the author's family in America that originated with a young black male in Africa who was kidnapped and sold as a slave. Today, whomever we are, it is appropriate to seek our own roots even if only to understand how and why our ancestors came to this country.

Our search for ancestors and heritage need not end, however, after a few generations with a list of family names and dates; it could be even more thorough. We could trace mankind's origin back beyond historical times. This clearly cannot be handled through written records or tradition or even mythology. It must be handled through scientific studies of the time before humans appeared. Though many approaches to human ancestry and heritage reflect a popular trend, there is a fundamental reason for seeking a scientific explanation to mankind's origins. We seek to understand how humans originated as a species because, from such knowledge, we may better understand what we are and why we are this way.

THE STUDY OF HUMAN ORIGINS

The study of humans is quite popular today in the public press as well as in our colleges and universities. School catalogues list many pertinent courses in various academic departments. We typically study our species academically in two ways: as a biological species and as a social animal. Psychology, sociology, history and anthropology are among the social academic disciplines that study humans and human behavior. Paleontology, the study of life of the past, plays a key role in the search for human ancestry. In addition, biology, the ultimate life science, has many subdivisions that can bear on human origins and some that specifically examine them. Courses in these disciplines range from introductory subjects for non majors to highly specialized courses for professionals. In the social sciences, the studies are primarily observational and interpretive, and are concerned with behavior patterns, the motives for our actions and about how and why we live in particular relationships with others. In the natural sciences, humans are studied in a more fundamental way to determine and understand the composition and nature of cells, tissues, blood, digestion, skeletal structure, heredity, disease, etc.,

especially the biochemical mechanisms by which all life processes are carried out and the evolutionary means by which changes have taken place.

Some of these disciplines study humans only as a species living today without giving much consideration given to our long evolutionary history. These disciplines tend to start with modern humans and do not consider how the events of our origin and evolution have shaped modern humans. There is much thought given to body cells, organs, skeletal structure, genetic matters, behavior patterns and social relationships of modern populations, but too little study is given to the formative events that operated during the long evolutionary history from primitive ancestral species to the complex people of today. These formative events and aspects are not simple to understand and easy to determine. They include complex patterns of adaptations in ancestral species that led to cumulative changes in structures, social behavior, survival tactics and more. Our very distant ancestors lived in a world without tools, without grocery stores and without police to protect them from faster and stronger predators who were always dangerous. The survival of these ancestral species demonstrates their success in changing.

Our species reflects the existence of literally millions of plant and animal species that lived in the past and are now extinct. The diversity of these millions of species, their distributions, their ways of life, their interactions, and their presence in different intervals of geologic time are all aspects that led to our own development. Most scientists accept that these species are the product of a process of change termed **evolution**. In this process, an existing species or a portion of it becomes modified by changes in its genetic makeup that respond to environmental pressures and becomes slightly different from the parent species. If enough changes occur, the population can become very different and be unable to interbreed with the parent stock. At that point a new species is said to have evolved. The rock record is clear that millions of organisms lived in the past and are now extinct, but we must never assume that they were unsuccessful. These species evolved and lived for many millions of years, even though eventually they all died out. We, too, as a species will likely suffer the same fate. Therefore, the study of evolution itself, its successes and its failures—if indeed there were any failures—is essential to understanding how and why humans inhabit Earth.

We should be concerned with any and all factors and aspects that offered advantages to our ancestral stock, because these helped shape them as species and early humans and enabled them to survive in a hostile world. We, their descendants, inherited the factors and aspects that they developed and now retain them as part of our genetic makeup. These have turned us into the scientific and biological species *Homo sapiens*. It is relatively simple to characterize our own species on the basis of bones and tissues derived or inherited from our genetic ancestors. But we have nonphysical aspects, too, that must be considered. We exhibit social relationships, moods and behaviors that contribute to making us human, and these must also have been inherited to some extent. Our ancestors were clearly able to survive in their world even though they could not compete physically with stronger and swifter predators. They must have developed combinations of physical, behavioral and social assets that somehow gave them advantages over other primates and other species that lacked these assets. Somewhere within our genetic systems can be found the traces of those same assets, the physical and social heritage that make us human. We must find, identify and learn about these aspects, if only to better understand ourselves.

The search for an explanation to human origins is basic to our understanding of ourselves and our place in nature. Most of us exclude this topic from much of our thinking and we may, thus, be limiting our potential to answer many questions about ourselves. All too often we are puzzled by antisocial, aggressive behavior of some individuals and commonly ascribe it to problems associated with their environments, perhaps problems acquired when they were children. While many experts accept this explanation as the sole cause, many others today have begun to inquire if this is indeed true. They ask if we could have inherited some of these behavioral aspects. Biological studies of structures and chemical activities within human brains now indicate that many behavioral aspects may be inherited. Whether aggression is inherited or learned is a legitimate question today, though some still disagree with and deny the possibility of its inheritability. We are more than just curious about these aspects and about our physical makeup as well. Why and when, for example, did we become bipedal primates? Why did we learn to use tools? When did we learn to make tools? Such questions may never be satisfactorily explained; however, these represent potentially sig-

nificant and fruitful areas of inquiry and should be pursued if we are to learn more about ourselves.

The interests we show in our species are not new, but experts today are looking at us in new ways with some fresh insights. To study humanity without considering such new ways is to study only a part of humans, and we may be making invalid assumptions about many human aspects. Clues to many questions of normal and abnormal behavior of modern humans may perhaps be found through a study of our past, a study beyond our written history. For these reasons alone, the study of human evolution is essential to our well being; we must seek the origin of humans to understand ourselves.

THE CONCEPT OF HUMANS

There are many different kinds of organisms on Earth and each taxon or category is recognizably different from the others. Such differences are also recognizable within any single taxon, any species. These differences between the individuals of any species on Earth, including humans (Fig. 1.1), are self-evident. We recognize different men and women, different dogs, cats, horses, trees and other creatures. That we cannot always distinguish between two individuals of any particular species is more often the result of our lack of experience than the lack of recognizable variability. Identical human twins, for example, are never so identical that they cannot be told apart. Some simple forms of life are obviously more difficult to differentiate, but experts regularly make such comparisons of multi-celled organisms with ease. With humans the differences can be quite obvious and are recognized by all of us—even slight differences between identical twins. We use various physical parameters of height, weight, obesity, skin color, shape, hair type and color, strength, swiftness, scars, mannerisms, behaviors, etc., as criteria to distinguish between individuals and between species. Some nonphysical characters like intelligence, leadership, charisma, etc., are less obvious than physical aspects but recognizable. This makes the work of social scientists difficult, but all these aspects are measurable to a degree. Researchers also question whether these differences are fundamental to the species or merely differences within a range of variability in any species and of no great significance. Are these characters perhaps accidentals due to environmental influences such as culture or some other human activity? Or are these differences biologically fundamental and due to

Figure 1.1 Recognizably different humans.

inheritance or perhaps to some underlying cause as yet undiscovered?

RACES

Scientists for years have classified people according to some recognizable physical differences thought to be inherent rather than accidental or acquired. Though allowing for individual variations among humans, scientists have, nonetheless, recognized and characterized several groups of humans into categories termed **races** (Fig. 1.2). We accept that each of these races of humans, or even races of frogs, are different from each other, yet they represent groups or populations that are basically similar and related. Clearly such organisms are subdivisions of the same species and differ in some physical characteristics from individuals of other subdivisions. But how and why are they different and yet similar? Are the differences very old in the species or are they recent? Though we have little difficulty accepting differences between races in less complex species, can we accept them for our own? Are mankind's racial differences basic to the species or are they ordinary variability due to place of origin? Can the differences be due to environmental causes, to culture, to evolution, or to some combination of these? Can they be identified?

Perceptive people and scientists for many years have been well aware of the differences between humans. Though there is variation among humans, like any other population of animals, it is not a random variation. Certain physical similarities are common in any relatively distinct population, and it is these kinds of general similarities within certain ranges that characterize populations as races. If there are physical differences between populations, why not biochemical differences, too? And perhaps psychological differences as well? If so, can we explain such variations?

Herodotus, the Greek philosopher, was aware of the clear differentiation between races of men and wrote of them. Egyptian art from 3000 B.C. portrays different races of humans with whom the Egyptians had contact. The earlier cave people left art showing different individuals, some fat, some slender, some bald. Today, many modern peoples consider other humans different if they are part of a different group, whether it be a different race, nationality, tribe or even family. Ritual warfare between stone-age tribes in New Guinea still takes place, though there are few real differences between the warring groups. Somehow each of these tribes consider itself different from the others and this leads to ritual warfare. But

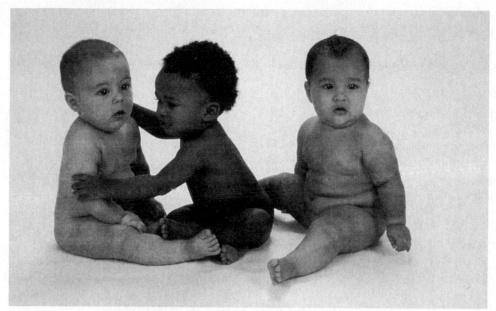

Figure 1.2 Different human races.

how much variation is necessary to be different has yet to be defined.

Today we can travel virtually anywhere in the world and see for ourselves the physical differences between peoples. As scientists we can interview people to learn about their behavior patterns, their customs, their psychological profiles and much more. We can obtain their physical and biochemical parameters. When we learn about them, we learn about ourselves. We do not doubt the existence of differences between races or types of humanity because we can see them, but we often question the significance of these differences.

Today we are much more concerned with the sociological and psychological attitudes of living peoples than their physical differences. The physical differences have largely been measured, but we do not know as much about the nonphysical attributes that characterize humans. We see examples of aggression and selfish behavior, of violence, of generosity, of warfare and altruism. Some experts believe that all of these are tied to our origins, that many of these behaviors are biological in origin and that they represent evolved sociological and psychological attitudes just as we have evolved physical structures and abilities. Yet we do not expend much effort in learning about our biological nature and origin, about how and why we behaved in the past, and about how and why we came to be

modern humans on Earth. Is it possible that we are afraid to inquire scientifically into human nature because we fear the answers? In the past, authorities imposed sanctions, including death, against individuals and groups who espoused unpopular beliefs, however correct they might have been. If we reject honest and serious inquiry into human nature, then we are examining only part of ourselves and can never develop a complete understanding of our species. We may be depriving ourselves of answers that may help us solve and correct many problems of today and those of the future.

HOW TO STUDY HUMAN ORIGIN

Whatever the reasons for studying the origin of humans, whether it be through formal academic disciplines or as a matter of curiosity, it is important that we seek objective answers. How then shall we conduct this scientific inquiry into the origin of humans? Is it fair to impose limitations, scientific or otherwise? Any objective data we obtain will have limitations that may preclude absolute certainty of many conclusions. We may be forced to rely on and accept the limitations of objective reasonable evidence rather than absolute proof. We may also be assured that some conclusions we derive will be unacceptable to many people because of their own personal bias about human origins.

The exact time and place of human origins are obviously lost in the past. Indeed, the crucial evidence may never even have been preserved in a form we could recognize. The record that is available includes relatively meager scraps and fragments of fossil bones, teeth, stone tools and other related evidence that paleontologists, geologists and others have dug from the ground. Yet, this fragmentary evidence has enabled us to draw many important conclusions about mankind's origin and history and to produce a good picture of how we evolved. Perhaps we may discover the branches of our family tree, perhaps even find our "hometown."

THE ORIGIN—LINES OF EVIDENCE

There is a universal belief that humans originated at some time in the past. All peoples of the world have developed explanations for this origin, typically as religious beliefs, myths and folklore traditions. These beliefs, however, do not provide us with a scientific explanation of how life appeared, how and why it changed and why humans are a product of this change. Evidence for the origin of humans, therefore, can be divided into several areas, traditions and myths, religions and science. All must be considered for their validity and reliability. How we evaluate these lines of evidence will depend upon our personal biases.

ORIGIN TRADITIONS

Among the earliest of traditions are those of an ancient people, the **Ubaidians**, who occupied the region corresponding to modern Iraq about 4000–5000 B.C. This area was invaded by a **semitic** people from Syria and Arabia about 1,000 years later and probably intermingled with the native Ubaidian people, adopting some of their traditions. Then the **Sumerians** moved into the area around 3500 B.C. and established the empire of **Sumer**, probably the first important civilization on Earth. They also developed explanations for the origin of the universe, heaven, Earth and much more. They believed that the heaven god **An** created the **Anunnaki** as his divine children. Two other gods created plants and animals to feed and clothe the Anunnaki. Then, these two other gods created humans out of clay and gave them breath (life) to serve these divine children, to tend their flocks, to weave their cloth and to ensure that the Anunnaki would not suffer. Later, different people migrated into this prosperous and wealthy land of Sumer

bringing their own customs, traditions and practices with them. Finally, **Amorites** from the Syrian desert successfully invaded the area and founded the first important dynasty of the Babylonian Empire around 1850 B.C. These **Babylonians** adopted the beliefs of the resident Sumerians, mixed them with their own and added to them.

The ancient Egyptians developed complex legends about the natural world and the gods that controlled it, some before the First Dynasty arose around 3100 B.C. The priests of Memphis believed that **Ptah**, the god of Memphis, created the world and everything in it simply by conceiving of the idea and commanding that it be done. Many of the Egyptian stories of creation, gods and other matters are traceable in theme to the Babylonians and the Sumerians. One particular theme explains the creation of the god of the air to separate the goddess of the sky from the continual embrace of the god of the Earth. The theme is taken virtually directly from the Babylonians, though the Egyptians improved it.

All of these people, whether of simple or complex culture, from the Sumerians to Egyptians and even to American Indians, shared a common belief in the finite nature of humanity. All believed that humans did not exist initially on Earth when it was formed but appeared sometime thereafter, just like the Creation account in the Bible's book of Genesis. This is not surprising, because people have always carried their traditions with them as they migrated from place to place. The present population of the Western Hemisphere, essentially an immigrant one, is comprised of peoples with their own sets of traditions derived from their ancestors and modified as they saw fit. Ultimately, many of these traditions came from very ancient cultures like those of Sumer, Babylon and Egypt. It may be unpopular to suggest that traditions have descended from a few simple stories and evolved over the millennia into complex stories, but the evidence of common themes supports such an interpretation.

RELIGIOUS EXPLANATIONS

Various religions have explanations for the origin of humans and these are probably related to older traditions. In particular, the Judeo-Christian faiths have expanded on some of the more ancient explanations of the Sumerians and Babylonians. These two religions hold that an all-powerful being, God, was responsible for the creation of everything—plants and animals,

Figure 1.3 Clarence Darrow addressing Scopes trial jury in the front two rows. Note the microphone from radio station WGN in the foreground.

rocks, water, air, mountains, stars and the universe. No special mechanism for this creation is provided in the book of Genesis, that portion of the Bible that contains the explanation, only that God willed that it be done (note the similarity to the acts of the Egyptian god Ptah). Though Genesis provides the story of the creation, it does so without details and does not particularly expand on the Sumerian and Babylonian creation myths which it parallels. In the Sumerian version, humans were created to serve the divine children of the heaven-god, An. In Genesis, humans were created to serve God directly and please Him.

Many modern individuals and groups still insist on a strict interpretation of the Bible and consider it historical fact. This view of the Koran is held by Muslem Shiites, some of whom are willing to go to extreme lengths to defend their beliefs. Advocates of essentially fundamen-

talist religious beliefs in the United States, too, have sought to establish their particular views as the correct ones. These groups have spread their religious-based views far and wide, even lobbying for laws that require the teaching of religious-based "creation science" in public schools.

The Scopes evolution or "monkey" trial in Dayton, Tennessee in 1925 marked the first of the great legal attempts to control teaching of science in the United States. In that famous trial, which challenged the Tennessee law against the teaching of evolution in public schools, a secondary school teacher, John Scopes, was charged with teaching evolution. He was defended by Clarence Darrow (Fig. 1.3) with William Jennings Bryan assisting the prosecution. He was found guilty, but the verdict was overturned because of a technicality.

In the 1970s, the California State Board of Education required all textbooks including evolution to devote equal time to opposing ideas, even those with religious bias. Fortunately, the California courts subsequently overturned that highly controversial act by ruling it unconstitutional because it required that a religious belief be taught in public schools—it violated the separation of church and state. In 1985, the same California State Board of Education, with more enlightened membership, rejected for public-school use those biology texts that provided virtually no coverage of evolution. The board ruled that to teach biology without serious coverage of evolution was hardly scientific. Similarly, anti-evolution laws passed by legislatures in Arkansas and Louisiana have been firmly ruled as unconstitutional and have been overturned by courts, including the U.S. Supreme Court.

Most scriptural scholars today do not consider Genesis a factual document. Many accept the belief that Genesis was written late in the history of the Hebrews, a considerable time after much of the post-Genesis Bible had been developed and written. Many workers have concluded that Genesis is a composite product representing four different time periods, each compiled by one or more authors and based upon the contemporary Hebrew traditions (Skehan, 1983, p. 307–314; Sandmel, 1978, p. 340–370). The Jewish scholar Sandmel (1978, p. 348) said,

> The traditions in Genesis are folk tales modified and embellished by religious belief. To seek to authenticate these as historically valid in the form in which Genesis relates them is to misapply a useful science.

Genesis can be separated into two distinct parts: an early portion (Gen. 1–11) of myths, legends, and folklore that discusses the creation, the origin of humans, the Fall in the Garden of Eden, the Flood of Noah, and genealogies and history leading to Abraham and the Hebrews, and a later portion (Gen. 12–50) of the patriarchal histories or sagas that document with great detail the history of the Hebrews as a people. Most modern scholars are willing to accept the historical data contained in the patriarchal sagas as the "facts" of the Hebrews, though they consider the early chapters to be less than factual.

A very important story in Genesis tells us of a massive worldwide flood of incredible size that was successfully survived by Noah and his family. This flood story is not unique to the Hebrews and seems to be universal among early peoples. Undoubtedly, the Hebrews obtained it from a Babylonian story known as the **Enuma Elish**. This Babylonian creation story dates from around 2000 B.C., at least 1,000 years prior to the earliest version of Genesis. The Babylonian account of the flood is in the **Gilgamesh Epic** and involves a hero named **Utnapishtim**. Apparently the Hebrews adopted the Genesis flood account from the Babylonians when they were held in Babylon as slaves and revised it in keeping with their own beliefs.

In the first part of Genesis we are informed that Adam and Eve, the first man and woman, gave rise to all humans. They had three sons, Abel, Cain and Seth, and, depending on the version, no daughters. Abel left no offspring but Cain and, presumably, Seth had children. Sandmel (1978, p. 350) notes a son of Cain and reports,

> It was Cain who first built one [city], for his son Enoch. (Where did Cain get his wife, if only Adam and Eve existed? . . .)

We must conclude that Adam and Eve had other children, daughters, or there were women nearby whom Cain and Seth took for wives. Ambiguities such as this create difficulties for theologians and laity alike and reflect the incomplete and nonfactual nature of Genesis.

SCIENTIFIC EXPLANATIONS

Biology is the principal science studying life of today but it is supported by **paleontology**, the study of life of the past. Paleontology is concerned with **fossils**, the remains or traces in rocks of organisms, plants and animals, that lived long ago. The record of fossils in the rocks of Earth is very well known today (Fig. 1.4). Virtually all scientists accept that life has been present on Earth for millions of years. These scientists recognize that these species of plants and animals have exhibited a general pattern of development throughout time, from the simple to the more complex, and that unknown millions of species have disappeared—they have become extinct. We know that strange creatures once walked Earth's surface, swam in the seas and flew in the skies. We know this because we have their fossilized bones, tracks and other evidence.

In general, the paleontologic record of life on Earth follows the pathway projected from the biological explanation of evolution. That explanation states that the simplest organisms should be present in the older rocks, but no advanced or complex multicelled creatures. The advanced types should appear progressively later in time in the younger rocks. The record in rocks, studied in **geology**, shows that Earth's oldest known strata do not contain any

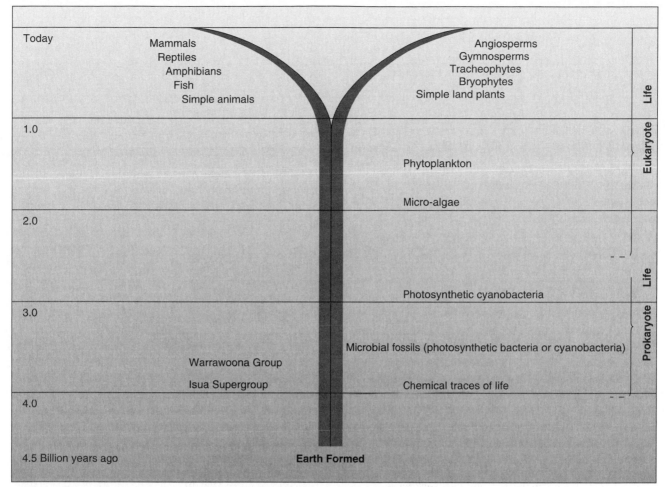

Today

Mammals
Reptiles
Amphibians
Fish
Simple animals

Angiosperms
Gymnosperms
Tracheophytes
Bryophytes
Simple land plants

Life

1.0

Phytoplankton

Micro-algae

Eukaryote

2.0

Photosynthetic cyanobacteria

Life

3.0

Microbial fossils (photosynthetic bacteria or cyanobacteria)

Warrawoona Group

Isua Supergroup

Chemical traces of life

Prokaryote

4.0

4.5 Billion years ago

Earth Formed

Figure 1.4 Generalized distribution of life through time.

recognizable fossils. Progressively younger strata contain easily recognizable fossils of increasing complexity, starting with single-cell organisms more than 3 billion years ago to large land and sea animals, including dinosaurs, of more recent times. Finally, only rocks formed in the last few tens of million years of Earth history contain fossils of primates, and humans are found in strata deposited only in the last few million years. This is precisely the general distribution to be expected if life has evolved from the simple, earliest forms to the complex creatures of today.

WHEN DID HUMANS APPEAR ON EARTH?
THE CREATION ACCOUNT

For centuries the Bible's book of Genesis provided the single authoritarian source of information about the cre-

ation of humans that was accepted by millions of people in the Western world. Genesis offered neither a mechanism for the origin nor a date when it took place. It simply stated that God created man, Adam, in His image and likeness from materials of the Earth and breathed life into him (from the Sumerian tradition?) on the sixth day of Creation. Countless theologians and philosophers have dealt with this topic, accepting the story and the "date" without considering when such acts could have taken place, or the mechanism by which the "creation" was accomplished.

In 1650, **Archbishop James Ussher** of Armagh, Ireland published a serious work on the chronology of the world (*Annals of the Old Covenant from the First Origin of the World*). In it he reported that the world had originated on the Saturday night preceding Sunday, October 23, 4004

B.C., nearly 6,000 years ago. He believed that this date could be easily arrived at by anyone using the evidence available from astronomical calculations, the Jewish calendar, nonbiblical records, and the Bible. Further emphasis was given to this date by the **Rev. John Lightfoot**, master of St. Catherine's College of Cambridge University, who said,

> . . . heaven and earth, centre and circumference, were created all together in the same instant, and clouds full of water. This work took place and man was created by the Trinity on October 23, 4004 B.C. at 9 o'clock in the morning.

There is little doubt what Archbishop Ussher and Rev. Lightfoot meant by this date. The year 4004 B.C. leaves no room for error or interpretation; it is an absolute point in time approximately 6,000 years ago. This date was largely accepted by the Christian world for more than 100 years; few people had reason to question it. But the seventeenth century when they offered this data was not as enlightened as the eighteenth century. By the eighteenth century early geologists had begun to study Earth's features and materials and they began to suspect that Earth's age was much greater than 6,000 years. As geologists examined the Earth to learn about its materials and operating processes, they demonstrated that Earth had a much greater age than thought, that Archbishop Ussher's age was much too short a time for the features of Earth to have formed.

With this belief in science, scientists began to question some of the strongly held views and teachings prevalent in the West. If the scientific studies indicated that previous ideas of Earth were often erroneous, what about other ideas? If the church views on the age of Earth were incorrect, could the explanation for the origin of humans as presented in Genesis be other than factual? As a result, some clerical leaders began to criticize scientific inquiry and some scientists. Despite the criticisms and even some condemnations, scientists moved freely to examine their respective spheres of interest.

GOALS AND METHODS OF SCIENCE

It is obvious that traditions and Genesis cannot and should not be used as either a factual document or as a scientific record. Genesis and science represent different areas of human endeavor and have different goals and different methods. Theological matters are based on faith, which can be defined as unreasoned belief. Science cannot be used to prove or disprove theological or even philosophical matters. Philosophy, as derived from man's internal and external perceptions of the world, attempts to explain the experience of human thought. It includes theology, a study of man's relationship to God. Science, as a form of philosophy, seeks to explain the phenomena of our material world and how it operates by studying the material world and its processes. Questions that seek answers to the nature of God and the relationship between humans and God are not pertinent to scientific inquiry, because philosophical and theological concepts are not tangible objects or processes and cannot be examined scientifically for their size, color, weight, electric potential, etc. Nonscientific concepts, however, are clearly appropriate to philosophy.

Scientists obtain their answers or explanations to questions of the material world in different ways than do philosophers. Typically, scientists solve questions using the **empirical method** whereby knowledge is derived from some kind of experience obtained through observations and experiments. Some use **inductive reasoning** by which the reasoning proceeds from the individual case to the general or universal principle. Scientists gather facts and from these cases develop general statements or explanations. As more facts or data are gathered, the general explanations may be modified or revised. Other scientists may use **deductive reasoning** which proceeds from the general to the specific. Certain principles or axioms are accepted initially and, from these, conclusions are derived by reason. A third method, hopefully shunned by scientists, is **intuitive reasoning** by which knowledge is known without any reasoning, sensory observation or other verifiable method. This is the weakest form of knowledge because it is simply opinion. It is considered subjective because it derives from within the individual or subject. As such it cannot compare with objective knowledge which comes from outside the individual and is not as subject to individual opinion or interpretation.

Science commonly requires that observations of phenomena be made and that tests and experiments be performed before any substantial conclusions be drawn. Furthermore, the observations, tests and experiments must be made with objectivity, without personal bias or interpretations, so that others can verify the explanations or conclusions using the same procedures. Scientific experimentation demands this aspect and, therefore, valid science must be **repeatable**. This is a prime requirement, that scientific tests, explanations and conclusions be completely documented, with numerical data and illustrations wherever possible, so that the objectivity of conclusions may be freely tested by others. In essence, science can only examine ideas that can be subject to tests intended to disprove them.

Science recognizes three kinds or degrees of general principles to explain the phenomena of nature: **hypothesis**, **theory** and **law**. These represent different levels of correctness of explanations or conclusions of the natural principles that "govern" natural things, listed in order of increasing validity. Such explanations seek to explain data or facts. **Facts** are pieces of information having objective reality. This is to differentiate them from **opinion**, which is subjective and relates to how an observer perceives reality; it constitutes a belief of less than positive knowledge. Opinions may be correct, but they may also be incorrect.

HYPOTHESIS

An hypothesis is the lowest ranked explanation and is a statement that seems to explain the facts of a case under consideration. As such, it must be amenable to testing and verification. The term hypothesis means below, under or less than a thesis or theory. In reality, several alternate hypotheses may all explain any one particular phenomenon. All may be equally likely to be true or correct, but tests or experiments can usually determine which particular hypothesis is more valid. If enough examples or cases are studied or tested, many or most hypotheses may be shown to be incorrect and one particular hypothesis may become the only logical one. Hypotheses, nonetheless, may have a low probability for being the correct explanation for any scientific phenomenon.

THEORY

If an hypothesis is successfully tested with many examples, it is considered to have withstood objective scrutiny and may be elevated to the status of a theory. A theory is a scientifically plausible or acceptable general principle determined by analysis of related facts. It is an explanation that holds true in all the cases tested or examined, and has,

accordingly, a much higher degree of predictability than an hypothesis. Many theories have excellent predictability but they cannot be raised to a higher ranking because of testing limitations. Thus, they may remain scientific theories until further testing allows their elevation to higher status. Often, some scientists may consider some theories to be more valid as scientific explanations than other theories, depending upon their personal judgments of the situations. Such acceptance is, however, dependent on the circumstances and the expertise of the scientists.

LAW

If a particular explanation works in all experiments or tests, and has a high degree of predictability, it may be considered a **scientific law**. A law is a statement of order or the relationship between natural phenomena presumed to be without variation under given circumstances. One such law is gravity. This law operates everywhere on Earth and beyond as well. Our astronauts have observed it working on the Moon, and because we believe it to be operational throughout our solar system, we can make predictions about circumstances on other planets. Using this concept we have been able to program space probes to perform certain maneuvers in outer space beyond the range of our own direct experience within Earth's gravitational field. Millions of miles from Earth, these spacecraft accomplish complicated maneuvers as programmed. Without the scientific principles and faith in the predictability of the law of gravity, we could not send spacecraft away from Earth and expect them to maneuver properly. We would certainly not send astronauts into space if laws such as gravity were mere hypotheses. Fortunately, the law of gravity works everywhere it has been tested and it is considered to be valid throughout the universe.

All the scientific understanding of natural phenomena, the hypotheses, theories and laws did not appear suddenly in textbooks. These explanations and predictions developed gradually as geological, biological, chemical and other scientific studies progressed from the few known facts into the unknown. Sciences developed from simple observations because some people asked the questions why, how and when, and began to discover the answers; science evolved from human curiosity and is a human invention.

SUMMARY

Today we face a dichotomy of explanations regarding human origins, one religious and the other scientific. Written records of religious-based ideas of the Sumerians and Babylonians have been in existence for more than 5,000 years, and successive peoples have developed their own particular versions of these origin explanations. Oral traditions of different people, the Ubaidians, for example, preceded these written ones by additional thousands of years. How these ideas came into existence cannot be ascertained at this distant date, yet some of them, particularly the later ones in the Hebrew and Christian Bibles, became the single most dominant influence on western civilization for the past 2,000 years. Few aspects of modern western life have been unaffected by this combined document, and the powerful ideas contained therein still dominate the lives of many individuals.

The scientific explanations, on the other hand, are not old ones; they have arisen only in the past two centuries from the accumulated work of numerous individuals in often unrelated fields. These curious workers were interested in discovering the workings of nature, the proper arena of science, and few, if any, were concerned with the religious implications of their discoveries. Many, in fact, were devoutly religious even as they carried out their scientific inquiries, and the same is true of many scientists today. Their studies provided explanations based on natural materials and processes, giving rise to great condemnations by those with religious viewpoints.

Many people, scientists and nonscientists alike, fail to understand that religion and science are different aspects of human endeavor, that each seeks to understand and explain entirely different worlds. Science attempts to determine the working principles of nature, its materials and its causes and effects. And to do so it must use objectivity that provides different levels of explanations. Religions attempt to understand the supernatural, something that is beyond the power of science to examine or determine. Scientific explanations, therefore, do not and cannot explain, justify or deny any aspect of religion, which is based in faith and not determined by nature. Religions, because they are not based on nature, should not be held to be in conflict with science.

It is clear that new views and, according to some, fallacious ideas of human origins have appeared in the past 150 years or so, ever since Charles Darwin caused a stir in the Victorian world with his concept of evolution and how humans originated. Though Darwinism became enmeshed in the religious disagreements regarding these origins, it was not by his choice. The survival of his explanation in the face of deeply rooted opposition is a tribute to his marshalling of the evidence. His explanation was the first that was based on objective evidence; it is self-evident and difficult to challenge because all can see it. Since his time, however, a widening rift between scientific and religious views has developed. In truth, the religious opposition tends to be fundamentalist in nature and is itself opposed by many religious groups. It seeks to attack and undermine the idea of evolution using whatever tools it can find. Today, Darwin's views are under constant reexamination by scientists and religious groups alike as they seek their versions of the truth. Many workers are continuing the search for human origins, but with new and radically different tools and methods than those used by Darwin. To date, their scientific evidence is impressive.

Though the philosophical, religious and scientific inquiries into human origins continue to this day and undoubtedly will throughout existence, we will always have difficulties obtaining answers to our questions. The principal events of human origin happened long ago and relatively meager evidence is available to us today. Perhaps the definitive evidence no longer exists; the rock record is not perfect and the important strata may have been destroyed by erosion or some other event. Perhaps we will not even recognize the evidence when it is found.

STUDY QUESTIONS

1. What evidence can we use to demonstrate that each human is unique?

2. What is meant by the *evolution* of species?

3. Why are the sciences of paleontology, geology and biology so important in studying human origins ?

4. What is a biological *race*?

5. What ancient peoples developed myths of human *origins* to explain how they came to be present on Earth?

6. Provide three lines of evidence to explain the presence of humans on Earth.

7 What did the state of Tennessee charge John Scopes with in his famous trial of 1925?

8. Why have the higher courts consistently overturned the so-called Creationist laws?

9. What ancient document served as source material for some parts of Genesis 1–11?

10. Why is Archbishop Ussher's date for the origin of Earth untenable?

11. How does science differ from religion?

12. What major events or discoveries in the past 150 years have introduced a new and revised awareness of human origins?

13. How do the three sciences of biology, geology and paleontology support the evolution of humans?

14. What are the three levels of scientific explanations and how do they differ?

15. How does a scientific explanation differ from a religious one?

2 Darwin's Idea

INTRODUCTION

All peoples have recognized the self-evident reality that there are many different types of animal and plant species living around us. Some of these species are closely related, others are very distant. Early naturalists often ascribed these similarities and differences to the acts of a perfect, all-powerful and eternal being, a **God** who could do anything, including creating the great variety of plants and animals all could see. No further explanations were necessary because the invoking of an all-powerful god as the source was sufficient in itself. Eventually, however, workers began to recognize similarities between organisms and arranged them into groups (genera and species) based on these similarities and differences. Some workers attempted to explain these relationships as changes whereby species were somehow modified into other species. This change or modification process known as **evolution**, as proposed by **Charles Darwin** (Fig. 2.1), was the first that provided objective evidence to support the change. He proposed that individuals changed slowly over long periods of time under the influence of environmental conditions. Darwin did not develop his concept in a vacuum. He capitalized to a great extent on the ideas and information produced by previous workers. These earlier naturalists or scientists did not always recognize the significance of their views or even attempt to verify them with evidence as did Darwin. They relied primarily on inductive reasoning without adequate proof. It remained for Darwin to put the ideas into one unique, working concept.

CHARLES DARWIN

The concept of evolution recognized by most scientists and accepted today was formally proposed in 1858 in papers by Charles Darwin and by Alfred Wallace. The concept was published by Darwin in 1859 as *The Origin of Species by Means of Natural Selection or the Preservation of Favored Races in the Struggle for Life*. Though the date of publication was 1859, Darwin had informally presented the concept as an essay to many of his friends and colleagues as early as 1844. The concept did not so much prove "evolution" as it did provide a great mass of supporting evidence from the disciplines of geology, paleontology, biology and others. In effect, Darwin left the world

Figure 2.1 Charles Darwin as a young man.

without a choice of viable scientific alternatives. He effectively proved the validity of evolution without proving it. In essence, therefore, evolution is the process by which the animal and plant species have been changed or modified from previous species. In the strictest sense it is the production of changes in the gene pool that appear as new forms of life.

Darwin's concept of evolution is not a complicated one, but it has had unbelievable influences on human affairs. Many people today have strong negative feelings about his concept and what they perceive to be its harmful effects on their world. All manner of problems of the modern world have been blamed on his concept, including rebellious youth, poverty, crime and a rejection of fundamentalist religious beliefs. Judge Braswell Deen of the Georgia State Court of Appeals said (1981)

> This monkey mythology of Darwin is the cause of permissiveness, promiscuity, pills, prophylactics, perversions, pregnancies, abortions, pornotherapy, pollution, poisoning, and proliferation of crimes of all types.

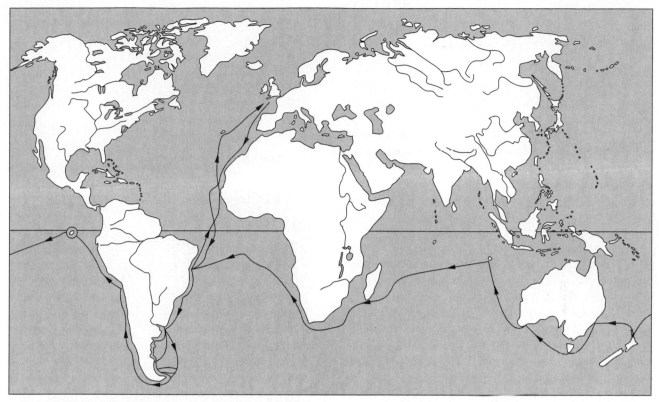

Figure 2.2 Map of the HMS *Beagle* voyage.

The fact that these problems existed long before Darwin and his concept is apparently irrelevant to such people. Darwin's purpose in seeking the origin of species was not to challenge any religious or other authorities. Rather he sought to find species' origin (1872, p. 373) because:

> To my mind it accords better with what we know of the laws impressed on matter by the Creator, that the production and extinction of the past and present inhabitants of the world should have been due to secondary causes. . .

Darwin was an educated man from a prominent family who lived during changing, even turbulent times. The intricate blending of factors that influenced his life and how he developed his concept has been recorded in the many volumes that have been written about him and his concept, including those by his family, relatives, friends, scientists, journalists, authors, clerics and by his own hand. Darwin had experienced many things in his lifetime that were unavailable to the average scientist or even the average person.

At the beginning of the eighteenth century, scientific knowledge was starting to accumulate at a startling rate and the sciences were becoming modern. Darwin was aware of many of these changes and developments because he knew or was in touch with many of the prominent scientists of the time, people who were making some of these changes and developments. After college, Darwin was recommended for the post of naturalist on **HMS Beagle** on a voyage that would last some 57 months (Fig. 2.2). During this time Darwin encountered geological and biological sights that were a continuing source of wonder to him. He was both puzzled and intrigued by the richness and variety of nature in different places. Finding fossil bones of many different and unknown animals, he realized that such creatures were no longer known on Earth, but were extinct species. Why, he asked, had some creatures disappeared from the face of Earth? Why had such creatures appeared at all if only to disappear in the future? How many other species had lived and become extinct? And, more importantly, what caused their extinction?

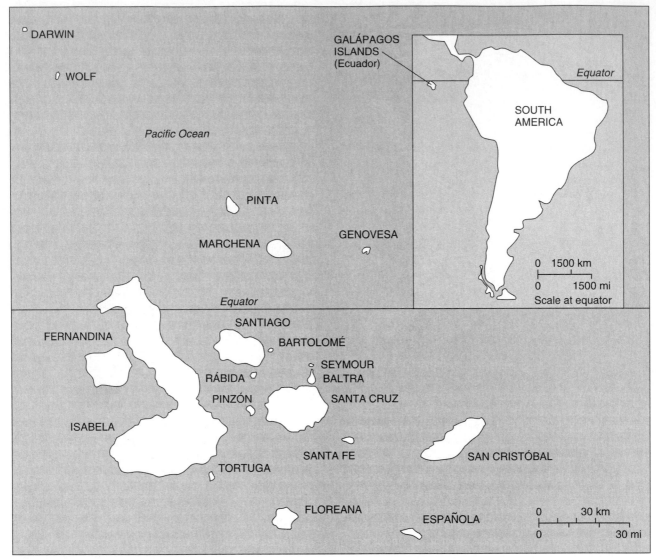

Figure 2.3 The Galápagos Islands.

Darwin saw, for example, that the Andes Mountains had been formed in the not too distant past. They were obviously an effective barrier to the spread of organisms from one side to the other. Could organisms have migrated or spread before the mountains had been built? Discovery of this and other kinds of geological phenomena in different parts of the world caused Darwin to speculate on the relationship between obvious geological changes and equally obvious changes in plant and animal life. He wondered if geological changes could have had enough significant influence on organisms to result in changes in species?

What Darwin saw, what he learned, what he concluded and what he explained generated such widespread interest in the evolution of plants and animals and in man's place in the universe that we cannot ignore it even today, more than 130 years after it was first published. Today many people unfairly condemn him and his work for the conflicts he unintentionally generated and, in fact, tried to avoid. Many of these people do not really understand what he

16

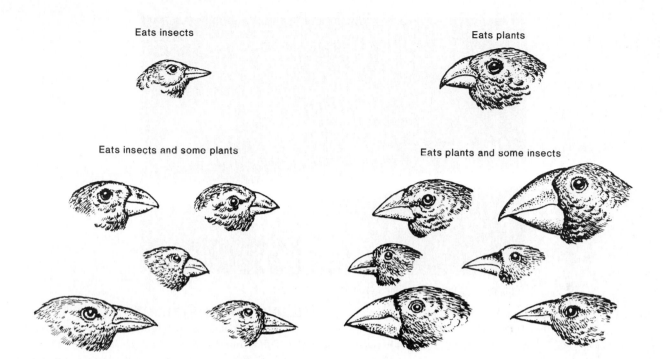

Eats insects

Eats plants

Eats insects and some plants

Eats plants and some insects

Figure 2.4 Finches of the Galápagos Islands.

said, what he meant and how his concept operates. Even many of his supporters did not fully understand the mechanism of evolution and how it operates to change species.

THE GALÁPAGOS ISLANDS

On the *Beagle* voyage Darwin visited the Galápagos Islands, a small group of rocky islands more than 500 miles west of South America (Fig. 2.3). These islands, all of volcanic origin, were formed when lava poured out onto the sea floor forming submarine volcanoes. The flows continued to accumulate until the volcanoes formed into islands. Obviously they were originally devoid of land life, but Darwin found them populated by varied and abundant plants and animals. These organisms could have arrived only after the islands had formed, after they had become hospitable to life.

These islands were named after the giant land tortoises (*Galápagos* in Spanish) that roamed freely, and Darwin had been told that many people could recognize the home island of any tortoise simply by looking at the shape of its shell and some other features; each tortoise type was unique to a particular island. Most of the other animals were also distinctive and restricted to specific islands.

Darwin recognized clearly that something was amiss in such unusual distributions of organisms. He concluded that the different species of plants and land animals on the islands had probably been introduced by accident, as adults or seeds and eggs that somehow crossed the open ocean from the South American continent. Once on the islands, these seeds and eggs hatched and, with the adults, became the founding populations that gave rise to the species that Darwin found during his visit. These species could not escape from these islands because of their remoteness and somehow each species became modified according to the demands of the islands they inhabited. This question of separate though similar species on the different islands troubled him. It was the study of birds on the Galápagos that finally provided him with the clue to the mystery, though it came many years later.

There are thirteen separate species of finches (Fig. 2.4) on the Galápagos, all resembling a single species on the mainland of South America; yet each is easily recognized by appearance and behavior, especially by shape of beaks. Darwin reasoned that the original population of birds was a single species that probably had been blown from the South American mainland to the different islands during a

17

Figure 2.5 Tool-using Galápagos finch.

storm. Once ashore the birds were restricted to the particular island on which they had landed. They were separated from those on the adjacent islands by the distances between them, and there was no exchange from one island to another. Because each of the islands varied slightly in environment, the separate populations of finches each had to react in its own way to cope with the subtle environmental differences. To survive in these slightly different ecological niches, the birds evolved slightly different behavior patterns based primarily on feeding habits and food supply (Fig. 2.5). In time the separated finch populations became so significantly different that they became different species; they had evolved.

Darwin chanced upon the most remarkable relationships between environments and sets of related creatures. He said,

> Facts such as these admit of no sort of explanation on the ordinary view of independent creation. . .

From the background of his training and experiences and the biological and geological phenomena he observed on the *Beagle* voyage, he began to look at these and other creatures in relation to each other and to their environments. What he saw were glimpses of something different that raised questions and challenged existing ideas and prevalent theories. He began to formulate his radical new concept.

THE DEVELOPMENT OF THE CONCEPT

Darwin did not develop his concept of evolution immediately following the *Beagle* voyage, but only after many years of careful study. It was not until 1837 that he began any serious writing about how species originated. This was recorded in a formal notebook in which he hoped to gather all the facts pertaining to the subject. By the middle of 1842 Darwin had written a "brief abstract of his theory" in 35 pages. By 1844 he had expanded this into a 230-page document which he showed to several associates. But he hesitated to introduce any radical concept that might cause more confusion than provide clear explanations. For more than twenty years he continued to add to this abstract, gathering facts and information from other workers and from his own experiences and experiments. Eventually he mustered a large document with more than ample evidence and examples to support his contention. In 1856 he wrote to Asa Gray, the American botanist,

> Either species have been independently created, or they have descended from other species. . . As an honest man, I must tell you that I have come to the heterodox conclusion that there are no such things as independently created species. . . I believe I see my way pretty clearly on the means used by nature to change her species and adapt them to the wondrous and exquisitely beautiful contingencies to which every living being is exposed.

Darwin envisioned the spread of species throughout geologic time by means of a series of gradational changes, an actual evolution of a species into one or more other species. He said that the plants and animals living today are the descendants of previous species from whom they evolved, and that their ancestral species had evolved from still earlier species and so on back through time. His process of evolution was one in which the effects of nature operated upon the different *individuals of a species*, or **variants** as they were known. This process of nature selected for survival those variants that were suitable for life in the existing environment from those that were unsuited. It was not a conscious process of selection by the environment or nature, but one that simply allowed variants to survive or die under the environmental conditions.

Nature creates far more individuals or variants in any species than could possibly survive. Most of the variants in any species are, thus, doomed to be eliminated before adulthood in some natural way to provide room for the others. Those that survived, Darwin emphasized, were somehow more fit or more suitable for their particular environment than others that failed to survive. The same was true of species. Many species were doomed because they were less fit or less suitable for survival in a given environment than other species. Though they had evolved in a given environment, this environment had somehow changed and made them unsuitable. Darwin termed this concept **natural selection** because the variants were selected for survival by undirected actions of nature. He said,

> As many more individuals of each species are born than can possibly survive; and as, consequently, there is a frequently recurring struggle for existence, it follows that any being, if it vary however slightly in any manner profitable to itself, under the complex and sometimes varying conditions of life, will have a better chance of surviving, and thus be *naturally selected*. From the strong principle of inheritance, any selected variety will tend to propagate its new and modified form.

> This preservation of favourable individual differences and variations, and the destruction of those which are injurious, I have called Natural Selection, or the Survival of the Fittest.

Darwin emphasized the mechanism of change by which variants and species evolved by stating that it was an undirected rather than a directed process in which nature merely provided a set of conditions in which an individual could survive or perish. Dawkins (1986, p. 21) stated with bluntness,

> Natural selection is the blind watchmaker,
> blind because it does not see ahead, does not
> plan consequences, has no purpose in view.

Darwin's concept was formulated only after he had spent years coordinating the key principles that served as its foundation. These key principles were the five ideas that formed the essence of natural selection:

1. The geologic concept of Uniformitarianism
2. The great length of geologic time
3. The Malthusian idea of population growth and limitations
4. The great variety of individuals in any species
5. The principles of selective breeding

Charles Darwin completely understood each of the fundamental ideas relative to his concept of evolution. He read them in their original publications and communicated with prominent scientists who themselves were experts and well versed in the important scientific ideas of the time. Further, he personally experienced these ideas in different ways, ways in which he was forced to question them in order to learn how things changed.

UNIFORMITARIANISM

The first idea, **Uniformitarianism**, was proposed by James Hutton, an early geologist (Fig. 2.6). Basically, this concept holds that the study of processes working today to shape the Earth are the same processes that worked in the past. Knowledge of these processes enables us to learn how features formed in the past. This view is accepted universally as the foundation of modern geology and it is valid even beyond Earth, our astronauts who walked on the Moon used these same principles to understand lunar geology. Once the principle was understood and explained, geologists began to recognize its potential and grasp its applications. Because Uniformitarianism provided the basis for scientific explanations of Earth, the study of geology became truly scientific and organized. This concept enabled the reconstruction of Earth's past, and Darwin was able to use it in developing his great unifying concept of change.

Figure 2.6 James Hutton.

THE GREAT AGE FOR EARTH

Darwin could not reconcile himself to the then accepted age for Earth of about 6,000 years. He believed firmly in slow and gradual evolutionary changes in species, much like the gradual geological changes observable everywhere. He knew well that Earth's geological features were formed over long periods of time, and he knew from his studies of the life of the past, the fossil record, that Earth had to be very old for evolution to have accomplished changes in species. Scientific research in his time had produced a date for the age of Earth of only about 100 million years, and this gave Darwin grave concerns. When this date was revised to about 24 million years it created even more difficulties for Darwin and other evolutionists. In time, however, it was discovered that incorrect assumptions had been made and that Darwin was right; Earth had existed for an unbelievably long time, now known to be about 4.5 billion years. This was ample time for both geology and evolution to create great changes, evolution to work its magic with organisms and modify them into other types and geology to do the same with Earth features.

MALTHUS AND POPULATION CONTROLS

Thomas Malthus (1766–1839), an English clergyman, studied economics and developed a view of life published as *An Essay on The Principle of Population* (1798). He considered the consequences of overpopulation growth by humans and proposed that the world human population would increase at a geometric rate (1, 2, 4, 8, 16, 32...), doubling on a regular basis. He also realized that the food supplies of any nation were finite and could only increase at an arithmetic rate (1, 2, 3, 4, 5...). The two rates of increase were obviously in conflict. He predicted that Earth's human population would increase much faster than the available food supplies needed to support it and would eventually face starvation unless some natural calamity reduced the population to a size consistent with the food supply. Malthus identified pestilence, war and famine as the natural calamities that operated from time to time and reduced the human population locally or universally. Such reductions, he believed, merely deferred the final catastrophe to future dates. Of special import in Malthus' views was the idea that among all creatures, plants and animals, there is a violent competition for survival. This was a critical aspect and it offered Charles Darwin a hint in the search that led to his concept of evolution by natural selection.

Darwin understood Malthus' ideas and accepted them. But even more, he borrowed some ideas from Malthus' work, which he had "read for amusement" but which provided him with a vital clue, namely, the competition between individuals for survival. He knew that all species of plants and animals tend to produce more eggs than could ever survive to adulthood in the wild. Flies or frogs or maple trees could easily overpopulate our world if all their eggs or seeds hatched and grew to maturity, producing offspring in turn. The balance of nature is such, however, that many animals eat these eggs and seeds and those of other organisms, keeping the fly, frog and maple and other populations in check. At the same time many eggs and seeds do not survive because they are subjected to unsuitable environmental conditions. Of those that hatch, the weaker ones tend to die off quickly because they simply cannot cope with the stresses. In this way, nature's way, only a small percentage of eggs in any species can survive to become breeding adults. Those that survive tend to be the stronger or more capable individuals because the weaker or less perfect ones have died off. The natural balance, thus, is maintained.

Darwin credited Malthus' ideas with providing the real basis for his own concept. In his autobiography he said (1876, p. 120),

> Fifteen months after I had begun my systematic enquiry, I happened to read for amusement Malthus on Population, and being well prepared to appreciate the struggle for existence which everywhere goes on, from long-continued observation of the habits of animals and plants, it at once struck me that under these circumstances favourable variations would tend to be preserved, and unfavourable ones to be destroyed. The result of this would be the formation of new species. Here, then, I had at last got a theory by which to work.

THE VARIETY OF INDIVIDUALS

Darwin understood that plants and animals must struggle for survival each day of their lives. Clearly, some individuals in any species would be successful and survive, others would not. The question that Darwin asked was *why* some organisms survived and others did not. Darwin understood that all individuals in a species were not equal, that some possessed characters or traits that made them somehow slightly more fit for their way of life than others. He explained that those individuals that survived were somehow favored with these better traits for a given environment. Those individuals that failed to mature and reproduce were somehow less favored with the better traits and characters; they were **unfit**. In effect, a species would tend to be dominated with better-fit individuals as the lesser fit died out earlier and reproduced less. Darwin showed how the many different individuals that comprise any species relate to each other and their environment, some well and others poorly, and how these relationships influenced their very survival.

Darwin lived in the country most of his life. He knew that there were real differences between individuals in any species; it was a self-evident fact that could be seen by anyone. Such differences, subtle or great, are the variations inherent in any species. Each variant or individual differs somewhat and to some degree from all other individuals or

Figure 2.7 Gregor Mendel.

variants in the species. Darwin knew this and accepted it, but he wondered why it should be so. These variants and their origin were the puzzling key to evolution. How were they to be explained?

The puzzle of variety was never satisfactorily solved by Darwin. Unknown to him, it was solved by a contemporary, **Gregor Mendel**, an Augustinian monk from Bohemia (Fig. 2.7). Mendel discovered the secret of inheritance by breeding varieties of garden peas and, in so doing, laid the foundation for **genetics**. By the very nature of sexual reproduction, half of the inherited characters are derived from each parent. But, differences in these characters establish them as **dominant** or **recessive**. The characters are myriad to begin with and then are combined essentially at random in successive generations. This results in variety of genetic traits and produces the **individuals** or **variants**, each of which differs in response to environmental pressures. Later workers discovered random mistakes in replication of genetic codes. These are **mutations** and they, too, contribute to variety. In fact, mutations are the only source of new aspects. These were the variants Darwin needed for his concept to work.

SELECTIVE BREEDING

As a country gentleman Darwin associated with other gentlemen in discussions of the common practices and problems that affected their success at farming and raising animals. All serious farmers of the time were well acquainted with the techniques of **selective breeding** used to improve livestock and crops. In practice, the farmer selected only the best animals for breeding, culling out those with undesirable traits in the hope thereby to improve his stock strain. When preparing seeds for next year's crop, he chose seeds from only the best and healthiest plants. A skillful breeder was able to improve his stock or crop markedly by learning how to identify the best individuals in the different animal breeds or the best potatoes, beans, corn, etc. This may well be the origin of state and county fairs and other competitions in which judges determine the best of any breed or species; however, such practices today are not limited to farm animals and crops.

To capitalize on selective breeding, a farmer first had to acknowledge that some animals of a species are superior to or better than others in some ways. But this is a self-evident fact. Good farmers knew which of their milk cows produced the most milk, or the least. They knew which breeds of cows were better at milk production, or had milk with higher butterfat content, were better at withstanding cold winters, produced more calves, were more resistant to illnesses, etc. Profit and success demanded that practical people apply common sense approaches to their methods. These individuals learned such practices by trial and error, an expensive method, but one that provided the experience that led to success. On selective breeding Darwin wrote,

> At the present time, eminent breeders try methodical selection with a distinct object in view, to make a new strain or subbreed superior to anything of the kind.

THE SYNTHESIS OF IDEAS

From these five fundamental ideas, Darwin was able to produce the first concept of evolution based on objective evidence. It is in Darwin's 1859 book, *The Origin of Species*, that we find the culmination of his experience, discussions and conclusions about evolution. Though he spent more than twenty years gathering data and refining his arguments, his conclusions can be summarized in a few sentences:

1. All species are composed of individuals or variants that differ from each other.
2. Natural environments exert great survival pressure on any population.
3. The environmental pressures tend to favor for survival those variants that possess suitable traits.
4. Thus, the selective action of the environment tends to allow variants with suitable or favorable traits to survive and pass on those traits to their offspring. Ultimately, the population will tend to shift in the direction of favored variants.

Darwin's concept has been summarized more succinctly and Darwinian evolution is often considered to consist of only two parts:

1. Naturally occurring variants in the population
2. Natural selection

The idea of variants, different individuals, in any species is self-evident and the effects of natural selection or the selective actions of the environment had been recognized by others long before Darwin. But no one was able to coordinate these obvious aspects and develop a realistic concept until Darwin combined them. Even Thomas Huxley, the prominent biologist and proponent-defender of Darwin, said when he had read Darwin's work, "How extremely stupid not to have thought of that!" Despite the simplicity of the concept, none of Darwin's predecessors or contemporaries were able to master it on their own. He was unique!

PUNCTUATED EQUILIBRIA

Not all the arguments in opposition to Darwin's view of evolution arise from nonscientists who insist on a literal interpretation of Genesis in which all plants and animals on Earth were created directly by Divine Will. Some scientists, too, hold other views. Though virtually all scientists accept evolution as a fact, some argue that all evolution is not a gradual one—the so-called **phyletic gradualism** in which very slow changes in species over very long periods of time result in the conversion (or evolution) of one species into another. Another concept, known as **punctuated equilibria**, developed by Eldredge and Gould

(1972) proposes, instead, that sudden accelerations or bursts in the evolutionary process can take place in which species form quickly. This spurt is then followed by long periods of time, **stasis,** in which no obvious changes in the species occurs. This process produces relatively sudden evolution of one species from another in perhaps tens of thousands of years rather than much longer intervals normally identified. Eldredge and Gould present evidence from studies of specific fossils in which such sudden appearances are identified. The process is not restricted to one kind of fossil but has been identified in a variety of phyla and in different geologic times of the past.

Not all scientists, however, agree with Eldredge and Gould in their view of sudden accelerations in the evolutionary process. Dawkins (1986), for example, argues that they have misinterpreted the evidence. His view of evolution is that it results from very infrequent changes in the genetic materials that transfer information from parents to offspring; perhaps hundreds of thousands of years or more is required to accumulate enough genetic changes to form a new species. He argues, further, that such sudden accelerations would require massive changes in the genetic material in a relatively short period of time, and such changes (that control development and inheritance from one generation of organisms to their offspring) would create far more difficulties for survival of these offspring than they could manage.

E. O. Wilson also points out (1992, p. 89) that the swift evolution postulated by punctuated equilibrium was

> . . . already a cornerstone of traditional evolutionary theory and therefore in no sense a challenge to it. The models of population genetics, the foundation of quantitative theory, predict that evolution by natural selection can be so rapid as to seem nearly instantaneous in geological time. The models also allow for stasis, or long periods with little or no evolution of a kind detectable in fossils.

SUMMARY

In Darwin's concept, every species consists of numerous individuals each with a slightly different set of characteristics or traits. Each individual has, therefore, a slightly different potential for survival in any particular environment. The individuals who have more characters suitable for any particular environment are more likely to survive than those who lack such suitable characters. The unfavored ones have a handicap in one or more traits and this makes them less likely to survive, reproduce and pass on their traits to offspring. It is this **inherent variability** in any population that is acted on passively by the environment as natural selection.

The very nature of evolution of species as envisioned by Darwin implied a set of unspecified conditions in an environment or a change in these conditions. Variants unsuitable in one environment could easily be favored in another. As geologic environments changed, and they did again and again throughout time, some variants were unable to adjust to the new changes and died off. Others succeeded in the new environments and became, in many cases, new species. The competition for survival that Darwin envisioned was not directly between individuals fighting each other, but with survival in environments and the difficulties this survival presented. The environments did not determine which individuals would survive, they merely provided the conditions that changed from time to time. The variants either had the ability to survive under those conditions or they did not. To say that organisms are adapted to their environments is an understatement; they are interwoven with the environments that produced them.

Darwinian Evolution is the only explanation for the origin of species that is supported by evidence and by experimentation. Supporting data from fossils are overwhelming in magnitude, depth and scope. Experiments have also contributed data from a wide range of disciplines, primarily biological, that are supportive of each other and Darwinian Evolution. Most compelling is the artificial evolution commonly known as selective breeding, practiced throughout the world today. Although Darwinian Evolution is the cornerstone of modern biology, some people have challenged it with a concept known as punctuated equilibrium. However, this recent interpretation of paleontologic evidence that examines only one aspect of

Darwinian Evolution does not challenge the validity of evolution as a concept. More than one scientist has challenged punctuated equilibrium itself and identified it as a misinterpretation of the data or an idea that falls within modern evolutionary theory.

The fundamental idea, however, that scientists continue to reexamine Darwinian Evolution using scientific methodology and materials is a measure of its importance. These studies lie well within the realm of ordinary science and serve to accentuate the scientific nature of Darwin's concept. Darwinism has withstood the challenges to date despite great opposition and this lends it even greater credence as a valid concept.

STUDY QUESTIONS

1. Why did the publications of Darwin's book with his ideas on evolution result in so much controversy?

2. Explain the two aspects crucial to Darwin's idea of evolution.

3. How does selective breeding relate to Darwinian Evolution?

4. Uniformitarianism is the great unifying concept of geology. Explain why it was so important to Darwinian Evolution.

5. What crucial clue to evolution did Darwin discover in the work of Thomas Malthus?

6. What information did Mendel discover that was so vital to Darwinian Evolution?

7. How did Darwin's idea of evolution differ from the popular biblical view of creation of species?

8. What two kinds of creatures on the Galápagos Islands strongly influenced Darwin? Why and how?

9. Why did Darwin consider a great length of geologic time necessary for evolution to operate?

10. How does evolution by punctuated equilibria differ from the usual view of Darwinian Evolution?

3 The Primates

INTRODUCTION

Darwin believed that humans evolved from primates, either the apes or Old World monkeys. Accordingly, researchers have examined these groups for evidence of this ancestry. Their success was not due to a single discovery of fossils but was the result of cumulative and independent research by a great many individuals working in different disciplines. Their results have led to long debates and discussions about the relevance of each fossil discovered, often resulting in different interpretations. Nonetheless, each fossil provided clues to unlocking the puzzle of our ancestry and each subsequent interpretation enabled scientists to better fit these clues together into a coherent picture of descent.

The distribution of primate fossil evidence throughout the geologic record (Fig. 3.1) supports Darwin's view of evolution. This pattern of appearances and ranges of the different stocks of primates, especially anthropoids, allows us to readily explain their origins and relationships through evolution. The earliest primates, the prosimians, arose just before the Cenozoic Era, but there is little evidence available to develop any real understanding about the earliest forms. The more advanced types appeared later during the era and by Pleistocene time all of the others had appeared, including humans. These fossil primates exhibit generally progressive biological advances or complexity in their features, structures, characters, traits and behaviors through time. What is even more important is that these aspects show trends that foreshadow humans.

The prosimians have the most primitive characteristics of the primates and are the most evolutionarily distant of them from humans. They gave rise to the intermediate stage, the Anthropoidea, which is the stock of higher primates that consists of the monkeys, apes and humans. The oldest of these Anthropoidea are the Old World monkeys and New World monkeys, two separate stocks that evolved independently in parallel fashion in the Eastern and Western hemispheres some 35 million years ago. The higher primates appeared in relatively recent geologic time, about 20 million years ago. These higher types are also known as the hominoids, humanlike forms that include the living and fossil apes and related forms, and the hominids, the living and fossil humans. All of these forms, their appearances and their distributions offer ample evidence in support of the Darwinian pattern of evolution.

The distribution of primates shows that the transition from earlier types to humans occurred slowly with many evolutionary experiments. These experiments indicate that the more revolutionary trend leading to humans did not appear early in primate history. Rather, it occurred late in the Cenozoic Era after the prosimians had finally spread into various environments and had evolved into many arboreal types or models. Once these new types, the anthropoids, successfully invaded different environments, the potential existed for improved new stocks among them to evolve into hominoids and hominids. The dryopithecines, in particular, are representative of these stocks that appeared and spread throughout the Old World. This dryopithecine stock is particularly critical because, as cosmopolitan hominoids, they are the ancestral group that ultimately gave rise to the great apes and humans.

Most of the genera and species of living primates, particularly the monkeys, apes and humans, possess biological attributes, both physical and nonphysical (see Chapter 4), that link them closely to each other. These attributes are the result of evolution from common ancestry; other explanations simply fail to explain the pervasiveness of characteristics throughout these different stocks. Goodall said (1976, p. 82)

> Recent biochemical and anatomical research has revealed striking similarities in the biology of man and chimpanzee in relation to the number and form of the chromosomes, the blood proteins and immune responses, the structure of the DNA, and perhaps most importantly, the anatomy and circuitry of the brain. These similarities. . . suggest that man and chimpanzee shared a common ancestor at some point in the remote past. We may argue that behavior which is common to modern man and modern chimpanzee probably occurred also, in similar form in that common ancestor—and thus in Stone Age man as well.

It is highly probable, therefore, that many of these attributes were already present in the ancestral primate

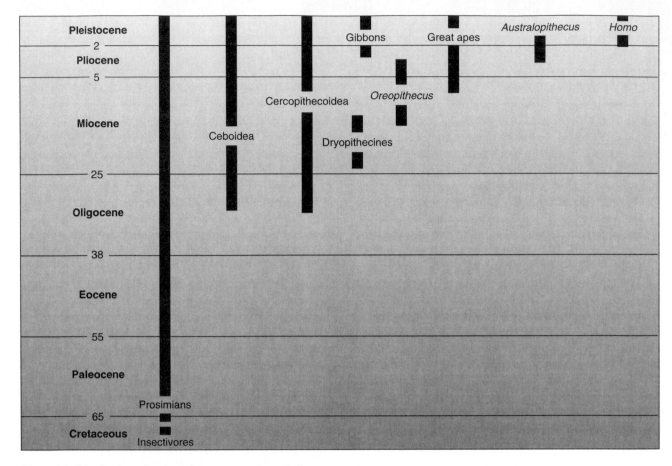

Figure 3.1 Distribution of major primate groups through time.

stock that eventually separated into monkeys and anthropoids, then into apes and humans. The most obvious of these attributes are the physical characteristics we can see and measure, though some workers have argued that these physical attributes are merely the result of parallel evolution. This is, however, too simple an appraisal. Fortunately, the nonphysical evidence also supports evolutionary relationships. Included in this category of evidence is the great range of animal behaviors that have been recognized through extensive field studies carried out over the past three decades. These studies and the evidence supporting the physical and biochemical data demonstrate forcefully the close relationships of the higher primates. As the most compelling of these evidences, the biochemical data is particularly definitive because it is not as open to interpretations as are the range of behaviors. Nevertheless, the significant and remarkable similarities between the behav-

iors of higher primates and humans provide us with critical clues to our own origin. Taken in total, therefore, the physical, biochemical and behavioral evidences support the path from the ancestral hominoids through the apes to humans by means of Darwinian Evolution.

Linnaeus introduced the name **primates** in his *Systema Natura* to include the highest placental mammals. He considered this group as prime, the first among animals. Included in it today are humans, apes, monkeys, prosimians, tree shrews and their extinct species. They are characterized in gross terms by **binocular vision, grasping hands with opposable thumbs**, **closed eye sockets**, **fingernails**, and the ability to **brachiate** (the ability to hang, swing, climb or pick objects with arms above the head). The latter is possible because the **clavicle** (collarbone) creates a sturdy shoulder that supports a pivotable arm. In addition there are several traits of major importance that provide

Figure 3.2 The tree shrew, *Tupaia*.

them with superior survival characteristics, notably enhanced brain ability, greater intelligence and strong social organization.

The oldest primates are known from Cretaceous strata and are represented by little more than jaws and teeth because of the difficulties of fossilization for small, arboreal creatures. The earliest primate, *Purgatorius*, apparently evolved from an insectivore ancestor and is known only by a few teeth from late Cretaceous strata in Montana. By the start of the Cenozoic Era, however, a variety of primates had appeared and are known from strata of the Paleocene Epoch. This stock was successful evolutionarily and underwent rapid adaptive radiation from Cretaceous ancestors. These early forms were about the size of living squirrels and are comparable in most respects with the modern tree shrew, *Tupaia* (Fig. 3.2). Their arboreal lifestyle, agility, intelligence and reproductive efficiency made them formidable as competitors, despite their small size. Life in trees contributed significantly to the rapid evolution of these traits and characters from which we have benefited by inheritance.

Today the primates are divided into two groups, the **prosimians** and the **anthropoids**. The prosimians, or the premonkey forms, include the lemurs and related types that are not as advanced as the monkeys. The anthropoids are the humanlike primates and include Old World monkeys, New World monkeys, apes and humans. Primates are predominantly arboreal animals characterized by flattened fingernails, well-developed clavicles or collar bones, five digits on front and hind limbs, opposable thumbs on hands

and usually on feet, closed orbits, two pectoral breasts, eyes usually toward the front of the head, an enlarged brain, and a strong social organization. Though most primate species spend most of their lives in trees, a few types are largely terrestrial. Humans are the only primate species totally terrestrial.

The appearance of the primates marked the final stages in the evolution of animals ancestral to humans. These highly intelligent, energetic animals, especially the apes, are amazingly like us in many physical structures and traits. Almost all higher primate species are highly social; only the orangutan is solitary. Males and females come together only to mate. Female orangutans care for their young for several years but otherwise form no groups beyond their own small family. The chimpanzees and gorillas have highly organized family systems with many features that directly parallel those of humans, and many workers have concluded that their behaviors are not independently derived but, like their physical structures, are the product of evolution.

PRIMATE CLASSIFICATION

There is no universal agreement on the classification, ranks or names of primates. Table 3.1 shows one classification of the most common or recognizable primates, living and extinct.

PROSIMIANS

The modern prosimians are generally small to cat-sized arboreal primates that range throughout the tropics of the Old World in Africa, Madagascar and southeastern Asia. Most are nocturnal, furry animals that feed mostly on fruit and insects. Based on the fossil record, the living species have evolved with little change from the earliest primates. Swiftness of motion, leaping and jumping are distinctive of most prosimians although the galago, potto and loris are relatively slow moving species. Excellent binocular vision and the ability to firmly grasp branches facilitates such active lifestyles in trees. Accordingly, their hands and feet developed opposable thumbs and toes that enabled such grasping ability. Their eyes also rotated to a more forward position to allow for stereoscopic vision and depth perception. Together these highly developed and specialized structural advances provided for a unique mobility in trees with upright posture known as **vertical- climbing-leaping ability** (**VCL**).

Table 3.1 Classification of the primates.

Order Primates

  Suborder Prosimii

      Family Tupaiidae*—Five living genera; *Tupaia*, the tree shrew; southeast Asia

      Family Lemuridae—Six living genera; lemurs and others; Madagascar

      Family Indriidae—Three living genera; Madagascar

      Family Daubentoniidae—*Daubentonia*; Madagascar

      Family Lorisidae—Four living genera; *Loris*, *Galago*, potto; Africa, southeast Asia

      Family Tarsiidae—*Tarsier*; Malaysian archipelago

  Suborder Anthropoidea

    Superfamily Ceboidea— Seventeen living genera; New World monkeys, marmosets

    Superfamily Cercopithecoidea—Eleven living genera; Old World monkeys

    Superfamily Oreopithicidea—*Oreopithecus*

    Superfamily Hominoidea

      Family Hylobatidae—Gibbons, siamangs

      Family Pongidae—Gorilla, chimpanzee, orangutan, gibbon, *Gigantopithecus*, dryopithecines

      Family Hominidae—*Australopithecus* and *Homo*

* placed in Order Insectivora by some workers

The advances in primate hand and eye development are also reflected in their ability to reach out to pick fruits and other foods and to catch insects, activities not ordinarily possible in nonprimate mammals that use their jaws for eating as well as for picking up objects. The structural and functional advances in primate hands, feet and eyes were accompanied by enlargement of their brains, particularly in those areas devoted to hand and foot control and in the temporal and occipital lobes, those areas devoted to visual and auditory stimuli, perception, integration and memory.

Such advances in vision and hearing were also essential to the development of social organization. Individuals in social groups must be able to recognize each other quickly and respond to visual signals and vocal communications for mutual interaction and defense. This is particularly true for lemurs which have a wide range of vocalizations for many situations and behaviors. Because most modern primates with these features and structures typically have highly organized societies, it is believed that some of the early primates probably also lived in well-organized social groups.

The living prosimians are mostly nocturnal animals of relatively small size and numbers. These adaptations enabled them to survive where they were not in direct competition for food and living space with their improved descendants, the monkeys, which occupy the same habitats. The prosimians are not diverse except in Madagascar where they have thrived without primate competition because of their isolation. As a result, prosimians, are not common animals, are usually difficult to observe and study, and occupy niches that provide maximum cover and protection.

FAMILY TUPAIIDAE. These are the tree shrews, small squirrel-like animals, the most primitive of the primates (see Fig. 3.1) and most closely related to the insectivores. They have some features that are primitive and typically insectivore in character and many others that are clearly more advanced and primate. They are diurnal, extremely aggressive and active omnivores that feed on insects, fruit and flesh. Tree shrews have the primitive mammalian dental formula of 2-1-3-3/3-1-3-3 (incisors-canines-premolars-molars). Their third incisor is much reduced in size, however, and suggests a trend toward the typical primate pattern. They use their lower incisors to groom their fur, a technique followed also by lemurs. Their hands are effective grasping instruments used for climbing but cannot efficiently hold food. They do not form large social groups.

Figure 3.3 Ring-tailed lemur, *Lemur catta*.

FAMILY LEMURIDAE. The largest group of prosimians are the lemurs (Fig. 3.3) which range in size from mouse to fox. They superficially resemble a fox with its large bushy tail. They are known only from the island of Madagascar where they are without competition from their monkey cousins and have become a diverse group of six genera and fifteen species. Lemurs typically have an elongate skull, much like that of a fox, with a proportionally larger brain than other prosimians and eyes arranged toward the front of the skull. As a result, binocular vision is well developed in the group and they are agile arborealists. Their grasping hands are used for little more than holding on to branches and eating. Active and highly mobile, they can move with swiftness using all fours in the trees or on the ground, where some species spend much time. Their diet appears to be exclusively vegetarian. They are highly social and live in troops of forty-five to sixty. Lemurs have the dental formula 2-1-3-3 (an exception is the genus

Lepilemur with 0-1-3-3/2-1-3-3). Their lower incisors and canines are procumbent, nearly straight from the jaw, and serve almost exclusively as grooming tools. Grooming is an essential part of their lives and has a social as well as sanitary function. They recognize alien lemurs with warning calls and threat gestures and so appear to be strongly territorially oriented.

FAMILY INDRIIDAE. Three genera comprise the indris which include the largest of the Madagascar primates. They are similar physically to lemurs, long-legged, extremely agile and able to leap with ease through the trees. They maintain an erect posture in the trees and usually walk bipedally when on the ground, though terrestrial locomotion is uncommon. This erectness has been accompanied by the development of lengthy arms and hands which, however, are used rarely in brachiation. They have been little studied, but generally live in family groups of three or four, are distinctly territorial and offer threat gestures and calls of remarkable loudness and song when they meet foreigners. In this respect they resemble gibbons, though their lack of brachiating behavior contrasts strongly with the gibbons superb brachiating abilities. Indris have the dental formula 2-1-2-3/1-1-2-3.

FAMILY DAUBENTONIIDAE. This taxon has a single species, *Daubentonia madagascariensis*, known also as the aye-aye. It is an aberrant, nocturnal, cat-sized species that is extremely shy and not well known. It has rodentlike teeth and an extremely elongated, thin middle finger which it uses for spearing grubs in holes in tree bark. The genus has the unusual dental formula 1-0-1-3/1-0-0-3.

FAMILY LORISIDAE. The members of this family, the lorises, potto and galagos are relatively slow moving, nocturnal, arboreal prosimians. Pottos are found in Africa and resemble small teddy bears. They have shortened index fingers that produce a powerful grip. The lorises, from southeast Asia and Africa, are either very slender animals or resemble pottos. Both types move slowly by creeping along branches or climbing up and down trees. Both are omnivorous in captivity. The galagos or bush babies, known only in Africa and some offshore islands, are small animals with long tails. Their second digit is specialized for grooming. In contrast to the lorises and pottos, the galagos can leap rapidly for long distances through trees. They are also omnivorous. All members of the Lorisidae have the typical prosimian dental formula 2-1-3-3.

FAMILY TARSIIDAE. The tarsiers consist of a single living species, the arboreal *Tarsius spectrum*, which inhabits the bamboo forests on some islands off southeast Asia. These are very small and chunky animals with enormous round eyes on a flattened face. Clearly, eyesight is their dominant sense. The ears are also large and very efficient. The hind limbs are very long as is the tail, which serves as a brace in its normally upright posture. Their feet are elongated because of large tarsus bones and provide extremely effective leverage for leaping. The fingers also have long bones and are specialized with large disk pads for better grip. Tarsiers normally perch or cling upright to tree branches and trunks as demonstrated by their foramen magnum which is located more underneath the skull than in any other prosimians. These animals are not well known except for information derived from captive individuals. They have the dental formula 2-1-3-3/1-1-3-3.

ANTHROPOIDS

The anthropoids arose during the early Eocene Epoch, presumably from plesiadapid stock by modifications to the skull and dentition. This stock is divided into four groups, one of which became extinct soon after its appearance. The three surviving groups are the best known and most abundant of primates, occurring in great numbers and diversity as monkeys, gibbons, chimpanzees, gorillas and orangutans, and in immense numbers as humans which count about 5 billion individuals. Buettner-Janusch said (1965, p. 183)

> . . . the Anthropoidea. . . is the only group of Primates clearly distinct from all other mammals. This is so for the fossil and the living forms.

Anthropoids are characterized by stereoscopic vision, closed back walls to the orbits, well-developed clavicles, two pectoral breasts and fingernails rather than claws. Diet in most anthropoids is onmivorous though examples of vegetarian, insectivorous, frugivorous and carnivorous behaviors are known. Two stocks of anthropoids are the Old World monkeys and the New World monkeys, the names referring to the different continents they inhabit. They are the oldest of the anthropoids and gave rise to the higher primates. Though the two monkey groups evolved separately and independently in different parts of the world, they likely descended from a common though unidentified ancestor.

The Old World monkeys evolved first, possibly in late Eocene time and several genera are known from Oligocene deposits of Egypt. This group represents a tight evolutionary lineage in which all members have a 2-1-2-3 dental formula. The New World monkeys appeared during middle Oligocene time. Their ancestral stock had apparently spread into the Americas and there evolved slightly differently than the ancestral stock that remained in the Old World.

The third anthropoid stock, the hominoids, includes the apes and humans. The apes differ from the monkeys in very obvious characters and features, although the gibbon, a lesser ape, is monkeylike. The living and extinct humans, the hominids, are also different from monkeys and in more ways than the apes.

SUPERFAMILY CEBOIDEA. These are the New World monkeys, so called because they inhabit only the Americas. They are characterized by a wide nasal septum that separates the nostrils and identifies them as the **platyrrhini** (flat nose). They separated from the main anthropoid stock probably during Oligocene time and evolved within the great variety of available arboreal ecological niches in the thick forests of the Americas. Today more than one hundred species of New World monkeys range from North America south of Mexico City through Central America and across South America. They exhibit parallel evolution to a marked degree with the Old World monkeys, though today they have been separated by at least 25 to 30 million years of independent life. All Ceboids are arboreal, most are onmivorous, and all but one are diurnal. Only a few do not have a prehensile tail, and they are particularly distinguished by widely separated nostrils. In contrast, Old World monkeys may or may not have tails, always lack prehensile tail abilities, may live on the ground and always have closely set nostrils.

Ceboids are divided into two families, Cebidae and Callithricidae (marmosets). The Cebidae have the dental formula 2-1-3-3; the Callithricidae have 2-1-3-2. Ceboids are a highly diverse collection of genera and species that range in size from very small to about three feet in length, exclusive of tail. They vary considerably in their feeding habits, size of social groups, territoriality, ecological niches, geographic ranges and more.

SUPERFAMILY CERCOPITHECOIDEA. These are the Old World monkeys and include the macaques, baboons, guernons, colobus monkeys and langurs. Also known as the **catarrhini** (hanging nose), they are generally recognized by their closely spaced nostrils. They range throughout Africa and across southern Asia. A single remnant of a wild population in Europe is maintained by the British government at Gibraltar. The cercopithecines are both arboreal and terrestrial forms; most species are arboreal but the terrestrial species have the greatest numbers. Two distinct groups are recognized, the omnivorous Cercopithecinae and the vegetarian Colobinae. A number of the arboreal species inhabit savanna environments along the forest edges where they feed on the ground and return to the trees at night for protection. Among these are the colobus monkeys of Africa and the langurs that are so diverse throughout Asia. Other terrestrial species, the baboons (*Papio*) of Africa and macaques (*Macaca*) of Asia, are among the best known and well studied of all monkeys. Both have strong social structures with dominance behavior controlling much of their way of life. Both

Figure 3.4 Male baboon's threat-display "yawn."

the baboons and the macaques exhibit social traits that could serve as models for primate adaptations thought to be present in the ancestral stock that eventually led to humans.

The baboons are highly social, territorial ground animals, although they sleep in trees as protection against predators, especially leopards that commonly hunt at night. Baboon troops range up to more than one hundred individuals and are bonded primarily by sexuality, according to several different researchers. Males and females exhibit several sexually dimorphic traits including some that are clearly related to mating. Males possess much larger canine teeth than females and use these as weapons and tools, and particularly in **threat displays** (Fig. 3.4). The latter are behavioral postures that threaten use of these powerful weapons toward other individuals or foreigners. Though these threats are most commonly used within a troop to establish and maintain dominance and order, the threatening act is itself generally sufficient and rarely is there a need

31

for actual combat. These threats are also used against members of an alien troop that approach or invade the resident troop's territory. Threat displays are used against predators such as leopards, and may actually dissuade such animals from an attack. Social bonding is so strong within baboon troops that adult male baboons, weighing about ninety pounds, will not hesitate to attack a hunting leopard of greater size stalking the troop. In such an encounter the leopard does not always win, and it rarely does when more than two baboons cooperate in defense of the troop.

Social behavior in baboons, especially dominance, has been well studied by many researchers; some postulate it to be the system likely used by early hominids. This behavioral pattern underlies a social system in which a dominant male is supported by a group of strong, slightly less dominant males of slightly lesser rank, much as a human leader is supported by strong aides. Some scientists have recognized, in addition to this hierarchy, a group of older but less physically powerful males. These support the dominant males and may be likened to wise and experienced advisors. It is unclear whether dominant males impregnate more females and produce more heirs than nondominant ones but it would make sense. This is precisely the situation in many nonprimate mammals and supports the argument that dominance is good for a species, because the more dominant individuals tend to produce more offspring having their better genetic materials.

Dominance conveys several advantages to lesser individuals within a troop, especially in food, protection, aggression, sexual relations, order and perhaps, mating. At the same time, dominance also imposes responsibilities. Dominant males serve this responsibility well, and the troop prospers accordingly. Single dominant males are particular protective of the troop and will unhesitatingly attack marauding leopards. This is identified as altruistic self-sacrifice and has often been compared to similar behavior in humans. Hierarchial dominance structure, thus, provides stability, social peace and protection against predators for the benefit of the entire troop. If dominant males do impregnate more females than subservient males, their acts of self-sacrifice make sense because they are protecting their own genetic descendants which make up most of the troop.

SUPERFAMILY OREOPITHECIDEA. The single extinct genus *Oreopithecus* occupies this enigmatic form which was discovered in lignite deposits of Miocene age in

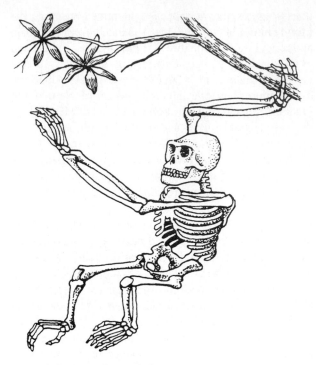

Figure 3.5 *Oreopithecus* skeleton.

northern Italy around 1870. More than one hundred specimens of this genus have been found since, and it is reasonably well characterized. It has been assigned various relationships to baboons, the great apes and to humans because of the inconclusive nature of some of its skeletal features. A nearly complete fossil skeleton of this genus (Fig. 3.5), the most complete nonhuman primate fossil, was found in 1958 in lignite beds in northern Italy where it had long been sought during the course of normal mining operations. Its possible relationship to hominids was emphasized at that time, and it was immediately dubbed the "Abominable Coal Man" by the news media. Its total skeletal features, however, show that it could be related to cercopithecoids and to hominoids, perhaps even to orangutans (Simons, 1972, p. 264). Accordingly, it has been given its own taxon equal in rank with the Hominoidea pending more definitive resolution of its status.

SUPERFAMILY HOMINOIDEA. The hominoids (manlike) are the primates that closely resemble humans and include the gorilla, chimpanzee, orangutan and gibbon, as well as living humans and a variety of extinct species known only through their fossils. One principal

Figure 3.6 Gibbon in brachiating mode.

distinguishing characteristic that separates this group from the other primates is the absence of a tail; none of the hominoids has one. Few people have difficulty recognizing the living apes, though the gibbon is probably the least known to the general public.

FAMILY HYLOBATIDAE. These are the lesser apes, the living gibbons and siamangs (*Hylobates* and *Symphalangus*), and several fossil genera, notably *Pliopithecus* and *Limnopithecus* of Miocene age. Gibbons (Fig. 3.6) and siamangs are the smallest living hominoids and inhabit the mainland of southeast Asia and many of the offshore islands. Like their larger cousins, the great apes or pongids, the lesser apes are characterized by **long canine teeth**, an **opposable great toe**, **lack of a chin** and **lack of a philtrum** (groove in the upper lip). Carpenter (1940) conducted pioneering field studies of gibbons and also did extensive early field work on different kinds of monkeys. He reported that gibbons live in small family groups typically consisting of an adult pair with one or two immature young. The young stay until they mature and then are driven out to start new families on their own.

Gibbons are gentle and normally unaggressive, but have a strong sense of territory and will fiercely defend it. For this purpose they use their efficient canine teeth and distinctive calls. Their vocalizations are powerful and consist of drawn-out hooting; they also seem to be useful in maintaining territorial spacing between families in the thick forest tops. These characteristic hoots can be heard up to one mile away in quiet air. Marshall and Marshall (1976, p. 235) report that an adult family pair will use such calls daily to broadcast to other gibbons various information including the extent of their territory, particular species, sex and offspring. Carpenter believed that territory was a dominating aspect of gibbons and other hominoids as well. He said,

> It would seem that possession and defense of territory which is found so widely among the vertebrates, including both the human and subhuman primates, may be a fundamental biologic need. Certain it is that this possession of territory motivates much primate behavior.

Gibbons are blessed with great skills in brachiating, the ability to swing or hang by arms. Their arms are extremely long and their hands have fingers that are elongated and curved inward which they use to hook onto branches rather than fully grasping them. Such structural adaptations allow them to swing easily and gracefully from branch to branch in virtually effortless fashion. Their looping curves through the trees can be so swift that watching their passage is difficult except at a distance. Specimens brachiating in zoos are truly spectacular to watch.

FAMILY PONGIDAE. The family Pongidae is a large taxon consisting of more than fifteen living and extinct species arranged in three subfamilies. Pongids are characterized by **long canine teeth**, **opposable great toes, no chin** and **no philtrum**. The family includes the ancestral genus, *Aegyptopithecus*, from the Fayum of Egypt that is believed by many to have been part of the ancestral stock that gave rise to the great apes and humans and to two other genera, *Propliopithecus* and *Oligopithecus*. The genus *Dryopithecus* with its six species, the living great apes, *Pongo*, *Pan* and *Gorilla*, and the large *Gigantopithecus*, complete the family.

Gorillas, chimps, and orangs have become fairly well known to science and the general public in recent years, largely because of the efforts of a few individuals and their assistants who study these animals in the wild. Jane Goodall has studied chimpanzees for more than nearly thirty years in Tanzania; George Schaller studied the orangutan in Sawarak and gorillas in the eastern Congo in the early 1960s; Dian Fossey studied the gorillas in Volcano Park, Rwanda and adjacent areas of Zaire and Uganda from 1967 until her murder in 1986; Biruté Galdikas has studied and rehabilitated orangutans in Borneo since 1971; and Carpenter studied gibbons in Thailand before World War II. All these workers reported in detail on the complex societies that these higher primates have developed, complete with examples of dominance, affection, frustration, gentle care, anger, cooperation and even murder.

Chimpanzees—The chimpanzee consists of two species, *Pan troglodytes* and *P. paniscus* (the pygmy chimpanzee), though the latter is not considered by some biologists sufficiently different to merit a separate species. Important data of chimpanzees provided over the past three decades by Jane Goodall and her colleagues and students from their observations in Gombe National Park, Tanzania have developed a much clearer picture of this primate, our nearest relative in the animal world. Though their structural features clearly separate us, they are a generalized model of arboreal types not very far evolutionarily from the ancestral stock or from us. In addition, they exhibit behaviors that are similar to many human manners; they are also nearly identical to us biochemically. These multiple aspects, structural features, behaviors and biochemistry support their close relationship to us, and the differences reflect the paths that we both have taken in diverging from a common ancestor. The chimpanzees remained conservative much like this common hominoid ancestor, whereas the hominid stock underwent changes and evolved into a more daring lineage.

The details of chimpanzee anatomy show many structural features that are superbly adapted to an arboreal way of life. Research has shown, however, that chimpanzees will spend much time on the ground in play and feeding, given appropriate habitats with open ground space. They can and do exhibit bipedal behavior on occasion, but it is an incomplete motion and without ease. They cannot stand fully erect because of their leg structure and other anatomical features, but they are able to use a bent-leg gait for a short distance at a slow pace. Nonetheless, these semiterrestrial animals have not completely adapted to a terrestrial lifestyle and at night commonly return to the trees where they build nests for sleeping.

Chimps are very active, noisy and exceptionally curious animals that live principally in forest habitats though some groups occupy savanna-woodland habitats at the edge of the forests. They are organized in bands of up to sixty individuals that usually split up into smaller groups of three to sixteen that fluctuate in size and membership as they roam at will throughout their home territory seeking food. They will defend this territory with vocalizations, threat gestures and physical force against other bands and even relatives. Goodall reported the ultimate aggression, murder, against former band members who left a parent group to settle into a new, adjacent home range.

Their societies are relatively complex, and they are capable of social behaviors that range from altruism to murder. They employ an open system of dominance hierarchy in which the dominant males exert their positions over all other less dominant members, generally in food, feeding position or some other way. Mating is not dependent on dominance but is probably related to it. Being related to, or friendly with, the dominant male has its advantages. Touching and grooming behaviors are important parts of their social life. They make tools that are unsophisticated compared to human-made tools but can use them for easily recognizable functions, just as humans do. Chimpanzees are largely vegetarian but have been reported to opportunistically hunt and kill for food some twenty species of mammals, including monkeys, young gazelles and others, along with birds and eggs. They are even known to attack, kill and eat infant chimpanzees, though their motives are unclear.

Gorillas—The gorilla (*Gorilla gorilla*) is the largest of the primates (Fig. 3.7); males weigh up to 450 pounds and females commonly are half that weight or less. Extremely shy, they are considered gentle by those who have studied them. They are terrestrial animals but often feed in trees and build nests there each night. Their movement on the ground is quadrupedal or tripedal using knuckle walking and they flee danger swiftly by moving on the ground, not in the trees. They are herbivorous and have never been observed to eat meat. Extremely powerful, they are able to tear apart almost any type of vegetation with their hands and their large canine teeth. These canine teeth are used effectively for threat displays, and they often charge swiftly with open jaws and an accompaniment of loud roars to emphasize their potential. Few animals will stand and face at close range such a roaring, gape-jawed giant in the

Figure 3.7 Dian Fossey with a gorilla.

normally thick underbrush of their habitats. Probably the only wild predator that normally enters these habitats is the leopard, but it is unknown whether or not they prey on gorillas. At present, therefore, humans are the only enemy of the gorillas.

Gorillas live in groups of up to twenty-five individuals led by a dominant silverback. Silverbacks are adult males whose back fur has turned from black to gray or silver as they reach full maturity. More than one silverback may inhabit a group, and their ranking may be determined by size. Silverbacks are dominant to all other males and females who, in turn, are dominant to the immature individuals. The dominant male leads the group to feeding areas, decides when to break for an afternoon nap, when to camp for the night, etc. The research by Schaller and by Fossey has provided remarkable and fascinating new information about their group and individual behavior, yet we still do not know enough.

Today gorillas are an endangered species. They are threatened by loss of their rain forest habitats that have been denuded by loggers and farmers who utilize the land for agriculture. This has created crisis situations that jeopardize the survival of gorillas in the wild. Sanctuaries have been created to protect some of the remaining habitats and the animals, but poachers have infiltrated these areas and killed gorillas despite penalties. A further complication in their survival is the seeming lack of sexual drive in the males noted in some zoos. They do not seem to actively attempt to regularly mate, either in the wild or in captivity. Perhaps their gentleness has become too influential and they may face serious decline in numbers through their reproductive behaviors. Fortunately, females tend to be good mothers provided they had good role models when young. Some zoos and wildlife parks have had excellent success in getting the gorillas to reproduce and successfully raise offspring. Observations of females with infants, both in captivity and in the wild, lead to the distinct impression that they are relatively little different from human mothers in giving extended care and attention to their offspring. Given reasonable habitat protection and security from poaching, these great animals may yet survive for many more years.

Figure 3.8 Female orangutan with young.

Orangutans—These red apes (*Pongo pygmaeus*) (Fig. 3.8) are in great danger of extinction in their native ranges of Borneo and Sumatra. At present, probably less than ten thousand are left in the wild, and their forest habitat, too, is rapidly being cut down for lumber and farming space. These apes are essentially solitary animals with contact between adults only at mating time. Almost totally arboreal, they rarely come down to the ground. On the ground they may use a bipedal gait, but it is awkward and ungainly. Their behaviors are not well known, primarily because of their solitary nature, the size of the range they occupy, the thickness of the forest habitat, and the fact that they live in trees high above the ground where they are difficult to observe and follow. However, all observations lead researchers to conclude that these animals, too, can be protected and survive if adequate measures are taken.

FAMILY HOMINIDAE. These are the living and fossil humans and are discussed in Chapters 7–12.

PRIMATE LIFESTYLES

The arboreal lifestyles of early primates played a major role in shaping their evolutionary development and that of their descendants in terms of structural adaptations and behavior patterns. Though arboreal behavior is a minor way of life among mammals, it is pervasive throughout the primates, almost exclusively so. Only humans have totally eliminated a tree-dwelling behavior, though a number of primates have become markedly terrestrial. Arboreal behavior, then, must be considered among those primate adaptations that ultimately led to the ground-dwelling forms and modern humans. Exactly how four-legged animals evolved into the arboreal primates and how these arboreal primates eventually gave rise to terrestrial humans is unknown, though several models have been suggested to explain how this giant step took place. Even these models tend to be only good guesses because of the lack of firm evidence. However, many factors that influenced the animals in this evolutionary pathway to humans have been proposed, including some that are very convincing.

This critical pathway was apparently followed only once by our primate ancestors several million years ago. The other primates failed to find this narrow trail and remained conservative animals living almost entirely in trees. Those nonhuman primates that have adopted a terrestrial lifestyle—baboons, macaques, gorillas—do so as a mixed adaptation; they commonly return to trees at night and for safety. Their changes of lifestyles were significant factors in producing particular adaptations of structure and behavior, and these adaptations led to the evolution of humans.

An arboreal lifestyle presents much greater difficulties for survival than typical mammalian life on the ground.

Accordingly, arboreal animals are more efficient in some ways than terrestrial animals, because they learned how to surmount the inherent difficulties for life above the ground. Tree dwellers usually carry out all their various activities, feeding, sleeping, resting, playing and reproducing, on narrow branches in leafy, three-dimensional tree habitats. They constantly run the risk of falling from moving and bending branches. To reduce this risk, these primates evolved certain modifications to the structures and behaviors of their originally terrestrial body pattern. In addition, their brain size and abilities increased so that they could successfully and efficiently operate these new structural modifications and behaviors. In all probability, these developed together by mutual feedback.

Arboreal habitats present more than simple difficulties of mobility in the trees. Visibility is restricted in the leafy treetops, and enemies and predators may be hidden from view until dangerously close at hand. Faced with such situations, primates had to become extremely wary and alert to ever possible dangers. Their only escape from such dangers was along branches and through the trees or downward to the ground. These primates had to become keen of eye, agile, swift and smarter to survive in environments where a moment of hesitation could spell doom. Those that could not meet these requirements did not survive for long, and their descendants with similar gene pools were also fated to short lives and eventual extinction. Though life in the trees was very demanding for these primates, the advantages gained through arboreal adaptations were significant in advancing them beyond the abilities of ground-dwelling mammals. These adaptations and advantages were the rewards that were ultimately reaped by their improved descendants, the new species that evolved into humans. Most of these advanced primate species, especially the apes and early men, added their own improved adaptations to provide for specific demands of their ecological niches.

PRIMATE SIZE

The ability of animals to move easily through trees is facilitated by small size. The early primates were typically quite small and are thought to have been similar to living tree shrews (see Fig. 3.2); their descendants tended to retain similar dimensions and only a few of the living arboreal forms are as large as a medium-sized dog. One very unusual fossil form was *Megaladapis*, a large, calf-sized lemuroid from Madagascar that became extinct very recently, perhaps within the last 1,000 years. Its skeletal anatomy suggests that it was not only arboreal but also probably a slow climber.

That such a large and probably slow-moving lemuroid primate evolved and lived until recently in Madagascar is a measure of that island's isolation, lack of predators and lack of competition from the more capable monkeys and apes which had not yet evolved or emigrated there. Large arboreal primates are unusual, because they cannot usually move their greater weight easily along slender branches. They also have much more difficulty keeping their size and weight in balance as they traverse branches that bend and sway; they are commonly restricted to the larger branches. To alleviate these difficulties, the early primates remained small, and evolved structures, features and behaviors to facilitate life in this specialized environment. Gorillas, orangutans, and chimpanzees are the living forms that violate these rules of size, but the gorilla is essentially a ground dweller, and the chimpanzee will utilize ground-dwelling behavior where possible. Only the orangutan spends virtually all its time in trees, probably because of the ease in moving above the ground through its thick forest habitats. In addition, the preferred foods of orangutans are found high in the trees. Gorillas climb trees easily for food, but they generally prefer to remain on the ground.

PRIMATE GRASPING HANDS AND FEET

The ancestral mammalian pattern was and is ideally suited for locomotion on the ground. It was based on an elongated body with a horizontal spine supported by four limbs placed directly underneath the body. Front and hind limbs became modified into paws or hooves with appropriate reductions of digits which provided direct contact with the ground. Forest animals, in particular, tended to utilize multiple digits to assist in quick turns as they ran in the underbrush. Grassland animals, however, found less need for sharp turns in underbrush but a great demand for speed. In most of these savanna animals, such multidigit "feet" tended to be less efficient, and these types gradually lost some of their digits and concentrated them into single or split hooves. Only the tapirs and rhinoceros of living mammals have three toes, other mammals have one, two, four or five.

Figure 3.9 Young baboon showing opposable thumbs and toes.

posable grip in both hands and feet to allow the firm grasping of branches, instead of simply walking and balancing atop them or clinging to them with claws.

These hand and foot adaptations required extensive modification of bone arrangements in the existing paws to provide for the opposition of the digits necessary for grasping hands and feet. Their thumbs and large toes rotated from a position parallel to the other digits to where they could bend toward and oppose the other digits, thus producing positive and powerful grip. This major modification, opposable thumb and toe, allowed these animals to hold tightly to any branch, limb or other object (Fig. 3.9). At the same time, these primates developed the muscles, nerves and brain ability to manipulate these hands and feet at will.

An added advantage of this grasping capability, perhaps a by-product, perhaps the impetus, was the ability to pick up food with the hands or feet instead of with the jaw. These modified paw structures allowed the primates to cling safely with one hand or foot, maintain their balance, and still be able to nonchalantly reach out and pick food with a free hand or foot. This ability to pick and grasp encouraged them to adopt an erect or upright posture in the trees. Such vertical posture with grasping hands and feet is much more efficient for arboreal activities than a four-footed walking one.

VERTICAL-CLIMBING-LEAPING ABILITY

As these arboreal primates became more vertical in their posture, they were better able to climb and leap through the trees. These aspects—vertical posture, climbing and leaping, the so-called vertical-climbing-leaping (VCL) ability—became the trademarks of arboreal primates and is a highly developed characteristic of the living forms, including those that are largely terrestrial. The terrestrial primates, excepting humans, however, have not completely lost their arboreal pattern, and they retain many of the arboreal structures and traits. Humans have emphasized the vertical posture even more than the other primates; they are the only primate so totally terrestrial that climbing and leaping behaviors are insignificant compared to vertical, bipedal walking abilities.

Arboreal primates also developed long, specialized bones and musculature in their legs and pelvic bones to accommodate their climbing and leaping behaviors. These provided the strength and mechanical leverage for leaping and climbing. Their feet and hands became flattened to

Horses carried this reduction of digits to an extreme by ultimately losing all their digits, save their index finger; they have retained two digits in the form of splints or cannon bones that flank their lowermost legs. Horses effectively stand on the tip of their index finger atop a once useful claw that is enlarged and modified into a hoof. Such reduction in digits, along with lengthened legs, increased their strength by providing additional leverage. This leverage ultimately translated into speed for escape.

The typical mammalian pattern of paws and hooves, however, was basically unsuitable in trees without the assistance of claws. Many nonprimate arboreal mammals use claws to assist them in the trees; squirrels, raccoons, felines and others are quite effective climbers. The tree shrews still retain this primitive condition of claws and live very successfully in trees. They are, however, very small and do not have to manage and support significant body weight. In contrast, the other primates developed the op-

allow for better grip on branch surfaces. Claws, which characterize so many mammals, were less suitable in their grasping hands and feet, and they became flattened into finger nails which provided broad support for the fleshy ends of the digits to assist the grasping ability. Even so, humans have emphasized and advanced the holding and grasping abilities beyond those of these primate ancestors.

The VCL ability (Fig. 3.10) adopted by primates resulted in a vertically oriented spine that developed from the typical horizontal spinal column of their terrestrial ancestors. Animals with horizontal spinal columns have their heads occupying a forward position at the anterior end of the body. This places the opening for the spinal cord (**foramen magnum**) at the rear of the skull where it connects directly to the brain. The face with mouth, nose and eyes is situated on the forward end of the skull opposite the foramen magnum. In living primates that fully exploit the VCL posture, especially hominids, the foramen magnum is located directly underneath the skull which thus orients it directly atop a vertical spine. The shift of this opening from the rear of the skull to underneath can be traced in varying degrees in species of primates and effectively records the evolutionary advance from early primates to humans. Tree shrews have their foramen magnum sited at the rear of their skulls, because these animals use a four-legged gait and have a horizontal spine. Lemuroid primates generally have their foramen magnum placed more underneath the skull than behind it to accomodate their degree of VCL posture.

As the anthropoid primates evolved, the different species adopted more vertical posture, and several accompanying structural modifications took place concurrently. Their hands became fine grasping instruments capable of reaching out and picking foods from all around them, and erect posture gave them greater reach. Standing upright also enabled them to move easily through trees, but to satisfy the demands of this new posture, they had to keep their heads upright. Their foramen magnum continued to shift from behind the skull to a position underneath the skull thus placing the head directly atop the upright or vertical spinal column.

This orientation is prominently developed in fossil primate skulls. Both the foramen magnum and the **occipital condyles** demonstrate this upright posture (see Fig. 4.2). The occipital condyles are bony projections that flank the foramen magnum at the base of the skull. They articulate

Figure 3.10 Prosimian (potto) in a VCL posture.

with the vertebral column to allow the skull to pivot. The position of a foramen magnum and occipital condyles in a skull thus identifies an animal's posture.

When these skulls were reorienting on the vertical spinal columns because of posture changes, the faces shifted to retain their forward position. The eyes moved from the sides of the head to the front where they provided overlapping or stereoscopic vision. This eye shift, with accompanying optic nerve modifications, provided sight with depth perception, a significant advantage to any creature climbing and leaping through trees from branch to branch to escape predators. Obviously, poor depth perception in arboreal animals was a serious handicap and could have led to misjudgment and a potentially fatal fall. Early primate eyes also developed another adaptation; they began to employ color vision. Experiments with many living primates indicate that the early primates very likely could also distinguish colors. It is unknown, however, whether living primates possess the ability to perceive colors to the same degree as humans. However, the acquisition of any degree of color perception significantly increased the early primate's ability to recognize objects in greater detail. These two visual aspects then, depth perception and color vision, clearly have conferred great advantages for fitness in arboreal habitats.

Such specialized adaptations conferred survival benefits on those individuals that developed these changes, however slight, in their structures, organs and behaviors. Wilson (1975, p. 23) pointed out that the cause of changes in species

> ... consists of the necessities created by the environment: the pressures imposed by weather, predators, and other stressors, and such opportunities as are presented by unfilled living space, new food sources, and accessible mates. The species responds to environmental exigencies by genetic evolution through natural selection, inadvertently shaping the anatomy, physiology, and behavior of the individual organisms.

Once a selected behavioral or structural advantage appeared in an individual primate or other animal, it was likely to be selected again and again in the offspring. Those individuals that lacked such advantages were not as efficient at survival and were more likely to die before the advantaged ones, thereby reducing the percentage of their lesser-fit genes in the gene pool. Though such advantages at first conferred only slight benefits for survival on the recipients, they would have accumulated rapidly in the populations where selective pressures of the environment favored individuals with advantageous traits. These populations would have gradually shifted to those with traits producing greater fitness, and in the primates, these advantages and features led toward humans.

THE PRIMATE BRAIN

All of the structural, functional and behavioral improvements that took place in the evolving primates were accompanied by related developments in the brain, especially in those portions that provided for the necessary sensory and motor function responses and control. A significant portion of the enlarged cerebral cortex of the primate brain is devoted to the functions of the grasping hand, a multipurpose instrument without equal in the world of machines. It is extremely versatile, sensitive and powerful, and primates would not be so successful without it. Significant advances also took place in those portions of the brain devoted to visual reception and memory as these areas increased in size and capability. Primates simply had to learn more to survive, and to learn more they had to know more and retain this knowledge. To accomplish this learning they also had to be able to see things, recognize them, put them into perspective in various situations and decide on the best courses of action. The improvements to their eye systems involving stereoscopic and color vision were interconnected with the ability to differentiate food from nonfood, and to recognize friends, foes, individuals from rival troops and potential mates. All of these powers, which are functions of the brain, had to expand and improve to accomplish these necessary decisions.

The ability to identify friends or enemies, for example, is essential to all animals for survival, but arboreal primates had to make such identifications instantly and correctly in order to survive in the trees. An animal might appear suddenly before a primate through the leafy curtains of a treetop at any moment. The primate had to decide immediately if the new animal was friend or foe. If an enemy, what kind of danger did it represent? Was it a slow-moving snake or a swift leopard? What was the best way to escape—perhaps turn and flee to another branch, go up or down or move across through the trees? Perhaps the new animal was not a predator but another primate. Was it a primate of a different species or the same species, possibly a member of a rival band from an adjacent territory? An individual from a rival group could be an enemy if their societal structures and territorial behaviors were so developed and organized. These questions became routine parts of their daily life, and individual survival depended on the correct answers and decisions.

These kinds of quick decisions by individuals were of value to the group, because each individual had responsibilities for the safety of the group. Whatever affected the individual could have great impact on the group, its own genetic kin. In times of danger from a predator could an individual alert the troop and still save itself? Could it alert the other animals to the source of danger, perhaps even the type and direction of danger? Could it make some noise or gesture, a visual signal or vocal one that would provide a message? Obviously, various danger warnings were ultimately evolved and used, because they are commonplace in primate life today. But all decision making did not involve danger. Perhaps the animal that suddenly appeared through the trees was not an enemy but a relative, a friend, or simply a member of the same troop, and there would be no real danger. Perhaps the friendly primate was the opposite sex and if so, might he or she be receptive to mating? All these decisions and judgments had to be made quickly within the confines of a small brain. These primates obviously learned to make such split-second decisions, because they are now routine behavior for living primates.

Primates learned to make vital, quick decisions routinely in their everyday life and, as a result, their brains and eyesight improved markedly. The decisions needed for survival required increases in brain power, which implies both brain size and brain quality. Improvements in their optical sensory systems required more enhanced visual receptors and optic nerves to send the signals to sites in the brain where all these signals were coordinated; the brain improved accordingly. As their visual abilities improved and the corresponding portions of their brains enlarged, they relied less frequently on their olfactory senses. These senses then declined in relative importance and, with less demand, corresponding areas of the brain reduced in size. Modern prosimians, for example, rely to a great extent on their visual senses and have smaller olfactory lobe areas in the brain than most other mammals, especially those who rely heavily on olfactory senses. Simply put, primates are more dependent on their eyesight than their sense of smell.

We humans use eyesight primarily for our information gathering and rely little on our ability to detect aromas. Fragrances still play a useful and important part in our lives, much of it with subtleties not well known. But our olfactory senses are not particularly critical for our everyday survival. We are not usually at great risk from predators and most of our olfactory senses have become reduced in capability through time. The anthropoids, as advanced primates, have smaller olfactory receptor areas than the prosimians, as well as smaller olfactory lobes in their brains.

The prosimian dependence on the sense of smell lessened as reliance on the other senses increased. The snout, where the olfactory receptors are located, became shortened as these receptors declined in importance and size. With the lessened need for the snout or jaws for picking up objects, primate faces became more shortened and, ultimately, flat. This condition is evident in all the advanced primates, which show considerable reduction in snout condition. In several forms, particularly humans, this condition resulted in a flat face.

The many decisions made by primates for survival in the trees taxed their brains to keep up with such demands. Having a larger brain was advantageous because it had more cells to allow for more memory and quicker decisions. Those individuals that randomly developed larger or better brains gained distinct advantages. They had more memory, learned more easily, made more complex decisions and were capable of more sophisticated behavior; they were smarter and better. Advanced primates were more able to survive because they had larger brains with better memories that gave them an enhanced ability to make better decisions; those that failed to make better decisions, or did not learn how to make them, were less likely to survive. The more intelligent ones tended to survive more than the less intelligent ones and were able to produce similar offspring that also survived and, in turn, produced intelligent or clever offspring. In time, with more to learn, remember and decide, such better brains became the rule rather than the exception. The primate way of life resulted in many obvious structural changes, but those aspects that were associated with the brain, especially the functional, behavioral and intellectual aspects, were of major importance. These developments, too, were eventually passed on to humans.

PRIMATE BEHAVIOR

The behaviors that we observe in primates and other animals are typically defined on the basis of human behaviors; we measure their behaviors by ours. For many years such animal behaviors were considered unrelated to human behaviors because of the generally accepted belief that animals could not be compared with humans, who were unique and possessed special abilities, especially behaviors. Many scientists, however, now accept that some human behaviors are genetically influenced and inheritable. Primate behavior has been extensively studied in recent years by several workers. C. R. Carpenter, Adrian Kortlandt, Jane Goodall, Dian Fossey, George Schaller and Biruté Galdikas have studied gorilla, chimpanzee, orangutan, gibbon, and monkey behaviors in the field over the past fifty years, and markedly so in the past three decades. They report remarkable activities that resemble and seem to parallel humans behaviors leading to the view that we inherited these from our primate ancestors.

Behavioral scientists and others have recorded and filmed higher primates displaying what are considered to be clear examples of jealously, frustration, temper tantrums, anger, love, affection, amusement, play, thoughtfulness, patience and murder. Our previous failure to accept these behaviors and emotions as possible and normal in nonhuman primates reflects an a priori belief that we are different from other animals who cannot possibly be capable of having such "advanced" feelings and behaviors as humans. Goodall's expressions of amazement at having witnessed the systematic "genocide" of one chimpanzee group by the males of a rival group provided ample evidence of what the "highest" primate considers to be normal behaviors in these "lesser" primates. Today many scientists have expressed the view that such emotions, behaviors and activities found in humans are to be expected in nonhuman primates. They add that these are inherent in humans where they represent an evolutionary heritage that was transmitted genetically. In effect, these experts maintain that many human behaviors originated in our primate ancestors who transmitted these to successive generations and descendant species, eventually to us. Some of our behaviors obviously were learned, others were probably inherited.

Most primate species are highly social, that is, they live in organized groups for mutual support and benefit. Some of the prosimians are solitary but all of the monkeys are social. Of the higher primates only the orangutan follows a solitary type behavior; all others live in groups. An essential component of these primate societies is the close physical contact (**cuddling**) that they enjoy from the very start (Fig. 3.11). Infants require this familiarity in order to adequately develop their social position in the troop and to prepare them to be caring parents and adults. Numerous studies show that without this close contact, these animals mature into adulthood with a variety of problems and generally produce offspring that do not survive or that are misfits in the troop. Such close physical contact is continued in adulthood with the common practice of **grooming** (Fig. 3.12). This behavior ensures that individuals within the troop recognize and trust each other and has a component of pleasure and sanitation derived from the activity.

The other higher primate species have such recognizable highly organized social systems that it is difficult to imagine that these systems are not the same as those of humans, who have the most complex societies and social

Figure 3.11 Baboon mother cuddling offspring.

systems of all animals. Some researchers who have recognized parallel behaviors in apes and humans have characterized the nature of social structure and organizations (societies) of different animals as clear examples and evidence of evolution. The study of the biological basis of all social behavior is termed **sociobiology** and Wilson (1975, p. 32) stated that

> Social evolution is the outcome of the genetic response of populations to ecological pressure within the constraints imposed by phylogenetic inertia. Typically the adaptation defined by the pressure is narrow in extent. It may be the exploitation of a new kind of food, or the fuller use of an old one, superior competitive ability against perhaps one formidable species, a stronger defense against a particularly effective predator, the ability to penetrate a new, difficult habitat, and so on.

To understand how some of the human behaviors relate to those of our anthropoid and ape cousins, we can

Figure 3.12 Grooming activity in baboons.

examine and characterize the behaviors of living anthropoids and apes. For convenience, we should probably examine those primates whose way of life is most similar to our own. These are primarily the terrestrial or near-terrestrial forms, especially the savanna baboons, chimpanzees and gorillas. The reason for this restriction is simple; it would make little sense to study more distantly related forms, especially arboreal types like the orangutans, because we would expect them to have fewer behaviors in common. The more closely related forms are more likely to have behaviors that are similar, especially if they were the source of human behaviors and evolution.

TERRITORIALITY

Territoriality is an important behavior pattern of most higher animals. Wilson (1975, p. 256) states,

> . . . nearly all vertebrates and a large number of the behaviorally most advanced invertebrates, conduct their lives according to precise rules of land tenure, spacing, and dominance. These

rules mediate the struggle for competitive superiority and may be considered enabling devices or behaviors that raise personal or inclusive genetic fitness.

He defines territory (p. 261) as

> . . . an area occupied more or less exclusively by animals or groups of animals by means of repulsion through overt aggression or advertisement.

Ardrey (1966, p. 3) is more blunt about the concept and states that

> A territory is an area of space. . . which an animal or group of animals defends as an exclusive preserve.

The concept of territory has been recognized by countless animal behavior experts who generally agree that the territory must provide some kind of resource that is worth defending. It is usually defended against competing members of the same species or against different species that are similar and use the same resources. At specific times of the year, territory may be defended almost blindly against any intruder, however innocent of base intent. This concept permeates human societies where it provides the basic premise for nations or countries, complete with boundaries that must not be violated. Unfortunately, even racial, religious and other groups claim rights to otherwise undefinable circumstances or turfs. To these areas and boundaries must be added the intangible rights to trade and deal with other countries or entities. To protect our territories and rights, humans establish diplomatic relationships. We also raise armed forces and are not reluctant to engage in war to preserve these rights and territories.

This concept is equally valid for various animals and animal societies. If the animal species is organized into social groups with unique territories, each group will defend its territory with aggressive behavior against intrusion by other members of the same species. If the animals are essentially solitary or are paired, they may also aggressively defend their territory. In bird societies, territories serve different functions depending upon the species. Bird territories provide shelter, space for courtship, mating, nesting, and food gathering. Territories might be fixed, or they might be transitory ones occupied by animals that normally range over broad areas in search of food. In the birds and some other animals, males establish territories that are almost exclusively used for breeding. In fact, the males of many bird species establish and defend such territories before any females arrive. To advertise these territories, male birds commonly sing and are often brightly colored for easy recognition. Females do not sing, so as to hide their presence, and are typically drab colored to blend into the foliage where they can raise their young without attracting attention, especially from predators.

DOMINANCE AND AGGRESSION

The purpose of territorial defense is obviously to restrict exploitation of an area's resources by others to reserve it for one's mate, offspring or group and to provide some type of sheltering range. In this way, animals can exploit the available food for themselves and their offspring, and ensure the successful maturation of their young and the continuation of their genetic line. To establish this space and security they may engage in some form of aggressive activity that establishes dominance against another group or another individual, whether by direct aggression, a threat display or some other behavior. It may

well be that such aggressive behaviors evolved initially as protective mechanisms to ensure survival against predators. A prey animal, for example, when cornered by a predator of another species and fighting for its life can present such aggressive defense that the predator might be deterred. The continuations of such aggressive defensive actions against others of the same species may actually serve the same general purpose, but with slightly different goals. Though a competing member of the same species is probably not interested in the resident animal as food, it may be interested in acquiring its territory for its own needs, especially as they relate to survival and reproduction.

Most workers who study animal behavior agree that defensive activities of animals are generally much more forceful than acts by invaders. Such behavior can become a group activity if local inhabitants combine their defensive behaviors against invaders. Though such defenses might be uncoordinated initially, success with their combined actions eventually could become typical group responses to intrusions as they learn to recognize the benefits of combined acts. In time, they would come to depend more and more on group responses for protection and security. Bernstein and Gordon (1974, p. 307) state,

> Concert group aggression against an outside threat thus may serve to preserve the group, and any mechanisms making the expression of such directed aggression more effective would have positive value. In this case at least, it is the expression of outward directed aggression, rather than its inhibition which unites and preserves the society.

Aggressive behavior by individuals against outside threats may thus have been very important initially in the development of social organization in primates as well as other animals. Such cooperation provides distinct benefits in many situations and some workers believe it is important enough to be the principal driving force behind human evolution.

Fighting between individuals of the same primate group to establish dominance would be counterproductive if serious harm befell either individual or if it reduced the well-being of the group. Instead of providing for better conditions, aggression could result in disintegration of the society with eventual loss of group actions and benefits against outside threats. To reduce the possibilities of harm or fatalities in such encounters, animals have evolved **inhibitory mechanisms**. These consist of two distinct aspects, the dominant behavior of one individual over another and the submission by the second, the acceptance of a subservient role. Together, these provide the basis for dominance structure by reducing bloodshed, ensuring better access to food, mating or some other advantage, and still provide for survival and well-being of the individuals and the group. Those animals that are successful in these dominance behaviors will likely have their particular behavior patterns passed on to their offspring. Those that fail to develop inhibitory behaviors may cause their genes to become predominant in the group's gene pool. Fighting to establish dominance that results in bloodshed, and perhaps fatalities, will usually weaken such a group and reduce the ability of the dominated individuals to survive against outside threats.

Dominance ranking is not as absolute in some species as in others and may vary somewhat with the situation. Some dominance orders are highly complex. Many of the anthropoids, the higher monkeys and apes, produce "committees" or "coalitions of peers" that assist the dominant male in ruling the group. The advantage of such dominance to the society is essentially its increased survivability. The advantages to the dominant individual are also great. Wilson (1975, p. 287) states,

> In the language of sociobiology, to dominate is
> to possess priority of access to the necessities
> of life and reproduction. With rare exceptions,
> the aggressively superior animal displaces the
> subordinate from food, from mates, and from
> nest sites.

Such actions would seem to give clear advantages to the dominant individual or individuals in spreading their genetic materials to the next generation. If the society constantly shifts in the direction of the dominant males, then the very characters that produce the dominant male will tend to become more common within the group by inheritance, and the society will undergo constant genetic improvement.

Aggression within primate societies serves important functions when it is directed at preserving order and providing for the well-being of the group. Jane Goodall reported aggression in chimpanzees that resulted in systematic murder by males from the parent group in an apparently senseless behavior. She observed a small group of expatriate chimpanzees that had left the parent-group and were murdered by males from the parent group. Though the reason for such acts was not established, she thought that the parent-group males sought to prevent the expatriates from settling nearby and setting up a rival territory. Whether the murdered chimpanzee expatriates were a real threat to the original group's territory or not, they may have been perceived as such by the killers. The territory and perhaps dominance status of the parent group was at potential risk from the newly established band, even though not directly invaded or threatened. They reacted to reduce this risk before it became an actual threat, and to accomplish this they simply attacked and killed the "threats." The parallels with human societies and activities are frighteningly close. Nations are not above engaging in "preemptive strikes" if perceived threats are recognized.

BABOON SOCIETY

Savanna baboons are highly social, terrestrial, territorial monkeys that live mostly in grassy areas with scattered trees, though they also inhabit forests and mixed environments. They spend most of their lives moving and feeding on the ground as a group during the day, though they sleep in trees at night as protection against predators. They are dependent on vegetation for their primary foods, especially high-energy seeds, seed pods, roots, blossoms and grasses, but they will eat insects picked opportunistically. They are also known to eat birds and small animals on occasion.

Baboon behaviors are so recognizably similar to human ones that some workers have identified them as models for the earliest type of human societies. Strum (1975) and Harding and Strum (1976) reported that olive baboons from Kenya have advanced sufficiently to become cooperative hunters. They noted that these hunters, observed over several years, gradually increased their predatory nature and the quantity of meat they consume with improvement in hunting skills (Fig. 3.13). They ascribed this increase to a decrease in the usual predators in the area by farmers protecting their livestock. The baboons, noting the decrease in predators and resulting increase in prey, have taken to eating some of these easily killed animals. Their behaviors show some similarities to those of very primitive peoples living a marginal existence as hunter-gatherers.

Figure 3.13 Male baboon eating a young gazelle after the kill.

SEXUAL DIMORPHISM

Baboon troops consist of up to one hundred or more individuals bonded primarily by sexuality, according to several researchers. Males and females are sexually dimorphic, that is, they are easily distinguished sexually by their features. Some of these features are clearly related to mating and other adaptations that may be due to their grassland environment where they are more vulnerable to predators. In these savannas, therefore, males are larger than females and possess much larger canine teeth that are useful as weapons. Their large body size and large canine teeth are important for their survival. Weiss and Mann (1985, p. 238) point out that as these canine teeth have enlarged, they have produced correspondingly larger roots to secure them in the jaw. This results in an enlarged face with larger muscles to operate the bigger jaw. The larger face needs larger and stronger neck muscles to balance the head. As canines enlarged, a variety of features also became larger, and thus helped distinguish males from females.

DOMINANCE

The baboons commonly use their large canine teeth for threat displays, those behavioral postures that threaten actions with these powerful weapons. The threats are often directed against other adult males as they establish their ranking within their **dominance hierarchy** (Fig. 3.14). Actual fighting between competing adult males to establish dominance does not often result in serious injury. Both combatants usually accept the situation realistically after sparring a few rounds. Mere threats are generally sufficient thereafter to maintain their dominance position until another male challenges the status quo. Dominance behavior among baboons provides an orderly social system and is very effective.

Threat behaviors by the dominant males are commonly used to establish order within a troop and are directed against others, adults or juveniles that create disturbances. There is usually little need for actual combat; the threat is adequate. Threats are also used against members of an alien baboon troop nearing or invading their territory. Threat displays may even be sufficient to scare some predators away from the troop. An adult male baboon weighs seventy-five pounds or more, has impressive canine teeth, two inches or more in length, and will not hesitate to defend the troop from a hunting leopard. But threats against leopards and other adult male baboons may not always be sufficient for the purpose, and actual combat can take place.

Figure 3.14 Aggressive behavior in a baboon troop.

The dominance behavior used by baboons is a key part of their society and must serve some vital purpose. It is likely in these sexually bonded animals that dominance confers some privileges related to mating. The dominant males seem to be more acceptable to females at times of ovulation than lesser males. The hamadryas baboons of Ethiopia live in harsher environments than the olive baboons of Kenya. They have smaller troops that consist of a few females and young led by a single male that tends to be much more aggressive in keeping his females away from other males. In both olive and hamadryas baboon societies, however, the dominant male fathers most of the offspring, which is good for the troop and the species. Dominance requires certain, often indeterminate qualities in the leader—generally strength, size, stamina, intelligence, and prowess in mating, among others. These qualities are good for the group and tend to be passed on by the dominant males to their offspring. If these qualities of dominance and leadership are the result of inheritance of an appropriately superior genome, then this type of social system enables the group to better survive. The leaders have developed from genes that confer some measure of superiority. Dominant males mate more frequently and produce more offspring than males with fewer leadership qualities. The troop improves in the direction of those leadership traits that are clearly advantageous for its survival.

As dominance conveys distinct advantages to the baboon troop leaders, especially in food and probably in mating, it also imposes responsibilities. Dominant males are charged with the defense of the troop, particularly of females and young when threatened by a predator. This defense will be taken to the point of suicidal self-sacrifice in apparent altruistic behavior if the need arises. By their mutually supporting actions, the hierarchial structure of these dominant males provides leadership, cohesion, order and stability throughout the troop, and protection against predators, all for the benefit of the entire baboon troop. This is true of all such animal societies that utilize dominance, not just baboons. Achieving a position of dominance or leadership in one of these groups is not an easy task or one that guarantees success, yet all baboons utilize this social plan.

CHIMPANZEE SOCIETY

Chimpanzee societies share several basic behaviors with baboons, including dominance systems and the hunting of meat on occasion. Chimpanzees are found generally in forests though they also live in the savanna-forest interface where they can take advantage of the better aspects of both environments. Goodall's studies of chimpanzees (*Pan troglodytes*) for over thirty years at Gombe National Park in Tanzania represent the greatest single observational episode and achievement in the study of nonhuman primates. Like Dian Fossey and Biruté Galdikas studying the gorillas and orangutans, Jane Goodall was recruited by Louis Leakey and began her studies at the Gombe Stream Chimpanzee Preserve in 1960. There she gradually developed the project to study approximately one hundred and fifty chimpanzees that had established territories in the over thirty square miles of heavily wooded valleys and ridges around Gombe Stream. Eventually she concentrated on about forty-five identifiable individuals in several groups of transitory composition in the ten-square-mile area along the Kakombe Valley. Her project over the years has attracted numerous researchers who helped contribute to its overall success.

All researchers at Gombe National Park recognized a strong social hierarchy in the different chimpanzee bands which were each led by a single adult male who exerted dominance over all other individuals, usually several females with their young of varying ages. This contrasted with the olive baboon society in which a number of dominant males supported the most dominant one. The social structure of the Gombe chimpanzee bands parallel that of hamadryas baboons whose small societies are composed of small family groups, each led by a single male. The hamadryas single-male groups are, however, probably small because they live in harsher habitats that would not support larger groups. Chimpanzees live in relative abundance in their forests and savanna-woodland environments; such areas can support large populations. Among chimpanzees the dominant male usually leads the group for a number of years until displaced by a younger male who better demonstrates the indefinable characters of dominance.

Researchers have carefully documented the complex relationships that exist between individuals in these Gombe chimpanzee bands and their behaviors, especially the making and use of simple tools, grooming, parental care, cooperation, murder, the hunting of small animals, eating of meat, communicating via gestures and much more. Many of these behaviors have been filmed and provide evidence that is available for all to examine. Some of these behaviors are not unexpected, but others are very surprising, especially murder, the use of gestures which are

recognizably similar, if not identical, to those of humans either by parallel or common origin, and the making and use of simple tools.

The identification of murder by chimpanzees (Goodall, 1979, p. 592) was perhaps her most shocking discovery. This action, long considered by many to be an exclusively human failing, has now been identified in nonhuman primates that are considered to be evolutionarily near to humans (chimpanzees) and in other non-primate animals. Though unexpected, murder by lions has also been observed by Schaller (1972) on the Serengeti Plain of East Africa. He reported a relatively high incidence (perhaps the rate was considered high only because it had not been previously considered) during his hours of observations. The rate of murders—number per observational time—vastly exceeds those of human murders by many factors. Most humans, in fact, have never observed such a horrifying event in their entire lifetimes of many tens of thousands of hours dealing with other humans. Among lions, however, a newly dominant male in a pride will kill the cubs of its predecessor. This will initiate sexual receptivity of the mother and allow the new pride leader to produce his own offspring.

GORILLA SOCIETY

Gorillas (*Gorilla gorilla*) are the largest and most peaceful of the hominoids. Two types inhabit different areas in Africa. The lowland gorilla inhabits the lowland forests of West Africa and the mountain gorillas live in the volcanic mountain forests along the borders of Rwanda, Uganda and Zaire. The two types probably represent remnants of a once more widespread population that ranged from east to west across central Africa, but which became separated as the climate modified. On the basis of destruction of habitat by farmers of today, it seems likely that the once wider range of gorillas was also reduced by influx of humans and their destruction of the gorilla's forest habitat. Certainly gorillas have little to fear from any other animal.

Gorillas are decidely terrestrial vegetarians and live on the ground most of their lives. Their great strength and agility enables them to climb into trees when there is some incentive, such as a desirable food. They generally nest in trees at night but always return to the ground in the morning for all their other activities. They commonly sprawl on the ground in relaxed fashion when resting at midday or in a resting-feeding mode. Their typical posture is upright when either sitting, squatting, knuckle walking or in active feeding. When alarmed they can quickly rise and are able to flee in great haste along the ground through the underbrush. As further evidence of their preference for a terrestrial lifestyle, it is noted that if alarmed while in trees they quickly return to the ground and flee to escape the source of danger.

Gorillas live in family groups of five to twenty-five individuals of varying ages dominated by a large adult male, a silverback, who is the absolute leader. He is easily identified by the distinctive color of his graying or silvering back fur and is typically the oldest male gorilla. The dominance structure affects all the members of the troop—adults, juveniles and infants. Social play is common among the juveniles or teenagers and often involves the older infants and the adults, including the silverback leader. Such play has the function of allowing the youngsters to learn relationships and attitudes, and develops their ranking in the dominance hierarchy. Females are normally protective and become capable mothers when raised with appropriate role models. Their care of infants is probably equivalent to that provided by human mothers. Other females in the group commonly behave as overzealous "aunts" who are eager to assist in raising an infant, a pattern recognized in many anthropoids, including chimpanzees and orangutans. Such behavior may frequently become excessive and will be dissuaded by the mother when an aunt shows too much interest in an infant.

Gorillas have been raised in captivity, though females must experience group rearing behavior to avoid serious problems when they become mothers. In recent years, zoos have significantly improved the quality of artificial habitats for gorillas, and their birth rate and success at rearing young have increased. Some zoos have specialized in one type of primate or another to such a degree that other zoos regularly send their specimens for breeding, birthing and rearing. Unfortunately, gorillas, especially gorillas (and orangutans) in the wild, do not seem to be increasing their numbers naturally. Human predation and the destruction of the forest habitat may be part of of the reason for this problem. Male gorillas in zoos also seem to suffer from infertility or lack of interest in mating. It may be that the gorilla is such a highly specialized animal that it cannot adapt to changing or different conditions, or that it requires virtually complete isolation, and any pressure from other animals or activities affects its reproductive drives. Such a slowly destructive process is not well known in the biological world.

Virtually all of the organisms that have ever lived, perhaps numbering in the hundreds of millions of species, are now extinct. This is a natural consequence of all life and such is the fate of all organisms. Environments are being modified instantly in geologic terms by humans all around the world. Gorillas simply may be following the path of all extinct animals, even though they may be assisted inadvertently by humans.

ORANGUTAN SOCIETY

The orangutan (*Pongo pygmaeus*) is the "old man of the forest" in the Malay language. They are rare animals, perhaps only ten thousand are left in the wild, and are found only in the forests on the islands of Borneo and Sumatra. Though some workers have reported them as living in small family groups, it seems clear now that all are solitary except at mating time and when females are raising their dependent young. These red-haired apes are in serious danger of extinction, in part because of illegal captures for various purposes, though strict conservation laws have reduced the illegal traffic somewhat. The greatest threat to their survival is loss of forest habitat. As a measure to help keep their numbers high, three centers were established in Malaysia and Indonesia during the 1970s to rehabilitate confiscated captive youngsters to life in the wild. At one such center, Biruté Galdikas was able to record exceptional details of their activities and behaviors. Films taken at one rehabilitation center show young orangs engaging in behaviors that could be seen in any human day-care center. The young orangs behave like typical young children, showing temper tantrums, petulance, frustration, greed, bullying, affectionate play, camaraderie, curiosity, perseverance and a host of other traits easily recognizable and previously thought to be unique to our own species. Galdikas also documented (1980, p. 830) an apparent case of murder by orangutans. These animals otherwise show little in the way of social structure that would lead us to believe that we are closely related to them.

SUMMARY

According to most scientifically accepted ideas, humans evolved from the primates, a diverse group of advanced mammals. Primates have captured our interests and our affections, perhaps because of our close relationship with them, perhaps because many are so ingenious, and perhaps because we can identify in them so many human qualities. Recognition of these and other aspects, especially when some are in danger of extinction, has led to important studies of their biology and their lifestyles. What we have learned about them has radically changed many previously established ideas. It has also added immeasurably to our understanding of human behavior and evolution.

Of all animals, humans have the most complex societies complete with intricate social relationships and behaviors. After the work of Goodall, Fossey, Galdikas and others, we now know that primates, too, have complex social organizations and behaviors that are similar to those of humans, even as many anatomical similarities between the two can be demonstrated. The premise that humans have descended from other primates by evolution is based in scientific fact and explanations. The demonstrated anatomical and biochemical relationships between humans and nonhuman primates cannot easily be explained in any other way. The development of anatomical and behavioral similarities between humans and the other primates argues for a very low probability for an independent origin. Conversely, explaining these similarities of our social organizations and behaviors along with our physical anatomy as products of evolutionary inheritance makes much sense.

Primates apparently descended from the insectivores. The most primitive living primate is the tree shrew, a type that has characters of both insectivores and primates. The combination of insectivore and primate affinities has led to their being classified by some workers as insectivores and by others as primates. Some species are arboreal and others live primarily on the ground. They are models of the earliest primates whose descendants today are found in great numbers but who remain essentially arboreal types; a very few species of primates utilize terrestrial behavior and only humans are totally terrestrial.

Arboreal lifestyles shaped the primates' anatomies and behaviors. Size, VCL ability and posture, brachiation, color vision, brain and opposable digits are all aspects of primates that relate directly to their arborealism. So, too, are many of their social behaviors. Life for primates, whether extinct or living, reflected these aspects which were undoubtedly enhanced by feedback from their arboreal habitats. Stereoscopic vision and hand-eye coordination are but two examples of modifications and abilities that enhanced their lives. At the same time their brains became much more efficient and larger organs; these primates simply became more intelligent. Humans inherited and exhibit most of these aspects, some to a greater or lesser degree than nonhuman primates. Some primates have descended from the trees to spend more time on the ground where they have found a greater variety of foodstuffs. Baboons and gorillas typically spend much time on the ground, though both are equally at ease in trees. Chimpanzees are primarily arboreal but also exhibit terrestrial behavior where the opportunity exists. Yet even today, baboons, gorillas and chimpanzees retain the opposable digits on their feet; they have not been down on the ground long enough to have evolved a walking foot.

In the study of primate behavior there is much support for the argument that humans evolved from ape ancestors. The human anatomy is very similar in many respects to that of living apes (see Chapter 4). Human behaviors, like our anatomy, reflect those of our nearest primate relatives. Chimpanzees, gorillas, baboons and many other primates live in social groups like humans. They establish dominance hierarchies and do not hesitate to use aggression to keep their rankings. Both chimpanzees and baboons have learned to hunt and eat meat, a practice undreamt of just a few short decades ago. All join in group defense, depending on their ranking and position in their societies. According to some workers, this cooperative ability is the key to why and how early humans were able to survive in environments where predators and scavengers possessed superior physical attributes. All these behaviors are well known in humans. Whether it was through cooperation, the bipedal posture, some other aspect, or a combination of features and behaviors, humans made the transition from apes to thinking and reasoning creatures.

It would be very satisfying to consider humans as separate and distinct from all other types of animals, a nonscientific practice readily accepted a few short decades ago. Certainly this would satisfy many individuals with particular viewpoints. The evidence, however, does not support such an absolute conclusion. Many of the very aspects that make us human have been recognized in other primates, particularly the great apes. Tool use and tool-making, self-awareness, murder and much more have been identified by highly qualified researchers. We humans, then, are the beneficiaries of evolutionary changes that began several million years ago in a primate stock.

STUDY QUESTIONS

1. Explain why stereoscopic vision is important in arboreal animals.
2. Why are opposable digits ideal for arboreal lifestyles?
3. What modifications were necessary for the early primates to develop opposable digits on their hands and feet?
4. Why is hand-eye coordination a key aspect of primates?
5. Why did the primates develop better brains than other animals?
6. VCL posture is a universal aspect of primate lifestyles. Why should this be so?
7. What is the purpose of territory?
8. What is the significance of hunting in baboon societies? In chimpanzee societies?

9. What is brachiation?
10. List four characters that separate primates from other mammals.
11. Distinguish between Old World and New World monkeys.
12. What is the function of a dominance hierarchy?
13. Offer an hypothesis to explain why orangutans do not live in adult groups.
14. Name three scientists whose work on ape behavior has changed many of our preconceptions about these primates and their evolutionary relationship with humans. Name their respective subjects of study.

4 Humans and Apes Compared

INTRODUCTION

We have a great affection for the primates. It may be no accident that most humans, especially children, are more attracted to monkeys and apes than to snakes and spiders. Perhaps we are simply acknowledging a kinship with these animals that resemble us in so many ways. Dogs, for example, are social animals with traits that we consider admirable—loyalty, affection, and generosity. Dogs began to work with humans in cooperative hunting many centuries ago. They were able to easily adapt to this relationship because it paralleled their own cooperative social structure. Humans and dogs gained from this mutual relationship; both worked much better together than either did alone, and both were able to be more successful in hunting. Few of us today keep dogs solely to hunt food—instead we retain them as pets and keep alive the bonds of a long-standing mutual friendship. Cats, on the other hand, are generally solitary hunters and are less clearly friends of ours. It is uncertain to many people whether we keep cats as pets or they keep us; many people love cats, many others dislike them.

We do not always have good reasons for liking particular animals. Deer and cattle are among the many gentle animals that have large, expressive eyes that appeal to us. This is true, too, of koalas and pandas, very popular animals in zoos. Such animals are thought of as cuddly and are often given as stuffed animal toys to infants and children to hold, drag around and even take to sleep. There is little inherent fear of such animals among humans, though some argue that we have conditioned children to accept these kinds of animals and encouraged their distrust of other types. These other, distrustful types are animals farther removed from us physically—grubs, worms, snakes and spiders—animals not commonly kept as pets. Many people are revolted by the prospect of baiting a fishing hook with a worm, and many others would never consider holding a snake. Such, then, is the affinity of humans for other animals. Those animals nearer to us physically and behaviorally are attractive, those farther removed are usually not attractive but are commonly repulsive. Primates are particularly attractive, but that may be my own bias.

Mere resemblances between humans and apes are not, however, adequate to demonstrate the evolutionary rela-

tionships. Fortunately, biochemical evidence of different types support evolutionary relationships. Various types of analyses have been performed on biochemical materials from humans and primates to determine any relationships. DNA divergence data, albumin immunological distance units, clotting factors in blood, serum antibody reactions, protein analyses and more have all been carefully analyzed and clearly demonstrate this close relationship of humans with the apes. Washburn and Moore (1974, p. 21) noted with regard to evolutionary nature of these types of evidence, "In no instance is the biological data contradictory. . ." The tests clearly demonstrate that humans are more closely related to chimpanzees and gorillas than to any other primates. Humans are less closely related to the monkeys, related still less to prosimians, and distantly related to other mammals, all in direct relationship to their respective positions on our family tree. These family trees were based at first primarily on physical similarity or morphology that, in turn, inferred evolutionary relationships, but now such tree models are supported by sophisticated biochemical analyses.

There is little argument among experts that humans share an evolutionary relationship with the other advanced primates, though some workers have challenged this on various grounds that range from religious to pseudoscientific to scientific. Kurtén (1984), for example, is one scientist who argues that humans did not evolve from the apes. There is less disagreement that such an evolutionary connection exists than where, when and how the branching of the hominid line from the hominoids took place. To determine when and how humans made the jump from the anthropoid ancestors, we must compare humans with our nearest primate relatives, the chimpanzees and gorillas, both typical higher anthropoids, and see how they differ from humans and how they are similar.

PHYSICAL COMPARISONS BETWEEN APES AND HUMANS

Humans and apes differ in many ways, including physical size, shape, intelligence, behavior, strength, diet and habitat. The physical comparisons between humans and the higher primates, especially the pongids, are a simple

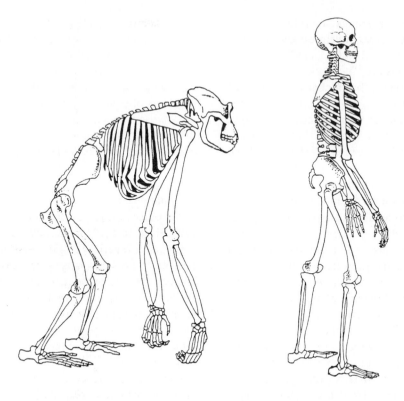

Figure 4.1 Gorilla and modern human skeletons in upright posture.

matter of measurements and direct comparisons of anatomy, skeletal materials, body systems and biochemistry, among others. Skulls and teeth, for example, are particularly informative in identifying the human or ape status of a fossil. Nonphysical or behavioral comparisons (Chapter 5) are also possible, but they are more difficult to study and interpret. The late Dian Fossey studied wild gorillas extensively in Africa after earlier work by George Schaller. Jane Goodall has studied the wild chimpanzees in Tanzania for nearly thirty years. And Biruté Galdikas has worked the orangutans in Borneo since 1973. All have provided us with valuable new data and interpretations about the behavioral patterns of great apes and their striking similarities to human behaviors.

SKELETAL DIFFERENCES

Humans and all other animals may be defined biologically and the nature of our physical structure examined in biological terms. The most obvious physical differences and similarities we have with the apes can be seen in any comparison with the gorilla or the chimpanzee (Fig. 4.1).

These physical features reflect adaptations, those responses of an organism to a particular environment, and these features and structures provide us with clues to the origin of lifestyles. The legs of humans are, for example, ideally designed for bipedalism compared with those of the apes. Human legs have elongated, straight long bones, rearranged articulation and bone ends, reorientation of some bones in the feet, revision of the pelvis, and repositioning of the muscles that control all these—all structural adaptations that resulted from our bipedalism. Java man, for example, received the name *Pithecanthropus erectus*, because its straight-shaft femur suggested an erect or upright posture. The legs and musculature of apes in contrast have a slightly curved shape and, with the pelvis shape, do not permit a fully erect walking posture. These differences reflect the selection of a bipedal means of locomotion by our ancestors and the resultant modifications to their structures. The ancestors of apes, however, did not select such a basic means of locomotion and their descendants, the living apes, retained a primitive curved leg and a knuckle-walking gait.

It is apparent, therefore, that the skeletal characteristics of living apes and humans offer ample grounds for differentiation between them even as behaviors indicate a close but less clear-cut relationship. The skeletal differences become less obvious, however, further back in time when the apes and humans had not yet evolved very far from the parent stock and the two groups were more similar. If apes and humans share a common ancestry but have since evolved in different directions because of changes related to habitats, environmental influences, mutations, selective pressures, etc., then we should expect to find divergences in many adaptive features, characters and traits, particularly in the skeletal structure. These adaptive divergences are precisely what we find.

Ape and human skeletons are distinctive and easily recognized and separated; the ape skeleton reflects adaptation to behavioral patterns in specific environments, whereas the human skeleton argues for adaptation to another environment. Fortunately, skeletons are hard materials suitable for fossilization and provide much information when preserved. Most primate fossils are recovered from water-laid deposits of streams and lakes, places where these animals met their death or into which they were subsequently washed after death. Though such deposits often contain recoverable fossil remains, we are not likely to recover many of them without extensive field exploration and a great deal of luck. Despite this, substantial numbers of important hominoid and hominid fossils have been found.

The most apparent differences between the bodies of humans and gorillas are their size and shape. Most humans tend to cluster about the same size and shape, even allowing for individual variations. Young (1971, p. 561) believes that

> . . . there must be selective factors that have
> kept men of about the size and shape that they
> are.

Gorillas, too, seem to be about the same size and shape. Humans have a relatively slender shape compared to the gorilla, and our rather delicate skeletons have different shaped skulls, jaws, teeth and feet that reflect our posture, our gait and our much less physical way of life. Our musculature, in keeping with our smaller skeletal structure, is much less powerful than a gorilla's which provides great strength; gorillas and other apes are exceptionally strong for their sizes. Adult male chimpanzees are about three times as strong as adult human males and the larger orangutans and gorillas are much stronger. The more delicate human skeleton and smaller muscles obviously reflect a different lifestyle, presumably emphasizing intelligence and reliance on tool use rather than brute strength. This same slender human skeleton is also present in our ancestors, the australopithecines, and suggests that they, too, did not depend on sheer strength for survival, but presumably relied largely on their intelligence and social organization. Our Neandertal cousins had much more robust skeletons than either the australopithecines or us, suggesting that their way of life depended greatly on strength.

The great apes are typically covered with a thick pelt of hair, black in the gorilla and chimpanzees, and red in orangutans. Morris (1967) referred to humans as the "naked ape," but we are not truly naked or hairless. Our bodies are covered with a fine, downy hair that is readily apparent on close examination. We are normally hairy on the head (at least some of us) and in other parts of the body, the legs, arms, armpits and pubic region. Some human males have quite heavy hair growth on parts of their arms and legs, a few others are rather hairy overall, especially on the back, but we all have given up the thick pelt of our primate ancestors for a relatively hairless body.

The skeletal features of humans, too, show numerous changes from those of apes. Not only do we differ in general skeletal aspect from the apes but also in details. In virtually all bones and teeth we find major and minor differences that can be used to identify the remains as either human or ape. These detailed differences provide us with a starting point from which we can assess the human or pongid nature of most pertinent fossil discoveries. Though the chimpanzee is our closest relative by any scientific standard, comparisons with gorillas are equally valid.

SKULL DIFFERENCES

The skulls of humans differ significantly from those of apes (Fig. 4.2). Gorilla skulls are larger and more massively constructed. This difference in size and other aspects reflect the lifestyles that each utilizes. These features include:

BROW. Apes have an extremely prominent **supraorbital crest (torus)** or **brow ridge**, the portion of the frontal bone that protects the eyes, serves as anchor areas for muscles, and contains sinus openings. It is large and distinctive in the pongids, pronounced but relatively smaller

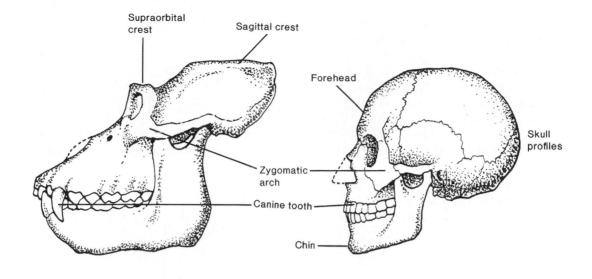

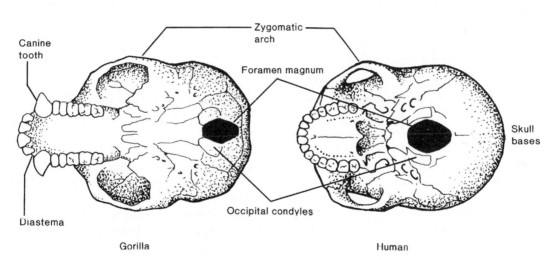

Figure 4.2 Gorilla and modern human skulls.

in the early hominids, and significantly reduced in living hominids. This brow ridge varies greatly in living humans from virtually nonexistent in some to noticeable in others, but is never apelike. Its reduction is an excellent example of evolutionary change through the hominoids and into the hominids.

FOREHEAD. Humans have a forehead that is generally vertical or somewhat sloping as a result of upward expansion of the **calvarium** to accommodate the greatly enlarged brain. In apes, the forehead slopes at a very low angle directly into the calvarium, which contains a relatively small brain.

FACE. The apes exhibit **prognathism**, a snoutlike condition that contrasts markedly with the **flattened face** of humans. Two evolutionary modifications affected the hominid skull shape to produce this facial shortening as it retreated from this prognathous condition. These were the loss of the very large olfactory receptor areas in the face as this sense became less important and the reduction in size of jaws and teeth as the hominids began to use different, less harsh foods. In both of these cases, smaller was better.

NOSE. Humans have a distinctive nose that is lacking in apes. Primates increased their dependence on eyesight in

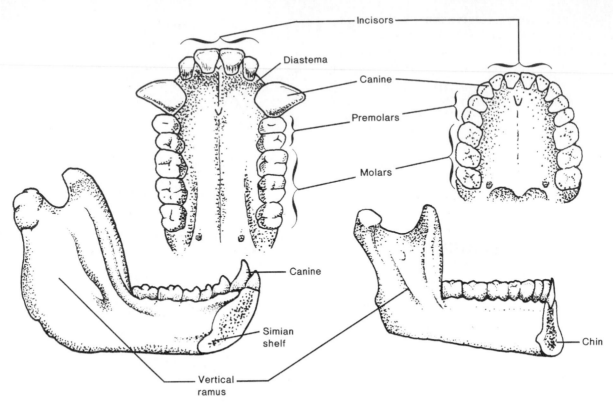

Figure 4.3 Gorilla and modern human jaws with teeth.

contrast to other mammalian species that relied on scent and the olfactory receptors. These olfactory receptors occupied much space in and around the nasal passages, and when olfactory abilities became less necessary, the receptors became smaller, especially in the hominoids, until they attained their present state in humans. We are not particularly dependent on our ability to scent various odors for survival. Therefore, the olfactory senses and structures were less essential to anthropoids for survival and consequently became reduced in function and size. The ultimate reduction is seen in the hominids, where the face receded and flattened more than in any other group. However, the external nose arose to provide a site for the remaining olfactory receptors and as a device to entrap inhaled foreign particles before they can reach the lungs and cause problems. The nose also warms air before it is breathed into the lungs.

JAW. As the diet of humans changed markedly from that of the hominoids, there was little necessity to preserve the very large **jaw** of ape ancestry (Fig. 4.3). The massive jawbone and large teeth of apes are quite capable of masticating very harsh plant fibers. The hominids, however, chose their foodstuffs from less harsh, higher energy foods. As a result, the hominids had no need for the massive chewing apparatus of the apes and began to evolve the smaller, more delicate jaws and teeth. When meats were included in their diet and provided significantly more energy, there was even less necessity to eat the sturdier and rougher plants that required large teeth set in strong jaws powered by large and powerful muscles. The massive jaws, teeth and accompanying musculature of the hominoids therefore were unnecessary for the hominids with a different diet. The hominid jaw, accordingly, became significantly reduced in size and retreated from the prognathous state as the teeth became smaller. At the same time, there was no demand for massive jaw muscles in such smaller jaws and no need to anchor these smaller muscles to the prominent and characteristic sagittal crest retained by pongids. Without this demand, the jaw muscles became reduced and the sagittal crest disappeared.

The shape of the hominid jaws changed significantly, as well. Ape jaws have their teeth arranged in a distinctly parallel-sided **U-shape**. The dental arcade of humans is not parallel but is flared outward slightly to the rear, producing an arched **V-shape**. This shape resulted from a narrowing of the front of the face and jaw as they were reduced in size.

TEETH. Teeth in humans are much smaller than those of apes. This also reflects the dietary changes that took place in the transition from early hominoids to hominids. The canine teeth, in particular, were reduced in hominids to the size of the other teeth. This contrasts markedly with the large canines of apes which serve both as formidable weapons in threat displays and as efficient tools for stripping vegetation of its harsh outer covering. The reduction in human canine teeth size was accompanied by the loss of the **diastema**, the gap between the teeth of apes that provides space for the large canine teeth when jaws are closed. Small human canine teeth allow for side-to-side movement of the jaws, an effective means of chewing using the cheek teeth. The apes cannot chew in this fashion because of the interference of their large canines with the other teeth.

One aspect of dentition that clearly relates humans to the great apes and the dryopithecines is the **5-Y pattern** of cusps in our molars (Fig. 4.4). This pattern originated in our dryopithecine ancestors some 20 million years ago and has been found in all primates descended from this stock. This pattern is found today in all modern and fossil hominids and pongids, the gorillas, chimpanzees, orangutans, gibbons, our ancestral hominid stock and many living humans. It is most commonly found in Asian humans, though not exclusively.

CHIN. Humans have chins, apes do not. The reduction in size of the hominoid jaw to the hominid condition weakened the strength of the joint (symphysis) between the two halves of the **mandible** or lower jaw. The harsh diet of gorillas requires a very sturdy mandible capable of withstanding the great pressures developed in tearing and chewing harsh vegetation. For this purpose the ape has a structural support on the inside of the point of the jaw in the form of a horizontal flange or shelf known as the **simian shelf**. When hominids changed diets and the jaw reduced in size, the jaw point receded and the face lost the prognathous condition. As these bones became thinner, the simian shelf effectively disappeared. But loss of this structural member in early hominids would have significantly weak-

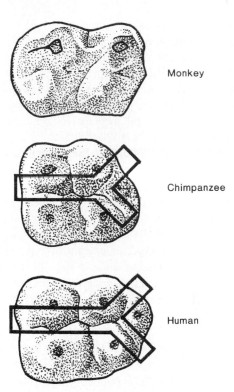

Figure 4.4 The 5-Y dental cusp pattern.

ened the symphysis and the entire mandible, probably creating difficulties in chewing tough foods. The evolutionary solution was to develop an alternative supporting structure, the external thickening called the **chin**. This feature produces the necessary strengthening of the jaw at its weakest point, but on the outside of the jaw rather than on the inside. Thus, the apes retained the simian shelf inside their massive jaws, and humans developed more delicate jaws with an external chin.

SAGITTAL CREST. The **sagittal crest** is a vertical ridge of bone atop the skull of gorillas. This provides the anchor attachment for the powerful jaw muscles. Smaller jaws in humans and hominids did not require the very large **temporalis** and **masseter** muscles that operated the effective and powerful jaw apparatus that apes used to chew tough vegetation. The human diet changed, however, and the massive jaw bones and jaw muscles of the apes were unnecessary. The jaw muscles became smaller to accommodate the reduction in size of the jaw bone. This also resulted in a reduction in size of the **zygomatic arch** or **cheekbone** and shifted it closer to the side of the face. The temporalis muscle passes under this arch and a smaller muscle obviated the need for such a large arch. Simultaneously, the reduction of chewing muscles negated the need for the sagittal crest, large supraorbital ridge and large cheek, all attachment sites for large temporalis and masseter muscles. Therefore, the sagittal crest disappeared altogether in humans, and the surpraorbital crest and broad cheek bones became much reduced in size.

FORAMEN MAGNUM. The **foramen magnum** is the opening at the base of the skull to allow passage of the spinal cord to the brain. Its location indicates the position of the skull on the spinal column and can identify posture. In humans, the opening is directly underneath the skull and provides for its vertical alignment atop the backbone. This orientation allows for erect or upright posture which is universal in hominids but unusual in apes. Gorillas have their foramen magnum placed underneath the skull at the rear to provide for a three-quarter erect posture, but not a fully erect one.

Fundamental to the placement of the skull on the backbone are the **occipital condyles**, knobs that allow the skull to articulate with the vertebral column. These knobs are situated laterally to the foramen magnum and slightly forward in humans . In apes they occupy the same position adjacent to the foramen magnum. But in apes, they indicate an in-between posture, neither erect nor horizontal.

TRAPEZIUS MUSCLES. Large and powerful neck muscles known as the trapezius muscles support the primate head atop its shoulders. These attach to a large sloping, flattened, triangular area at the back of the skull. In humans, these muscles are much reduced in size and attached principally to a much smaller area at the rear base of the skull.

POSTCRANIAL SKELETAL FEATURES

A number of skeletal features posterior to the skull can be used to identify the remains as ape or human. These are:

VERTEBRAL COLUMN. The ape vertebral column or backbone (Fig. 4.5) is gently curved into an arch that terminates at the base of the skull where it is attached by the powerful muscles that fit into the large triangular area at the rear of the skull. Human spines, in contrast, are recurved sinusoidally to distribute our weight evenly over our legs. The recurving of the human spine also increases the muscle leverage of the various structures of torso and legs. The modifications of shape that human vertebral columns have

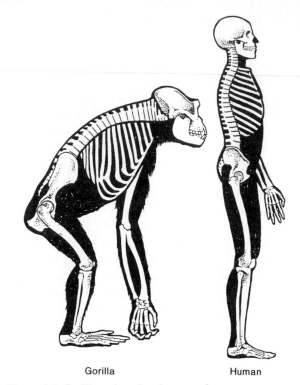

Gorilla   Human

Figure 4.5 Gorilla and modern human spines.

undergone, however, are not evolutionarily perfect. Rather, the shape is a compromise between needs of the various functions and of posture. The erect posture of humans is a highly desirable way of life, yet sitting and other positions can create problems. The result is the evolution of a flexible columnar structure that can be weak and is susceptible to damage and pain. Lower back pain, a common human ailment, often results from stresses on the spinal column that cause "pinching" of nerves, strained muscles and ligaments, and damaged cartilage and vertebrae.

PELVIS. Upright posture in hominids was possible because of modifications to the spine, pelvis, legs and feet. The human pelvis is shorter, broader, more flared and angled outward (Fig. 4.6) than the ape pelvis which is elongated and nearly straight. The human pelvis is also curved to the front to form a basin that assists in supporting the sagging internal organs that are awkwardly suspended from the now-vertical spine in our upright posture. This contrasts with the four-legged animals whose viscera are suspended from horizontal backbones that are parallel to the ground and who, therefore, have no need for a bowl-shaped pelvis.

58

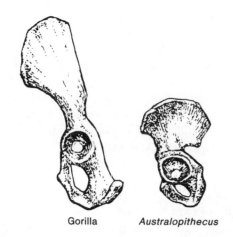

Gorilla   *Australopithecus*

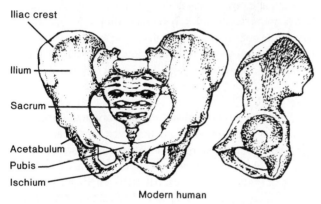

Iliac crest

Ilium

Sacrum

Acetabulum

Pubis

Ischium

Modern human

Figure 4.6 Pelvic bones of gorilla, *Australopithecus africanus*, and modern human. Adapted from "The Antiquity of Human Walking" by John Napier. Copyright © 1967 Scientific American, Inc. All rights reserved.

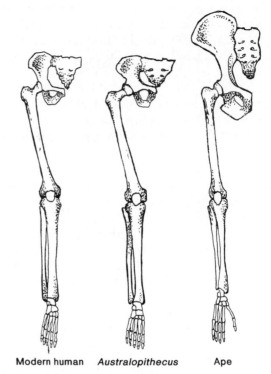

Modern human   *Australopithecus*   Ape

Figure 4.7 Legs of modern human, *Australopithecus africanus* and ape.

The pelvic leg sockets in humans are set in the central part of the pelvis to balance the torso over the legs. This also allows the leg a great degree of rotation and, accordingly, a large swing for a walking gait. The gluteus muscles that connect the leg and the pelvis are also attached differently in humans than in apes and function in a slightly different manner. These muscles provide human legs with strength for walking, running and climbing and for maintaining the upright trunk.

LEGS. Human legs are heavily muscled, essentially straight columns that are situated directly under the torso. They articulate exceptionally well with the pelvis to allow various bipedal activities. Human legs tend to converge toward the knees (Fig. 4.7) in a somewhat knock-kneed posture that is modified from the parallel thigh posture in apes. The knee joints of humans are exceptionally versatile and permit ease of motion and great range of movement, including the ability to completely straighten the leg, an ability lacking in apes. Muscles of human legs are equally specialized and well developed, even though they have counterparts in the apes. For example, the thigh muscles of humans are attached to the pelvis to provide for upright posture, walking and running motion, strength and endurance. In apes, the leg motion is without grace and smoothness because of the different manner in which the muscles attach and the bones articulate. The human buttocks are enlarged gluteus maximus muscles that attach the legs to the pelvis; apes lack such enlarged buttock muscles, and they cannot perform the erect posture and bipedal motion that is commonly and easily exhibited in humans. Further, the different shape of the human pelvis provides for the muscles to be attached directly downward to the femur. In apes the same muscle must make a curve around the

59

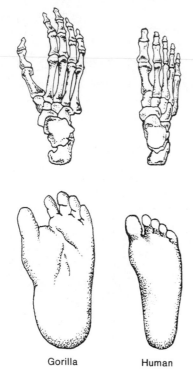

Gorilla   Human

Figure 4.8 Feet of gorilla and modern human.

straight pelvis, and this arrangement forces apes to take awkward bipedal steps that curve or arc outward rather than move straight forward and backward as in humans. In particular, apes also lack the enlarged calf muscles of humans that are so important in moving the feet on ankle pivot points. Living apes will never equal the ability to move, run and jump as gracefully as ballet dancers or even as clumsy humans.

FEET. The feet of apes are not feet in the human sense, rather they are "hands" placed at the distal ends of the legs. They have opposable large toes (Fig. 4.8), as do all other nonhuman primates, in contrast to the parallel toes in the walking feet of humans. Apes use these to grip branches when climbing or to pick up objects as desired. Their toes are long and reach back to the ankle. The ape ankle is small and designed for semierect posture which they maintain using their arms braced on their knuckles. An upright posture without such knuckle walking requires apes to stand on the sides of their feet and exert much muscular control of their legs and body to maintain balance for even a short time. Human feet are far more highly specialized

and are very long and semirigid with the digits or toes of approximately the same length and parallel to each other. Thus, they provide a firm base to support our entire body and allow us to remain upright or walk for hours with little effort.

The relationship between the size and robustness of the first toe in humans and the other toes largely determines how the foot functions. The ball of the foot is formed at the bases of the toes where they can articulate and flex in a straight row. The arrangement of the foot bones with the ankle forms a tripod, a very stable structure. A cross arch is formed behind the ball of the foot that provides two points of this tripod. A longitudinal arch is formed from the middle of the toes to the heel, which is an extended part of the foot, and provides the third point of support. The ankle bones are arranged directly underneath the body and distribute the body weight vertically to these points of the foot tripod with relatively little muscle effort. None of this is possible with the feet of apes.

THE BRAIN

The brain is a critical and complex organ of the central nervous system in animals. In the invertebrates and lower vertebrates, it is neither a prominent nor large structure nor is it capable of significant activities compared with the brains of higher vertebrates. All multicelled animals have their activities organized and coordinated by nerve cells. In invertebrate animals, which typically are unsophisticated compared to vertebrates, the brain is absent or essentially nonexistent and its functions are carried out by nerve cells and nerve cords. These animals also evolved sensory systems that relied on receptors or organs scattered throughout their bodies to provide them with information about their environments, especially about food and danger. Eventually these sensory organs and systems began to concentrate nerve cells in one place at their anterior end. This became the head, and the end of the enlarged nerve cord became the first brain.

In chordate animals, the nerve cord became dorsal and hollow within a specialized protective tube of cartilage or bone. The bony tube we possess consists of interlocking, articulating pieces of bone called **vertebrae**, hence the name, **Vertebrata**, for the group. The head end of the nerve cord in early vertebrates enlarged fully into a complete brain with three parts, **forebrain**, **midbrain** and **hindbrain**. Though these parts have undergone modifications

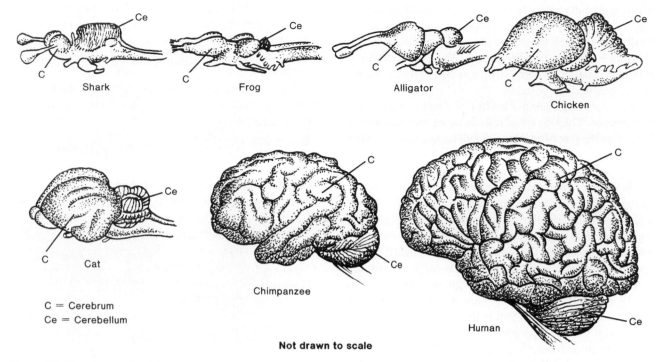

C = Cerebrum
Ce = Cerebellum

Not drawn to scale

Figure 4.9 Vertebrate animal brains.

in different types of vertebrates, the same three are recognized throughout (Fig. 4.9). In primitive vertebrates that rely on scent, like sharks, the forebrain or **telencephalon** has a pair of outgrowths, the **olfactory lobes**, that receive nerves from the nostrils and deal with the important sense of smell. Paired swellings behind the olfactory lobes on the upper forebrain make up the **cerebrum** and posterior to it is the **thalamus**. The midbrain or **mesencephalon** also has swellings, the **optic lobes**, that deal with sight. The hindbrain or **rhombencephalon** has a swelling, the **cerebellum**, adjacent to the midbrain which narrows and joins to the spinal cord. This narrowed portion is termed the **medulla oblongata**.

As different types of animals developed specialized sensory functions according to their ways of life and environments, different portions of their brains became enhanced. Fish, for example, rely more on smell than some of their other senses, and their olfactory lobes are well developed. Amphibians and birds, in contrast, are more involved with sight than smell, and their optic lobes and cerebrum developed at the expense of the olfactory lobes. Reptiles have much more to think about in the complex

land environments than fish or amphibians do in water. Accordingly, reptiles developed a much larger cerebrum and **cerebral cortex**, the gray matter outer layer of the cerebrum. Reptiles also developed a new portion of the cortex, the **neopallium**, which is concerned with receipt of sensations other than smell. In mammals the neopallium became even better developed than in reptiles and spread out to cover the top half of the cerebral cortex.

In modern mammals the neopallium is so enlarged that it essentially is the cerebral cortex. This is the center for coordination of many different stimuli and responses, and allows for far more complex behaviors than are possible with the simpler brains of lesser vertebrates. Simple brains allow for only simple behaviors, because they have fewer brain neurons to interconnect in different ways. The more complex the brain, the more complex the behavior that is possible, because more brains cells are available to connect with other brain cells. This principle is analogous to memory chips in computers, the smaller the memory bank, the less sophisticated are the tasks that a computer can perform. An increase in the size of the computer RAM (Random Access Memory) and ROM (Read Only Memory) will significantly affect the speed and potential for computing. This analogy between computers and brains is not totally true on the basis of size alone, because brain size is also related to the size of the animal. When the neopallium enlarged relative to body size, however, greater intelligence then became possible.

Theoretically, mammalian brains could evolve to immense sizes. They do not simply because the brain size is limited by the size of an infant's skull that can pass through the opening in the mother's pelvic bones during birth. These restrictions imposed by skull dimensions have limited brain sizes, though there is an evolutionary demand for increasingly larger brains and greater brainpower. Mammals met that particular challenge by developing brains with relatively larger areas of cerebral cortex, the place where these complex thoughts took place. When the skull limitations were reached in hominoids and hominids, the cerebral cortex continued to enlarge but became complexly infolded or wrinkled. This resulted in a greater cerebral cortex that fitted within the same-sized skull and, for a given volume, appropriately better brains. Humans' brains developed with the greatest infolding, i.e., the best brains. Although other primates have infolded brains, the total areas devoted to particular functions are significantly different.

ARBOREAL BEHAVIOR AND BRAIN WEIGHT

Life in the trees was a blessing for the early primates even though it created unique problems for them. These problems were emphasized by the requirement of mobility along swaying branches high above the ground. Any imbalance or false step could result in a fall that was likely to be harmful if not fatal. Agility, small size, grasping hands and feet, and stereoscopic vision were natural consequences of evolution as these animals adapted to this specialized arboreal lifestyle. The adaptations also included modifications in bone structure, changes in musculature to control these structures, the nerves to and from the brain to provide for control of these, and increases in appropriate portions of the brain to organize, coordinate and supervise everything. Above all, the improvements in the brain made all these other aspects possible.

Successful arboreal behaviors emphasize the need for great physical and intellectual abilities. The structural, muscular, neural and brain advances developed by these early primates were improved upon by their successors.

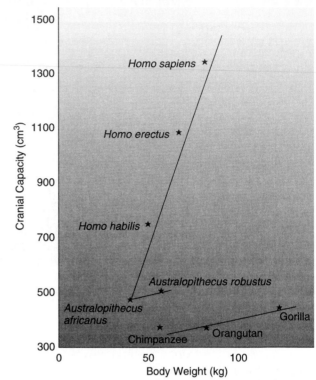

Figure 4.10 Brain size vs. body weight in advanced primates.

Azimov (1963, p. 152) points out that

> Of all the senses, sight delivers information to the brain at the highest rate of speed and in the most complex fashion. The use of a hand with the numerous delicate motions required to grasp, finger, and pluck requires complex muscular coordination beyond that necessary in almost any other situation. For an eye-and-hand animal to be really efficient there must be a sharp rise in brain mass.

With all these adaptations and advances, the primate's brains became larger and significantly better instruments capable of complex functions. Even so, the great apes tend to have relatively small brains, about 340 grams (12 oz.) in orangs, 380 grams (13.5 oz.) in chimpanzees, and 540 grams (19 oz.) in gorillas.

The chimpanzee is the smallest of the great apes with a weight of about 50 kilograms, but its relatively large brain gives it a brain mass:body-weight ratio of about 1:120 (Fig. 4.10). Adult male gorillas weigh about 250 kilograms but their brains give them a much smaller brain:body-weight

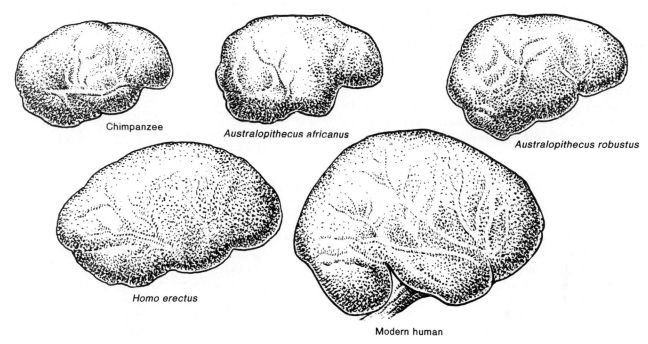

Chimpanzee

Australopithecus africanus

Australopithecus robustus

Homo erectus

Modern human

Figure 4.11 Fossil hominoid and hominid brain casts.

ratio of 1:420. Even the most ancient of hominids had much larger brains per body weight. These early hominids, the australopithecines, had brains that weighed about 600 grams (21.5 oz.) and a body weight of about 25 kilograms for a brain:body-weight ratio of 1:42. If everything we are led to believe about brains is correct, then the australopithecines were much more capable intellectually than their pongid cousins. Even these australopithecine brains were not far from modern humans in brain:body weight. Modern humans have ratios of about 1:47 based on brains averaging about 1300 grams (45+ oz.) in weight and range in volume from less than 1,000 cc to more than 2,000 cc. It is this large, highly sophisticated brain in humans that enabled the most significant steps in the long history of life on Earth. Fortunately for many of us, brain size has not been shown to correlate directly or indirectly with intelligence. Intelligence is still an intangible whose secrets are unknown, and people with large or small brains are fully capable of great intellectual achievements.

Evidence from endocranial casts of various fossil hominids shows that australopithecine brains differed from pongid brains in being more human (Fig. 4.11). Holloway (1974, p. 113) pointed out that

. . . both the direct evidence of neurological organization and the indirect evidence of comparative size appear to indicate that Australopithecus and at least one other African primate of the period from three million to one million years ago had brains that were essentially human. . .

BIOCHEMICAL RELATIONSHIPS

The biochemical similarities between humans and apes are perhaps the most compelling evidences to indicate a genetic and therefore evolutionary relationship. The basic premise is that if humans evolved along with the great apes from a common ancestor, then the living humans and apes should be demonstrably similar to each other in many respects, physically, behaviorally and biochemically. We have already seen how they are similar in physical terms and in many behaviors. Biochemically, a common ancestor would have had a relatively similar molecular makeup in its DNA, blood, proteins, etc. When some changes occurred in the genetics and biochemistry of a portion of the ancestral population by mutations, perhaps environmentally stimulated, enough differences were generated to turn them into a new species that became the human stock. The genetic identities of the ancestral species and this early human stock were nearly identical or probably very similar shortly after the separation. With time, as random mutations appeared in the genetic material of each, their biochemical and genetic compositions must have diverged. The appearance of the new species of apes from this ancestral stock also would have resulted from changes in the biochemical and genetic makeup of a portion of the population. Comparisons of the measurable differences in various biochemical aspects between living apes and humans should indicate how much these two groups have separated from the common ancestor and each other. Much similarity would indicate relatively close relationships and that they diverged only a short time ago. Little similarity would indicate there was no close relationship and that their separation would have taken place in the very distant past. More distantly related animals would be expected to show even greater differences in their biochemical makeup.

Nuttall (1904) proposed using molecular analysis tests of antibodies in sera to determine degrees of relationships between species of many vertebrate animals. Since then, various molecular biochemical tests—sophisticated analyses to identify similarities between humans and apes—have been conducted by different workers in many parts of the world. The results are the same wherever they have been performed; they demonstrate very close affinities between apes and humans. Immunological analysis, DNA divergence data and protein comparisons have all proven highly informative in verifying the nature of this relationship. Even more importantly, these tests provide data that demonstrate the degree of relationship much better than could any physical similarities. What they cannot show with certainty is when the humans and apes separated from their common ancestor or each other, though good estimates have been projected. Scientists simply have no absolute way yet to determine the rate of genetic change. Fortunately, however, such ages can be identified using sophisticated radioactive isotope analyses of rocks, and these identify approximately when these separations took place.

BLOOD PROTEINS

Goodman began molecular tests in the 1960s to obtain information on genetic relations in primates. These immunological tests compared blood proteins of humans with

Table 4.1 Primate relationships based on immunological comparisons.	
Man	0.0 distance units
Chimpanzee to man	0.2
Gorilla	1.0
Orangutan	2.2
Gibbon	3.4
Old World monkey	4.0
New World monkey	6.5
Prosimian	10.0
Tree shrew	13.0

those of apes. Results (1967) showed that humans and African apes were closely related, just as Darwin and Huxley had prediced many decades previously. Table 4.1 shows the results of these tests as distance relationship units.

SERUM ANTIBODY COMPARISONS

Wilson and Sarich (1969) studied proteins of humans and apes to determine similarities between the two and the degree by which they differed. They injected a human blood protein, **albumin**, into rabbits. The rabbits responded by making blood antibodies that fought this foreign protein from human serum. The researchers found that rabbit serum containing these antibodies (antihuman antibodies) could be injected into human serum where it would produce a quantity of precipitate that was directly due to the attack by the rabbit antibodies (antihuman) on the human serum. Antihuman serum from the rabbits was then tested against the sera of various other mammals, including gorilla, chimpanzee, orangutan, gibbon, Old World monkey and New World monkey. When the antihuman antibody serum from the rabbit was added to the serum of chimpanzees, it produced a precipitate somewhat less than that formed by reaction between the rabbit and human sera. This lesser precipitate from the chimpanzee-rabbit sera reaction was due to the fact that the chimpanzee serum was only slightly different in composition from the human serum. The rabbit antihuman antibodies reacted with this serum of slightly different composition and so produced less precipitate. The chimpanzee serum had less "human" serum composition and therefore could not produce as much precipitate. The chimpanzee serum, though very similar to human serum, was not identical to it and was

Table 4.2 Immunological distances in some higher primates.	
Man to chimpanzee	7 units
Man to gorilla	9 units
Man to orangutan	12 units
Man to gibbon	15 units
Man to Old World monkey	32 units
Man to New World monkey	58 units

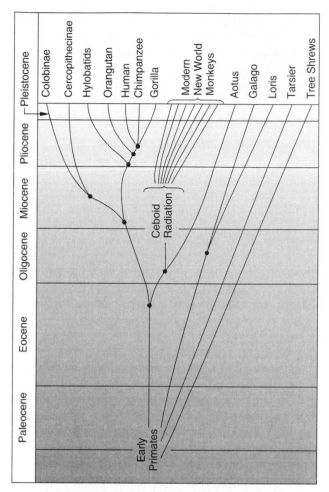

Figure 4.12 Primate lineages based on immunological distance units.

eventually calculated as being seven **immunological distance units** from human serum (Table 4.2). This same test applied to the serum of a gibbon resulted in an even smaller precipitate for the same reasons. These results were calculated and show that gibbon-human sera are fifteen immunological distance units apart.

Using this same test procedure other primates show progressively lesser precipitates in proportion to their distance evolutionarily from humans (Table 4.3). These relationships can be charted graphically (Fig. 4.12) to show when in time these primates separated from each other.

Another graph (Fig. 4.13) using such biochemical data shows the divergence of humans from the other apes and the times when such events took place. This branching pattern, using Old World monkeys as a baseline, measured genetic distances between these species. It shows that Asian apes diverged from humans and African apes between eight and ten million years ago. The pattern failed to identify the separation order between humans, gorillas and chimpanzees except to indicate that it happened about five million years ago.

Table 4.3 Albumin reactivity in some higher primates.

	Index of Dissimilarity		
Species of Albumin	Anti-*Homo*	Anti-*Pan*	Anti-*Hylobates*
Homo sapiens	1.0	1.09	1.29
Pan troglodytes	1.14	1.0	1.4
Pan paniscus	1.14	1.0	1.4
Gorilla gorilla	1.1	1.17	1.31
Pongo pygmaeus	1.22	1.24	1.29
Symphalangus syndactylus	1.27	1.25	1.07
Hylobates lar	1.27	1.25	1.0
Macaca mulatta	2.23	2.0	2.3

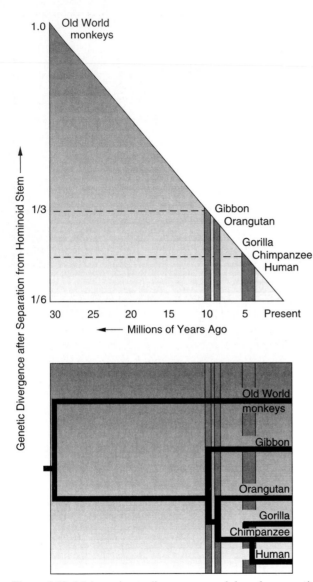

Figure 4.13 Living primate divergence graph based on genetic distance data.

DNA COMPARISONS

Another biochemical test involves comparative studies of DNA hybridization from humans and apes and produces averages of the differences of each. Results allow interpretations of evolutionary relationships. Such tests, developed in the 1960s, are useful in determining human-chimpanzee relations because the tests measure the materials fundamental to organisms, the genetic material of inheritance,

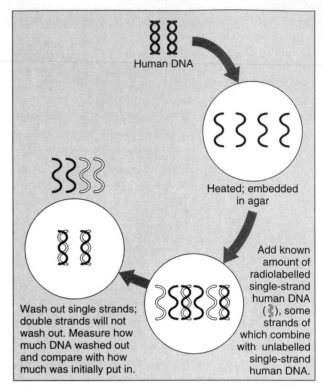

Figure 4.14 Hybridization of human and chimpanzee DNA.

the DNA from the chromosomes. During the process of meiosis, DNA strands separate naturally in the sex cells to produce the gametes. The separated strands from each parent are then free to combine randomly at fertilization to form zygotes that are new individuals. Researchers then can use these principles to test the DNA relationships between humans and apes.

The DNA strands can be made to separate in the laboratory, where they also can be recombined. A single strand of human DNA so treated can be combined with a similarly separated single strand of chimpanzee DNA to produce an artificial or hybrid strand pair. Examination of the hybrid strands show that a double strand composed of one human and one chimpanzee strand has a 97.5 percent fit; only 2.5 percent of the DNA amino acid sequences are different (Fig. 4.14). If the two strands were from the same species, the fit would be perfect. However, because the two strands are from different species that have descended from a common ancestor, this slight difference demonstrates that their DNA has undergone only 2.5 percent change since the time the two species separated. This DNA

Table 4.4 Partial amino acid sequences in primate hemoglobin chains.

Species	\multicolumn{11}{c}{Position}
	5	6	9	13	21	33	50	76	87	104	125
Homo-Pan	PRO	GLU	SER	ALA	ASP	VAL	THR	ALA	THR	ARG	PRO
Gorilla	PRO	GLU	SER	ALA	ASP	VAL	THR	ALA	THR	LYS	PRO
Gibbon	PRO	GLU	SER	ALA	ASP	VAL	THR	ALA	LYS	ARG	GIN
Rhesus monkey	PRO	GLU	ASN	THR	ASP	LEU	SER	ASN	GIN	LYS	GIN
Squirrel monkey	GLY	ASP	ALA	ALA	GLU	VAL	THR	THR	GIN	ARG	GIN

difference is due to mutations that affected the DNA strands of one species or the other, or both after they had separated. Other workers have verified this test method and demonstrate that humans are more closely related to chimpanzees than to other primates. Gorillas are shown to be only slightly more distantly related to humans than the chimpanzees. The other primates are progressively more distantly related.

HEMOGLOBIN AMINO ACID SEQUENCE COMPARISONS

Analysis of the amino acid sequences in hemoglobin protein of humans and other animals has almost identical results to the DNA comparisons and the human-animal sera methods. Hemoglobin is the vital molecule that transports oxygen in the blood in air-breathing animals. It consists of two molecular chains, an alpha unit with 141 amino acid units and a beta with 146 units, for a combined total of 287 amino acids. The two chains from human hemoglobin compared to chains from chimpanzee hemoglobin showed *no differences* in the sequence of 287 amino acids; they are identical. The chains of hemoglobin of gorillas compared to chains of human hemoglobin showed only one difference in each chain for a total of two. Similar comparisons with the gibbon and two monkeys produced the same general expected results. These results (Table 4.4) indicated that humans are more closely related to the chimpanzee than the other anthropoids, and the gorilla is a close second.

MITOCHONDRIAL DNA SEQUENCES

Ruvolo, et al., (1991) analyzed mitochondrial DNA sequences encoding the cytochrome oxidase subunit II gene from humans, apes, three monkeys and some non-primates tissues to establish whether a close relationship within the *Homo-Pan-Gorilla* grouping could be found. They discovered (p. 1571)

> Overall, *Homo* and *Pan* show 9.6% difference in COII sequences, compared to 13.1% for *Homo-Gorilla* and 12.1% for *Pan-Gorilla*. Three methods of phylogenetic analysis yeild congruent topologies containing a *Homo/Pan* clade. . .

They further state (p. 1572)

> . . .reanalysis of coding and noncoding regions of nuclear DNA and mtDNA provide statistically significant support for a *Homo/Pan* relationship. . .

and add (p. 1573)

> The COII gene sequence comparisons clearly favor a *Homo/Pan* clade. These mtDNA data add strong support to the results of both recent DNA sequence studies and DNA hybridizations studies. . . The mitochondrial COII gene trees suggest a separate ancestry for the common *Homo/Pan* clade longer than that seen in any other mtDNA analysis and comparable to that seen in DNA hybridization and nuclear rRNA sequence studies.

Evidence obtained from *all* the biochemical tests demonstrates that humans are more closely related to the great apes than any other group of animals and are closer to the chimpanzees than other primates. To expect such biochemical similarities for so many different aspects in non-related animals would require coincidences of high orders that are statistically unrealistic. We are faced with the inevitable idea that humans and apes are closely related biochemically by means of a common ancestry.

THE NONPHYSICAL COMPARISONS

In addition to the various physical and biochemical criteria that define and distinguish humans and apes, some nonphysical aspects are now being studied and accepted. Many of these are also difficult to recognize, understand and interpret objectively because they are primarily behavioral in nature. Until recent years, these were considered uniquely human aspects. With no standards to identify them or determine their usefulness in separating humans from the apes, such criteria are better as indicators than conclusions. Several such criteria, which include intelligence and thought, language, speech and social relationships, have in recent years been recognized in apes as a result of extensive field and laboratory studies. Unlike the physical differences between humans and apes, the nonphysical differences are not as obvious as might be expected. In fact, the evidence suggests that these criteria are nonunique to humans, can be studied in the apes, and positive comparisons can be made between them.

The apes appear to have nonphysical qualities and behaviors related to those of humans, though less advanced. These do not indicate a difference in type between those of humans and apes, but a difference in relative degree. There is much disagreement among experts from different disciplines about the significance of each nonphysical character and how they make us human, either individually or collectively. Nonetheless, we can examine many of these nonphysical aspects in apes and find support for the view that they are likely to have been derived genetically from our primate ancestors. The presence of similar behaviors in both humans and apes supports the argument that both were probably descended evolutionarily from a common primate ancestor. These anatomical and behavioral qualities and aspects are the same ones that define humanity and which we must examine if we are to determine when, where, how and why the humans split off from the conservative pongid stock (pre-chimpanzees?) that remained in the tropic forests and savanna-woodlands of Africa.

SOCIAL BEHAVIORS

Behaviors are patterns of actions or responses by individuals, groups or species to each other in their environment. The various factors of living space, food and water, shelter, safety and the influence of other organisms create circumstances that affect behaviors in any given species, and these behaviors influence how individuals survive and evolve. Weiss and Mann (1985, p. 223) state:

> Evolution operates through the mechanism of natural selection, in which individuals who possess traits of greater fitness within a particular environment have a better chance of reaching reproductive maturity and passing on their genes to the next generation. In this interaction, it is successful behavior that accounts for the ability of certain individuals to reach reproductive maturity and produce offspring carrying their genetic material.

Behaviors manifest themselves as ways of life and are the determinants of various lifestyles and nonphysical aspects of life. As such, they are largely determined by successes within environments. If a solitary behavior is more successful to a particular group of individuals than participation in a social group, then that particular population or species very probably will adopt the solitary lifestyle. If social group behavior is better for a particular group of individuals than a solitary lifestyle, then that population or species probably will evolve in that direction and adopt such a social structure. Of the apes, only the orangutans have opted for a solitary lifestyle. The fact that all other anthropoids have adopted a group social structure suggests that group structure offers more opportunity for success than solitary behavior. Clearly, populations tend to evolve toward successful behaviors; animals with successful behaviors tend to produce more successful offspring than other individuals. Interpretations of primate behaviors require rather detailed examinations of the animal's subtle ways by direct observations in the wild. Such research, unfortunately, has been slow to develop but has, in recent years, become increasingly significant. This research has shown that some higher primate behaviors are so similar to those of humans that we cannot help but interpret a close biological relationship.

The relatively new biological discipline of **sociobiology** has attracted our attention to the biological nature of behaviors, especially social behaviors, in many different kinds of animals, particularly the higher primates—humans and apes. The discipline attempts to explain many of these behaviors and social relationships as biological in nature. There are ample arguments to demonstrate that some of these behaviors are genetically derived and capable of being transmitted from generation to generation along with other, more obvious physical characteristics like eye color, height and body stature. Such group social and individual behaviors of close relatives like humans and apes are not likely to have been produced in parallel by unrelated animal stocks that followed independent evolutionary paths. The more likely origin is through an evolution from a common ancestor that discovered and adopted the benefits of cooperative behavior and began the trend for social organization with all its benefits. The idea that some social organization and behaviors are genetic and, therefore, inheritable is supported by many workers with impressive data. Though some individuals in various disciplines have objected to this concept, the supporting evidence is mounting. Wilson (1975, p. 22) states that research

> . . . convinced us that behavior and social
> structure, like all other biological phenomena,
> can be studied as "organs," extensions of the
> genes that exist because of their superior adaptive value.

The behavior patterns exhibited by animals, including humans, are the natural consequences of their responses to their environments. The sum total of our human way of life, for example, is very likely influenced by, or the result of, anatomical *and* behavioral changes that took place in some advanced hominoids sometime during or before the transition into early hominids just a few million years ago. The anatomical changes can be identified and traced through their fossils. What behaviors they might have used is far more difficult to determine. However, similarity of behaviors found in both living apes and humans, despite totally different lifestyles, supports an evolutionary relationship.

HIGHER PRIMATE SOCIAL ORGANIZATIONS

Both apes and humans live in social groups, societies, though human society, even the most primitive, is much further advanced than those of the apes. Culture, the cumulative acquisition of knowledge, achievements, and moral and ethical standards, is perhaps the only social attribute not identified in apes. But what level of achievement in a nonhuman species would we consider sufficiently advanced to be a culture? Is there a minimum level of attainment that marks a culture? If culture is a result of higher intellect, then we should not expect to find it in species that lack the necessary higher intellect. To expect to find culture in apes may be a search doomed to failure, much like the attempt to teach apes to talk; they lack the speech apparatus. However, we should keep in mind that experts once decreed that the minimum determinant of fossil hominids was a brain size of 700 cc. Today we believe otherwise.

Primate societies consist of varying numbers of individuals of the same species living together and acting together in cooperative fashion, a situation true of many types of animals, including insects. Among the prosimians, only the Madagascar species are all social and live in groups; the other prosimians are solitary. New and Old World monkeys all live in social groups (Fig. 4.15), as do all other anthropoids, except the orangutan. Thus, the commitment to social groups is virtually total in higher primates but less so in the prosimians where solitary behavior is more common. This pattern may well represent an evolutionary trend that originated in early prosimians when they were just starting to adopt social groupings as a desirable behavior that offered more advantages than solitary behavior. When the anthropoids appeared later, this social group behavior was firmly established in the primates as a nearly universal pattern. Both human and ape social behaviors show remarkable similarities, including both individual and group patterns that contribute to the success of their societies and the ways that these individuals interact.

Social behavior is so widespread among living hominoids that it probably appeared in an ancestral type millions of years ago. The "Dawn Ape of the Fayum," *Aegyptopithecus zeuxis,* is considered to have been either this ancestral species or part of the ancestral stock that lived some thirty million years ago. Simons and Kay (1980, p. 18) identified large canine teeth in some *Aegyptopithecus* jaws and small canines in others. They interpret the differences in the sizes of these teeth to be sexual, with the large canines in the jaws from males and the smaller ones from females. They explain,

> Sexual differences in canine size are usually
> found only in living primates which live in
> complex social groups with more adult females than males.

The differences in size of these canine teeth make sense only if *Aegyptopithecus* males and females lived together in social groups. These larger male canines would have

Figure 4.15 Baboon troop foraging.

served for a variety of sex-based behaviors, including threat displays as in living anthropoids. Threat displays would have been less likely if the animals were solitary because there would have been no regular contact and less need for males to exhibit threats with large canines.

The sizes of social groups and the nature of their social organizations are obviously related directly to their environment pressures; these animals have learned to survive by adopting techniques that work in these environments. Different species of animals follow different social organizations, depending on their needs and, probably, their evolutionary history. Some primates, like gorillas, hamadryas baboons and many Old World monkeys, have adopted the harem structure which consists of a single dominant male, several females, and their immature young. Other primates, like chimpanzees, use a large indefinite structure with membership that can shift into smaller unstable subgroups that change size and membership from time to time. Some others, like gibbons, some New World monkeys and most humans, form small family groups that consist of an adult male, an adult female and their immature

offspring. All individuals born into these groups learn how to live within the rules, customs, behaviors, etc. that are necessary for life within their society. These social "laws" serve to keep disorder within the group to a minimum and provide for its greatest benefit, whether this be safety, protection, care of the young, feeding or some other aspect.

Because there is so much to learn about how to live within such social groups, the anthropoids tend to have prolonged adolescent stages in which to allow adequate time for learning. Most nonprimate animals have, in contrast, very short spans of immaturity and adolescence. By their second or third year they are adult and can mate and raise offspring. Primates spend much longer periods of time before they achieve full adult status. Gorillas do not mature until about ten years or after, and male chimpanzee are socially mature at fifteen, females at thirteen. Humans in modern advanced civilizations are socially immature until about fifteen to twenty years, varying somewhat with the individual and the sex. Such prolonged immaturity is essential to the individual's social and physical development and reflects the complexities of their behaviors that must be learned for success in survival and as a member of a group.

INTELLIGENCE

The learning process is essential for survival in virtually all higher animals according to their lifestyles in a complex world. Intelligence of high level is required of animals that must learn complex social behavior in addition to the other basic skills needed for finding food and avoiding becoming food compared to those that have no real social system. It is no accident that the higher primates are among the most intelligent of animals. It is also deliberate that the first American astronaut into space was preceded by a chimpanzee who was able to follow learned acts under the uncertain influences of the space environment. Such higher primates are intelligent and capable of learning complex behaviors that are not natural to them; NASA considered it better to risk an ape than a human in this great test. Their great abilities to learn and remember are a direct consequence of their greater intelligence. They are simply smarter than all other animals, save one.

Intelligence is a condition or capacity that resides in the brain. Humans have larger and better brains than apes and this reflects our greater intelligence. It is one character clearly related to a particular body structure that has not yet been adequately defined or determined, even for humans. Research into nonhuman primate intelligence is today still in its infancy compared to that of humans. And even though studies of human intelligence have been ongoing for the past century, there has been remarkably little in definitive results achieved for the great amount of research.

COMMUNICATION

Vital to the learning process and directly related to intelligence is the ability to communicate with others, both friends and enemies. Though higher animals and the nonhuman primates lack what we might consider language, all, nevertheless, communicate to some degree. Communication is the ability to convey ideas to others who can react accordingly. Such communication ranges from simple **vocal** and **visual signals** in nonhuman primates to human languages that are virtually infinite and have few limitations. Visual signals are, of course, common and may be easily understood. They may involve the hands, arms or other parts of the body. **Facial expressions**, in particular, are very informative. **Sign language**, using the hands, is an effective means of visual communication and was once common between explorers and natives who spoke different languages. Modern researchers have used formalized sign language to communicate with both gorillas and chimpanzees. **Body language** is a type of visual signal that conveys specific meanings by means of postures of various parts of the body. The tilt of a shoulder or placement of crossed legs can often communicate more than words among humans, and it may be of great importance among the nonhuman primates. Vocalizations are a particular effective means of communication in many higher animals. Fish, amphibians and reptiles regularly communicate using sounds, and birds commonly sing to send messages to other birds. The mammals are a highly diverse group of advanced vertebrates that have many different kinds of vocalizations, growls, grunts, coughs, screams, howls, purrs and others. These comprise large repertoires that are capable of sending a variety of messages. Just imagine the effect that one such vocalization, the roaring of a tiger or lion, would have on most of us when heard nearby in the wild.

SPEECH AND LANGUAGE

Humans have highly developed **speech**, a unique form of vocal communication. Speech is the sophisticated use of vocalizations to convey all manner of messages and meanings. These vocalizations consist of sounds that are combined in various ways to accomplish the intent of the communication. The sounds can be varied by type, tone, volume, speed, inflection, emphasis and other ways to produce symbolic sounds or words as formalized **language**. Furthermore, the words can be united in various combinations depending upon the situation. During the nineteenth and early twentieth centuries several researchers attempted to train apes to speak. All attempts resulted in failure and it soon became apparent why apes could not and would not speak like humans. Unlike humans, these animals lack the necessary neurological centers in the brain and anatomical structures for language and will never be able to speak.

In the 1960s, however, several researchers recognized this restriction and began to teach hand sign language to some great apes. The substitution of the hand language for vocal language avoided the inevitable failure in attempting to teach them to speak. The hand language used was **Ameslan**, **Ame**rican **S**ign **Lan**guage for the deaf taught

Figure 4.16 Koko signing to Dr. Francine Patterson.

throughout the United States. The female gorilla, **Koko,** was reported (Patterson, 1978) to have learned more than 600 different signs during ten years of training. She regularly and appropriately uses some 375 of these (Fig. 4.16). Using these hand signs, Koko makes jokes, swears, teases, invents new combinations of signs for new objects and performs a variety of other human acts. She communicates easily with her trainers, even to the point of expressing her fears, telling lies, remembering events and emotions, understanding of past and future, and signing to a younger male gorilla. Some researchers have objected to this as a demonstration of language by arguing primarily about the interpretation of the results. These objectors voice the opinion that Koko is not really talking but has learned a behavior that provides some rewards. Their argument, however, is much weakened by observations that Koko is able to develop new words (signs) for new objects by combining previously learned though uncombined descriptive terms. For example, she referred to a zebra as a "white tiger." Similar novel constructs from uncombined signs have been reported from other pongids.

One determinant of human language is that it be able to combine and recombine sounds or noises for differing purposes. As such, Koko has fulfilled this determinant and is speaking a language. Different researchers have verified that chimpanzees have also learned signing ability and often use it to communicate among themselves, even when humans are not present. Even more amazing, they are also known to teach the ability to young individuals that have had no contact with human signing.

There is ample evidence for nonlanguage communication, often of a sophisticated form, between gorillas, chimpanzees, orangs, monkeys and a variety of lower animals. However, not one of these animals has demonstrated or developed a language-like form of communication in the wild. Their Ameslan speech and language is known only as a result of human intervention and training. Perhaps we are the ones who require the training in ape communication. Perhaps we have been missing much of their normal communication because we don't know how to read or understand it. A recent novel had humans using a signing gorilla to communicate with wild gorillas. What a marvelous opportunity it would be if brought to fruition. Someday we may be able to use our enlarged brains sufficiently to understand much more of the animal world and their smaller brains, and with it our own ancestry.

Hominid and, especially, human intellectual abilities developed as they evolved. Direct evidence of their intellectual capabilities are demonstrated by their tool cultures. Isaac argued that as species of humans evolved and developed better brains that produced better tools, their language abilities also must have evolved and become better.

GROOMING

One special type of communication and social contact among nonhuman primates is grooming, an activity using their very dexterous fingers (Fig. 4.17) to comb, stroke and clean the hair and body of another individual, commonly in an exchange. Such behavior develops and requires close contact and trust between the individuals of the group, and this promotes harmony, acceptance and cohesiveness that leads to group stability. It also enables the individuals to recognize each other in a very close contact that satisfies their psychological or social needs.

TOOL USE AND TOOLMAKING

Perhaps the most unusual behaviors in chimpanzees are those of tool use and toolmaking. These activities have become well documented in numerous field studies of the past three decades. Tools are any natural objects that are

Figure 4.17 Jane Goodall observing grooming activity of chimpanzees.

used to perform some act. But a great distinction must be made between tool use and toolmaking. Both orangutans and chimpanzees have been observed to use tools in the wild. Chimpanzees use sticks and branches as weapons to frighten other chimpanzees or even to ward off leopards. Kortlandt (1962, p. 138) observed a chimp with a club attacking a leopard and noted that "Apparently chimpanzees use clubs instinctively." Goodall filmed a staged encounter between a stuffed leopard and a band of chimpanzees in which the chimpanzees began to scream and run about after sighting the "predator." One chimpanzee began to thrash a branch around, eventually wielding it more directly against the leopard. In another case, one chimpanzee wielded a branch against another. In all such cases, however, the objects were picked up opportunistically and used as found. Galdikas was able to film (1975, p. 466) a young orangutan inmate at her rehabilitation camp wielding a stick against a young dog, perhaps in play. The dog, however, appeared to take the attack very seriously.

Though tool use is an important step toward humanity, it is not as crucial as toolmaking, which differs substantially. Tools can be natural objects picked up and used as they are found. In toolmaking, however, a natural object is somehow modified and used to fulfill some desired objective. This implies considerable thought as to the function of the tool before it can be modified and used. Chimpanzees often eat termites and commonly modify small branches to poke into the nest holes and remove them. This procedure has been filmed along with the teaching of this interesting behavior. Chimpanzees also drink water from depressions in trees by means of leaves which they have modified as sponges. Though chimpanzee toolmaking is very simple and clearly lacks the degree of sophistication that is found in modern humans, they do make tools.

THE ORIGIN OF BEHAVIORS

Behaviors are described as techniques or ways of living that animals adopt for survival in any given environment. In new or changing environments, animals with given behaviors often find themselves disadvantaged when the old behaviors are less suitable in the new conditions. Either these animals modify their old behaviors or adopt new, more successful ones, or they may well die out and the population, perhaps the entire species, may become extinct. New behavior patterns can arise in different ways. Some behaviors that produce increased fitness can be learned accidentally by an individual with sudden insight, as happened with Japanese macaques that were being studied (see Chapter 6). New structures can appear by random genetic changes in the population and these may generate new and useful behaviors. Hunger brought on by natural causes, such as drought, can cause animals to seek new foods, thus also initiating new behaviors. Similarly, the appearance of new predators into an area can generate new behaviors as the inhabitants avoid capture. Once a new behavior proves successful, it can be taught to others, especially to offspring. These are **learned behaviors**, and all higher species benefit from such learning.

Most higher vertebrates, particularly mammals, learn their ordinary behaviors during the long process of maturing. These animals commonly do not mature for many years and during this time remain with one or both parents until they learn enough to survive. In contrast, many lower vertebrates, species of fish, amphibians and some reptiles, provide no care at all for their young but simply lay their eggs and leave. Any behavior patterns that the young of these species follow are probably not learned from others but must be learned by themselves, or they are inherent. Learning by oneself can be painful if a mistake results in death. Behaviors that are not learned are instinctive.

Instinctive behaviors are inherited genetically. They probably result ultimately from the broad range of successful behaviors that all species develop. Such behaviors are commonly practices selected because of individual preferences; in effect, they are genetic preferences. Those individuals that practice poor or bad (for fitness) behaviors are less likely to survive than individuals that adopt good behaviors. Poor or bad behaviors probably will be removed quickly from the behavioral repertoire of the population and be lost to the species as the practitioners (and their genetic programming) are eliminated. Good behaviors, and their genetic instructions, on the other hand, are those that have proven to be successful by trial and error, and they tend to be continued by offspring. This process has been recognized in many different animals, particularly insects.

Fabre made landmark discoveries (1901, 1918) of numerous cases of parasitism as an instinctive behavior in wasps. Ichneumonid wasps are a very large group with many thousands of species. The females of many different types lay their eggs on other invertebrates, especially on the larvae of butterflies and moths. These hosts are paralyzed by the wasp to serve as foodstuffs while her eggs hatch and grow. The wasp larvae penetrate the host bodies and consume all the edible tissues. They save the essential organs to the last to prevent the deaths of their hosts before they are completely grown. These larvae cannot possibly learn this specialized behavior from their mothers, because the females depart immediately after laying the eggs and have no further contact. The larvae know to burrow into the host and what to eat in order to survive. They save the vital organs until the very last, otherwise the host dies, and without a supply of food they, too, will die. These larvae eventually mature into adults that go through the same process of reproduction by preparing hosts, laying eggs in these hosts, and also by knowing exactly what to do to survive. These wasp larvae have complete sets of instructions for this remarkable and sophisticated behavior contained within them; they do not learn it as they mature. The explanation is, of course, that this vital information is transmitted from the parents to the offspring through the genes in the gametes that produce the new individual.

One remarkable example of this behavior involves the psammocharid wasp that utilizes the trap-door spider as host for its eggs. These spiders are formidable predators with powerful jaws and are fully capable of killing wasps. Female wasps must attack these spiders, overpower them and paralyze them in preparation to lay their eggs. To immobilize the spiders, the female wasps must overturn them and then, dangerously near the spiders' jaws, inject the venom near a particular nerve center to paralyze it. If the venom is injected too far from the nerve center, the spider is not paralyzed and will likely kill the wasp. If injected too close to the center, the spider will die and the laid eggs will not survive. All the essential information for such procedures is not taught to these offspring by their parents, because there is no contact once the eggs are laid. The instinctive ability of these wasps to perform such behaviors must be contained within the genetic materials. It is, therefore, inheritable.

Insects are strongly oriented to such instinctive behaviors. In effect, their small brains are simple computers with relatively simple instructions. Push the right computer button, and you get a particular action. Provide the proper stimulus to an insect, and it behaves in a particular manner. Push another key or provide another stimulus, and the insect will behave in another way. Many insects react to **chemical stimuli**, scents that they detect, by performing particular actions. Ants, for example, will follow a trail of formic acid laid down by another ant to some goal. If such a scent is laid down in a circle, the following ants will blindly follow that circular path without deviation. Theirs is not a thinking behavior, but an instinctive one.

Tinbergen reported (1951) examples of instinctive behavior generated by **visual signals** or **stimuli**, particularly in stickleback fish during reproduction, a crucial time for any species. The male fish apparently recognize the enlarged belly of a female swollen with eggs as its visual signal. This prompts him to perform a zig-zag "dance" over a shallow nest on the water bottom that triggers a response in the female. She deposits her eggs in the male's nests whereupon he fertilizes them. Each of the actions are sign stimuli that trigger the sensory apparatus in the brains of male and female sticklebacks to induce specific motor responses, behaviors that are fixed action patterns. The male behaviors are sign stimuli recognized by the females which trigger further fixed action patterns in her, and so on. These feedback behaviors are clearly inherited.

Mammals have relatively few behaviors that are induced by sensory stimuli like insects. This is probably because mammals live in highly complex environments with situations that change from moment to moment. Each young mammal spends time learning how to survive in rapidly changing situations, and this length of time varies with each species and, perhaps, inversely with instinctive behavior abilities. Humans, in particular, have to survive in far more complex situations than wasps, ants or stickleback fish, and humans apparently have few instinctive behaviors. To survive we must learn prodigious amounts of material essential for survival. We must learn how to deal with the subtleties of the many different situations we experience, including our environment. There is no single solution to most situations but perhaps many different ones. Accordingly, we learn how to cope with driving automobiles, keeping our employers happy, finding mates and raising children who want to be treated as adults. We learn how to purchase the best insurance for our needs, provide for our retirements, submit income tax forms, send

birthday presents and cards to parents and friends, remember anniversaries, write books, go fishing and much more. There are no set paths to success in human behavioral activities in contrast to those of the animal world, and different humans are successful in different ways. These successes require noninstinctive or culturally learned abilities, because they are not the simple acts that most other species must cope with; they are also rarely life threatening. Accordingly, our brains must be very large to provide storage for all the information we need to deal with complex situations

There is still no instinctive behavior in humans for keeping away from hot objects and fire, yet humans have had close contact with fire for hundreds of thousands of years. We are all fascinated by flame and, as children, generally must be burned at one time or another to learn wariness. Simple admonitions, "don't touch it, it's hot," are not warning enough to most of us. We must actually experience pain from flame or heat before we become wary. The lack of an inhibitive behavior in humans is probably due to the short time that we have had fire and the fact that it is not crucial that we avoid it. In fact, we probably welcome our association with fire because of our dependence on it for warmth and for cooking. Such inhibitions against fire will never become instinctive in humans because of these contradictory aspects of value and fear. Many adults also, who should know better, are not repelled by "Wet Paint" signs but are compelled to touch such a posted surface.

Humans are probably born with only a few instinctive behaviors, though some workers list many. These are seen best in infants who have not yet learned to overcome their inhibitions. Of vital importance is the ability to carry out complex muscular movements of suckling which are necessary for infants to feed on their mother's milk. An infant's system is still too delicate at this stage to tolerate other foods and would starve without mother's milk or a substitute. Another instinctive behavior is the solid and strong grip of an infant's hands. Infants will cling with surprising strength to an object placed in their hands; they grab tightly and may not easily let go. This grasping reflex is probably a remnant of the critical hand grip exhibited by monkey and ape infants that cling to their mother's body hair. Their very survival depends on this holding ability as they are carried through trees high above the ground. In the arboreal orangutans, for example, the infant clings to its mother's hairy coat for the first twelve months of life and may never leave it during that entire time. Though humans do not move through the trees and subject infants to such danger, we have not yet been separated from the apes long enough to have lost this trait. Beyond these two instincts which are vital to infant survival, human behaviors are probably too complex to be reduced to purely instinctive reactions. Though we may tend to follow general patterns, instinctive behaviors are of relatively little value to us because we no longer live in purely wild environments where automatic reactions are desirable. We live in societies that present different dangers and stresses so new and complex that we have not yet evolved instinctive behaviors against them, or may not be able to evolve.

All of these views lead us to conclude that some of the behaviors in humans and other primates, primarily the social behaviors, are largely the result of or strongly influenced by genetic factors. The behavioral similarities between humans and apes and other primates are so striking that mere coincidence will not explain any of them. Primate behaviors that better fit an individual for survival within any given environment are more likely to be genetically selected than those that reduce fitness. This is because better fitness confers survival advantages over those behaviors that reduce fitness in the environment. The genetic systems ordinarily provide for a broad spectrum of behaviors because of randomness in the gene pool, but some behaviors develop greater fitness than others. Better fitness includes reproductive success, and individuals with these desired characteristics probably produce offspring with similar traits. The behaviors or traits that tend to reduce fitness are disadvantageous and tend to be self-eliminating. In time, advantageous behaviors permeate the gene pools and societies of primates as natural selection operates to eliminate unsuitable ones.

The survival advantages gained by any species or individual are not necessarily in terms of longevity of the individual, strength, intelligence, speed or some other aspect, though all contribute to better fitness. Any advantages gained must be related to successful reproduction, which is essential for the survival of the species. Successful strategies or behaviors that produce truly successful offspring tend to increase in the gene pool at the expense of unsuccessful behaviors. In this way species are "guided" by evolutionary chance that confers selective advantages to certain animals over others. Once such advantages appear, the recipients become locked into a pathway from which they may not easily escape. The conservative apes living in forests and jungles of Africa and Malaysia fit into this model, and their fate is beyond their control. They do not seem to have the potential to change but will survive or become extinct at the hands of humans. Different advantages appeared in the hominoid stock that became human when our ancestor hominids took a different pathway.

SUMMARY

Darwin had deduced that humans had descended from apes, not the living ones, but the stock that had given rise to them. Examination of the detailed anatomy of fossil primates, essentially unknown in Darwin's time, strongly supports this view. Today, scientists can demonstrate this relationship in many ways, physiologically, skeletally, structurally, molecularly and behaviorally, and this evidence is impressive in scope, content and specificity. The evidence shows that we are closely related to the living great apes, most closely to the chimpanzees and slightly less to the gorillas, and that we are less closely related to other primates according to the postulated lineage.

The view that humans were linked evolutionarily to other primates, especially the great apes, was based primarily upon anatomical features just a few decades ago. The anatomical features of humans and the great apes can easily be seen by all. Skull features, dental patterns, lack of a tail and postcranial bones with scars indicating musculature all supported this interpretation. Other features support this hypothesis, too. Digestive systems, blood, endocrine systems, reproductive systems, and more all bolster this relationship. In recent years, new techniques in biochemistry have been developed to allow for analyses that are even more conclusive in establishing evolutionary relationships. Humans and living apes share a great many biochemical similarities—blood types, DNA makeup, blood amino acid sequences, serum antibodies and more. To explain these similarities as other than evolutionary simply begs for coincidences beyond acceptability.

At the same time, impressive, long-term, field behavioral studies of the great apes by several dedicated scientists have more clearly identified their ways of life. These, too, support an evolutionary relationship between humans and the apes. Social behaviors of chimpanzees and gorillas have close similarities to those of living humans, so close that it is impossible not to see the parallels. The very nature of chimpanzee and gorilla group behaviors reflects what we see every day in our own lives. Hugging, touching, comforting, kissing, playing, disciplining, teaching and more are recognizable in these apes, particularly the chimpanzees. For years experts drew distinctions between humans and the other animals, because only humans used tools. Then workers noted a number of species using tools, among them chimpanzees, orangutans, several species of birds and sea otters. The argument separating humans and other animals then became one in which only humans made tools. Now this behavior, too, has been observed in chimpanzees who have learned to make several different kinds of tools and have passed this practice to their offspring. It seems that as we continue to study our primate relatives, we find more ways in which we are similar and more evidence that we are closely related.

The pathway that humans followed in evolving from other primates was not very clearly defined until just a few short years ago. All the lines of evidence that have been examined demonstrate a close relationship in advanced primates in no uncertain terms. Today the arguments against human evolution from earlier primates have dwindled mostly to nonscientific clichés.

STUDY QUESTIONS

1. Describe the differences between human jaws and teeth and those of a gorilla.

2. What is the function of the supraorbital crest in gorillas? Of the sagittal crest?

3. Describe the evolutionary pathway in primates of the 5-Y dental pattern.

4. What is the function of the diastema?

5. What is a simian shelf? How do humans compensate for the lack of this feature?

6. What is the significance of the position of the foramen magnum in apes and humans?

7. If chimpanzees and humans are different species, explain why hybrid chimpanzee and human DNA can have a 97.5 percent fit in the amino acid sequences.

8. How do brains of humans differ from those of the great apes?

9. Why is absolute brain size not a reliable tool in separating species of primates?

10. Given the fact that the alpha and beta hemoglobin amino acid sequences in chimpanzees and humans consist of 287 amino acids and are identical, does such a pattern suggest randomness or some other relationship?

11. How are instinctive behaviors different from learned behaviors?

12. What evidence supports the belief that instinctive behaviors are genetically transmitted?

13. What is the distinction between toolmaking and tool use?

5 The Primate Pathway

INTRODUCTION

Hominoids (manlike) are advanced primates, living and fossil, that include the great and lesser apes and humans. Species of humans and members of the genus *Australopithecus* are termed **hominids** (man, the family of man). Prior to 1940, the fossil evidence of hominids was relatively sparse and poorly known from scattered finds in Europe, Asia and Africa. Workers speculated on the relationships between these few forms and their significance, sometimes correctly in light of later information, often inaccurately. Family trees were developed and proposed, but they were not particularly illuminating because of the inadequacy of the meager fossil record. Too few fossil finds were available to accurately portray the populations of these species as we know them today and the construction of family trees suggested relationships that were speculative and not satisfactory. However, this early muddled situation provided adequate incentives for researchers to seek and discover "missing" fossils that many believed were yet to be found and that would clarify these relationships. Some scientists devoted their careers to finding these "missing" fossils but only a few were successful.

Since 1945 numerous fossil primate specimens—hominoids, hominids and other fauna that shared the same environments—have been found, including new genera and species. The numbers of these diverse and well-preserved specimens have allowed reconstructions that permit clearer understanding of these populations, their nature and variability. Where previously a few known specimens allowed some limited interpretations of the species and their relationships, the newly discovered specimens included much associated supporting evidence. These new specimens provided a much better understanding of these species and have allowed us to refine and expand their definitions, especially in relation to their variability. Furthermore, we have now been able to draw the evolutionary family tree of primates and humans with bolder and sharper lines and with much greater detail. We now understand more about the environments in which they lived, their way of life, the evolutionary pathways that they followed and their relationships to each other and to modern humans.

THE ANCESTRAL PRIMATES

The ancestral primates evolved slowly (see Fig. 3.1) throughout the Paleocene and Eocene Epochs, remaining as relatively small arboreal animals thriving in the security of their tree habitats. By Oligocene time a great variety of monkeys had appeared and spread throughout the Old World. During this time also appeared *Aegyptopithecus*, which is generally regarded as the ancestor of higher primates or an important member of that line. This species, or a very close relative, gave rise to the most important anthropoid types of the Miocene, that group of ancestral apes known as the **dryopithecines** (Fig. 5.1). These apes became widespread during the Miocene and have been found in Europe, Africa and Asia. Their widespread distribution in different environments resulted in the appearance of population differences as they responded to the various pressures imposed on them by the local environmental conditions. These populations then continued to be shaped by these pressures and evolved into several species, all part of a great cosmopolitan stock. Today these dryopithecines are highly regarded as the ancestors of the living hominoids, the apes and those forms that eventually became humans (Fig. 5.2).

FOSSILS OF NONHUMAN PRIMATES

Fossils of primates are far from common. Despite the fact that primates have lived for at least 65 to 70 million years of Earth history, they are much rarer than Paleozoic organisms that lived more than 250 million years ago. This paradox is due primarily to the relatively small numbers of primate species and individuals and the likelihood of preservation in the environments they inhabited. Birds are probably equally unlikely to be found as fossils because of their habitat and way of life and the poor prospects that they will be quickly buried after death. Birds are especially vulnerable to destruction because their delicate bones are thin and hollow to conserve weight; they are much less resistant to weathering, bacteria and scavengers than the bones of other, nonflying animals. At least the bones of primates are amenable to fossilization and their teeth, which are very resistant, are exceptionally well suited to it.

The primate way of life in trees and, uncommonly, on the ground offers less chance for preservation and fossilization than does life in the seas, where burial by shifting sediments is a common occurrence. Primates, furthermore, are pursued by predators that quickly consume the soft

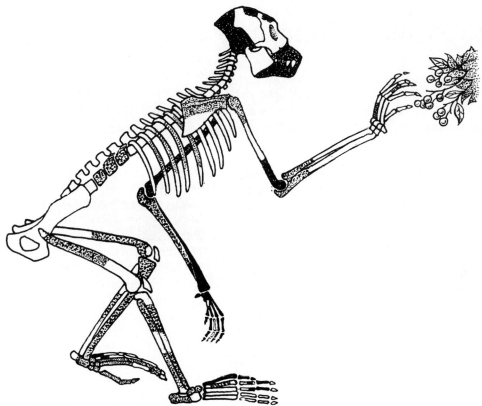

Figure 5.1 Reconstruction of *Dryopithecus africanus* skeleton.

tissues and discard the broken and disarticulated bones on the surface of the ground where they are commonly destroyed by natural causes. Those primates that die of natural causes typically also come to rest on the ground where they are likely to be consumed by scavengers, including hyenas, jackals, wild dogs, lions, leopards, vultures, among others, and also by weathering and bacteria. Only rarely will primates suffer quick burial so that they stand a fair chance of being fossilized.

Streams and lakes offer the best sites for such random burials. There primates may drown and then be buried in sediments; perhaps they wash down streams into lakes. But all such burial sites are temporary in geology, because they are subject to later events that may destroy them. These deposits may be eroded as rainwaters fall and as streams shift their courses, or the strata may be tilted, overturned and fractured as the landscape is uplifted in crustal movements. Such burial sites may also be covered by other layers as volcanic eruptions spread ash and lava across the countryside. Many fossilized primate remains may not be uncovered by subsequent erosion for millions of years, another element of chance in the long line of events.

Because primate fossils are relatively rare, they have not yet provided us with a complete picture of their evolutionary history. Numerous gaps in specific lineages exist without any clues to intermediate forms. Slowly but surely, however, these gaps are being filled as workers continue to make finds. A classic case is that of the deposits in the Fayum of Egypt. For many years only a few fragmentary primate fossils were found in these important beds of critical age. Then, with concerted effort, Elwyn Simons and his team found a great many primates fossils that significantly increased our knowledge of their life in this area during Oligocene time. One difficulty in reconstructing such primate lineages is that the earliest primates were usually small. Finding them is a problem.

A second difficulty in dealing with early primates is the diversity of opinion among experts as to the validity of

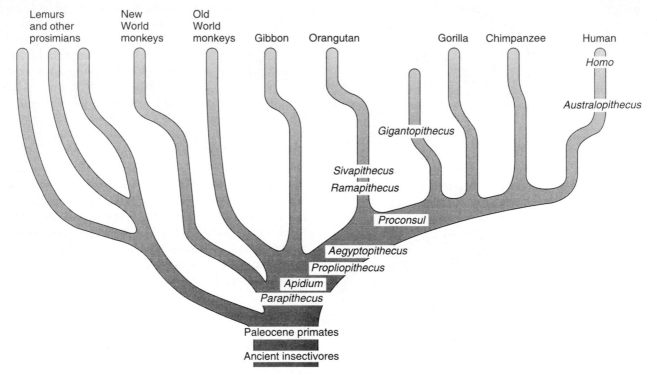

Figure 5.2 Hominoid and hominid lineages.

particular fossil characters that allow assignment to one species or another. Different workers will emphasize one trait over another. When the fossils consist of incomplete remains and broken fragments, decisions are sometimes arbitrary. The subsequent discovery of more complete and better-preserved fossils often results in major shifts of emphasis and interpretations. One particular form, *Pronycticebus gaudryi* (Fig. 5.3), was originally (Buettner-Janusch, p. 113) considered a tarsiod species. An expert subsequently decided it was a lemuroid. Still later, a third worker associated it with the lorises. Such inconclusiveness hampers adequate reconstruction of evolutionary paths and relationships but must be accepted and understood until such time as adequate fossil remains allow for acceptable conclusions.

PALEOENVIRONMENTS AND PRIMATE FOSSILS

The life of the past has been preserved as fossils in strata that record ancient environments. Such environments can be identified through the features that are preserved in the sedimentary rocks. Ancient deserts, swamps, river systems, ocean shores, beaches, sand dunes and other environments have all been recognized in strata. Within each of these environments are features and characters that reflect the natural dynamic forces that initiated the deposition and created the sedimentary features. Rivers, for example, have meandering and straight paths, sand and gravel bars, shallows and deep channels, eroding banks, depositional features, and more; all were carved and deposited by the energies of the waters that flowed across the land surface to the seas. Each part of any environment was produced by slightly different conditions of energy and directive forces that formed, eroded and deposited slightly different sedimentary beds and features. Subtle differences in sediment grain sizes, mineralogy and sedimentary features are the principal clues. In addition, animals and plants lived in these environments and were fossilized. These fossils may be characteristic of particular microenvironments if they were fossilized where they lived and have not been moved by some agency after death.

The available geologic evidence indicates that the Cretaceous Period was a time when great changes affected

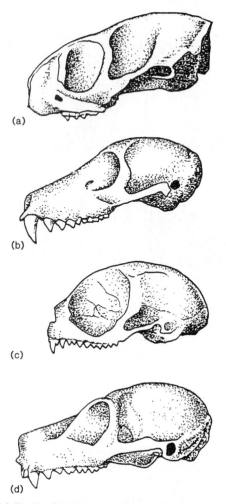

Figure 5.3 Skulls of (a) *Pronycticebus gaudryi* (X1.25), (b) Ring-tailed lemur, *Lemur catta* (X1), (c) Tarsier, *Tarsius spectrum* (X2) and (d) Loris, *Galago garnetti* (X1.3).

Earth's surface. The first half of that period was relatively peaceful with stable conditions that had existed for tens of millions of years since the Jurassic Period. Great shallow seas stretched across western parts of the United States from the Arctic Ocean nearly to Mexico and washed along shores that varied in topography. The eastern portion of this sea ranged through the midcontinent region depositing great limestone beds. The Appalachian Mountains in the east had been worn down to a low range by streams that eroded sediments and deposited them both eastward and westward along ocean shores. Forests covered the land-

scapes in the east and west and herbivorous reptiles fed on plants that filled these regions. Carnivorous dinosaurs, some of great size, stalked the animals that abounded, including smaller reptiles, amphibians and the great herbivorous dinosaurs. Birds found safety in the trees where predators were less able to catch them. Mammals, which had appeared during Triassic time some 60 to 75 million years previously, were still small creatures that survived, perhaps because of their size and agility, but they had not yet experienced great success as a group.

During the latter half of Cretaceous time, profound geologic changes in the landscape began to take place. In the western part of North America the Laramide Orogeny forced rock layers into contortions, and uplifting took place from Alaska southward into Mexico as the Rocky Mountains were formed. The seas that had covered the middle of the continent began to retreat as the sea floor rose. The older, more primitive plants which covered the landscape were being replaced by a newly evolved type of plant, the more efficient angiosperms, those flowering plants so familiar to us today. The precipitation that had nourished vast forests was reduced when the rising Rockies began to block off the moisture-laden, prevailing westerly winds. These effects continued into the Cenozoic Era and the forested regions of the midcontinent east of the Rockies gradually became the almost treeless plains we know today as the semiarid western interior of the continent, the Great Plains.

In other parts of the world, similar changes were taking place at the start of Cretaceous time. The continents were still largely interconnected but this soon changed as the forces of plate tectonics moved the crustal plates apart to eventually produce the global pattern we see today. The North Atlantic basin began to open with these plate movements and delineated the present continental shapes of North America and Europe, though the two continents were still connected at Greenland. Europe was largely above sea level during early Cretaceous time, but shallow seas slowly advanced and covered almost all of it. These shallow seas contained great quantities of organisms whose limy hard parts, after death, accumulated as massive limestone deposits that characterized the Cretaceous Period. In Asia, shallow seas began to flood the landmass from the north, but this flooding was incomplete. In Africa, the northern half was flooded by shallow-water seas, though the southern half remained above sea level.

In this global setting mammals underwent significant evolutionary changes into great variety as they responded to the major modifications affecting the landscape. These variants were evolutionary experiments in survival in the changing environmental conditions; some succeeded but many failed. Dinosaurs experienced a gradual decline until they finally disappeared around the end of the period. Whether by a great calamity following one or more great meteorite impacts or by some other mechanisms, the dinosaurs died out at the end of the period. Mammals expanded into this void as different experimental types found some favorable environments. The dominant mammals of today, the placental mammals (**Eutheria**), had not yet overshadowed the marsupials (**Metatheria**) and the monotremes (**Protheria**). It was in these late Cretaceous environments that the important newcomers of the animal world, the primates, appeared.

CRETACEOUS PRIMATE FOSSILS—135–65 million years ago (mya). The Cretaceous Period ended about 65 million years ago with a changed landscape on Earth. The new landscapes that developed before the end of the Cretaceous became the habitats for the primitive primates. These were the new results of evolution of existing mammalian stock as they "experimented" with ways to become better adapted and to survive in the changing environments. Despite their obvious evolution at this time, the fossil evidence for Cretaceous primates is very sparse. *Purgatorius ceratops*, the oldest known primate, is represented by a single tooth from late Cretaceous beds of Montana. Another species of *Purgatorius*, present in slightly younger beds of Paleocene age in the same area, is much better known but consists only of some fifty teeth.

PALEOCENE PRIMATE FOSSILS—65–54 mya. The evidence from Paleocene fossil plants strongly suggests that the environments of the northern hemisphere were subtropical to temperate with extensive forests. Only thirteen primate genera are known to have inhabited these environments during this epoch. Six genera are known from middle Paleocene strata, in addition to *Purgatorius*, which survived from the late Cretaceous. These six genera are all found in the United States in a belt extending through New Mexico, Colorado, Wyoming and Montana. One of these middle Paleocene primates, the arboreal form *Plesiadapis* (Fig. 5.4), is found in both France and Wyoming and demonstrates a land connection between Europe and North America, presumably through Greenland. This

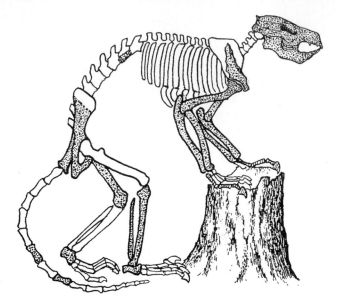

Figure 5.4 Skeleton of *Plesiadapis*.

genus had flattened claws like those of the living tree shrews rather than the fingernails of typical primates.

By late Paleocene time several more primate genera appeared in North America, Asia and Europe, some from all three continents, and these demonstrated a spread across the Northern Hemisphere. The faunas of South America are totally different from those of the Northern Hemisphere. These distributions suggest that primates arose in the latter. No primates of this time are known from Africa, Australia or southern Asia.

EOCENE PRIMATE FOSSILS—54–35 mya. The environments of the Eocene Epoch were similar to those of Paleocene time, though they were warmer and wetter as shown by the fossil plant evidence. The similarity between numerous mammals in both Europe and North America indicates that the Greenland land connection probably continued. The Paleocene primates had died out and were replaced by prosimians that underwent a maximum radiation into a variety of habitats. In some respects, this adaptive radiation provided hints that foreshadowed human evolution some 50 million years later; several types of prosimians developed primate traits that became more emphasized in the anthropoids. These included enlargement of the brain, shortening of the snout and erect posture.

In comparison to the brains of earlier Paleocene primates, the brains of Eocene primates were enlarging and

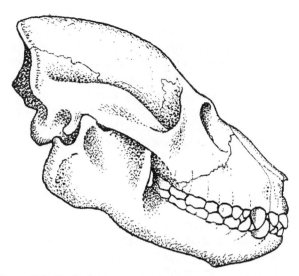

Figure 5.5 Skull of *Adapis parisiensis*.

lengthening as different sensory and motor areas were emphasized. Their snouts began to shorten as they developed greater reliance on hand, eye and brain ability and less reliance on the olfactory senses, which depended on the numerous receptors located in the snout. The foramen magnum in their skulls was shifting from the rear area to a position more forward and underneath the skull. This indicated a change to an erect posture with the head atop a more vertical spine similar to that of living prosimians.

More than forty genera of primates are found in Eocene strata. They are probably not representative of the primate faunas that existed during this nine-million-year time span, but they do indicate the difficulties of finding fossilized primates. These finds are scattered mostly through North America, but seventeen genera have been found in Europe, China, and Burma. The early Eocene has but one non-North American primate genus of the thirteen known. By middle Eocene time, however, thirteen genera are known from North America, Europe has eight genera and China has one. During late Eocene time, only five genera are known from North America, four from Europe, one from China, and two from Burma.

The North American and European Eocene prosimians represent an important collection of types that show affinities with the living lemurs, lorises and tarsier. One group, the Adapidae, appear to be the ancestral stock for the modern lemurs and lorises. A member of this group is *Adapis* (Fig. 5.5) from the gypsum deposits of Montmartre

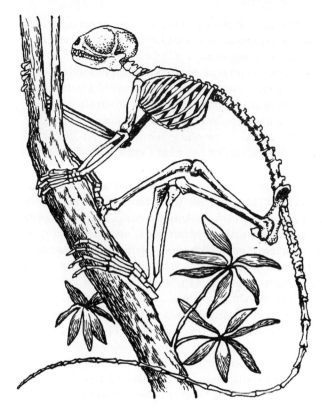

Figure 5.6 Skeleton of *Smilodectes*.

in Paris, the first discovered genus of primates and described by Cuvier in 1821. Two other genera, *Notharctus* and *Smilodectes* show affinities to modern lemurs in grasping hands and feet with opposable fingers, elongated snout, orbits at the front of the skull and other skull details. *Notharctus*, a well-studied species from Wyoming, was interpreted by Gregory (1920) and others as having advanced skeletal specializations of hands for grasping and legs for leaping. Simons (1972, p. 138) associated these and other characters with the vertical-climbing-leaping traits of living species. *Smilodectes* (Fig. 5.6), also from Wyoming and very similar in many skeletal features to *Notharctus*, is perhaps the most completely known of all pre-Pleistocene primates. Another group, the Anaptomorphidae, are known only from North America and appear to have been ancestral to tarsiers. They are characterized by shortened snouts, reduced size of front teeth and cheek teeth that have tarsiod traits. *Tetonius*, an abundant and well-known genus from Wyoming has important cranial and dental features that support its tarsiod tendencies.

Radinsky (1967) emphasized the importance of the early appearances of some characters found in these two stocks by noting that

> The occurrence of enlarged occipital and temporal cortical areas in such old tarsiod and lemuroid prosimians as *Tetonius* and *Smilodectes* suggests that expansion of those regions . . . may have been one of the critical adaptations responsible for the Early Eocene radiation of the Primates.

Two primates from the latest Eocene or earliest Oligocene of Burma were for a time considered by some experts as the earliest Anthropoidea. These two species, *Pondaungia* and *Amphipithecus*, are known only from fragmentary jaws with some teeth. *Amphipithecus* was originally found in 1923 and *Pondaungia* in 1927. More specimens of both genera were found from 1977 to 1978 and clarified the position of these creatures. Ciochon studied both genera and believes (1985, p. 35) that

> Both fossil forms exhibit a combination of lower and higher primate features, with the latter considerably more predominant, indicating that they were at or across the evolutionary transition from prosimian to anthropoid.

Amphipithecus has jaw and teeth features that are similar to those in all anthropoids, but not those in prosimians.

OLIGOCENE PRIMATE FOSSILS—35–25 mya. General cooling began to affect the climates of North America and Eurasia during Oligocene time and primate fossils were sharply reduced in numbers, though rich mammalian faunas thrived. The primates may well have retreated southward in search of warmer climates, reflecting current primate distributions. Only two Oligocene primate genera are known from North America, each from a single specimen; none are known from Europe or Asia. Extensive temperate forests developed and covered northern Europe, and the Mediterranean Sea suffered serious short-term reduction in size. To the south in Africa, tropic conditions continued and a very important fossil primate locality developed in Egypt in the Fayum. The Fayum is a low region in the desert about 65 miles southwest of Cairo and 120 miles south of the Mediterranean shore. Much of the region is irrigated by a canal from the Nile River and highly cultivated today. During Oligocene time the Mediterranean Sea stretched south beyond the Fayum and deposited marine strata throughout this region. These seas periodically advanced and retreated, alternatively inundating the region and forming marine beds, then retreating from it and allowing rivers to flow across the former sea floors and deposit terrestrial sediments. These alternating marine and terrestrial strata record the shifts of the Mediterranean Sea levels in a complex sequence.

Tropical rain forests formed in the area of the Fayum as rivers meandered sluggishly through the region during low sea levels to deposit six hundred feet of sediments. There is no evidence to indicate these forests had any open areas, and they are believed to have been quite thick. The extent of the forests is indicated by the fossil tree trunks found there, some of which are nearly one hundred feet long. Other plant fossils, mangrove trees and water lilies, indicate the area had abundant water; these required wet conditions, either in sluggish waters along the streams or in ponds and swamps. A rich fauna of mammals and other animals lived in the streams and surrounding forests, and some were preserved as fossils. Various aquatic forms are known, including lungfish, catfish, crocodiles, gavials and turtles. The land dwellers include tiny carnivores about the size of weasels, elephants of both small and large sizes, and a rhinoceros-sized herbivorous species. Bats, rodents and four primate genera are also known in these beds that are dated at about 24 to 27 million years.

The first fossil of a primate found in the Fayum was a mandible with teeth found by Richard Markgraf in 1908 and described by Schlosser as *Parapithecus*. Seven or eight fragments from the Fayum represented the total of Oligocene primate fossils from the entire world until Simons found many more during the Yale expeditions that began in 1960. In 1964 a single primate skull, *Rooneyia*, was found in Texas, doubling the number of species from North America. Today more than one hundred specimens of numerous species are known worldwide from the Oligocene. The oldest of these is *Oligopithecus*, found in 1961 and dated at about 32 million years ago.

The primate fossils present in the Fayum, including *Parapithecus*, *Apidium* and *Oligopithecus*, show that the primate stock was starting to diverge into different lineages. *Parapithecus* and *Apidium* demonstrate several characteristics that place them close to the ancestry of New World monkeys. They have the same dental formula (2-1-3-3) as the Eocene genus *Amphipithecus*, a form acknowledged to be near the ancestry of these anthropoids. *Oligopithecus*, however, has a different dental

Figure 5.7 Skull of *Aegyptopithecus zeuxis*.

formula (2-1-2-3), the same as Old World monkeys and humans, and may be nearer the hominoid ancestry than *Parapithecus* and *Apidium*. The best-known primate from these strata, however, is *Propliopithecus*, originally thought to be ancestral to gibbons, but now considered to be a more generalized type. Simons (1972, p. 214) believes that another Fayum species, *Aeolopithecus*, may be even more closely related to the gibbon and that *Propliopithecus* may be closer to the dryopithecine stock that followed.

The most important species from the Fayum beds is *Aegyptopithecus zeuxis* (Fig. 5.7), found as a wind-damaged jaw in 1906 but undescribed until 1963 (Simons, 1965) and not really well known until 1964, when Simons found several more specimens. This is the largest Fayum primate and is dated at 33 to 34 million years ago. About the size of a cat, *Aegyptopithecus* could have evolved from *Propliopithecus* on the basis of several features. Several jaws have been found in two sizes suggesting sexual dimorphism, with the larger jaws probably representing males. Limb bones show that it was an arboreal quadruped. Skulls found in the early 1980s show that the species had a small brain, a short face and a prominent sagittal crest to anchor powerful jaw muscles. In these respects Simons reported (1984, p. 20) that it resembles numerous members of the later important hominoid group, the dryopithecines, particularly *Proconsul* of Miocene age, the hominoid *Sivapithecus* of Miocene age which descended from the dryopithecines, the robust Pliocene hominid *Australopithecus*, two great apes, gorilla and orangutan, and macaques.

These Oligocene primates from the Fayum clearly demonstrate that during Oligocene time an important branching was developing within the anthropoid stock. The main stock continued in a conservative fashion and led to the monkeys. Another branch may have given rise to New World monkeys at this time or earlier. And a third became more innovative and led through *Aegyptopithecus* to the apes and ultimately to humans. The same features that tie these various advanced primates with *Aegyptopithecus* also demonstrate the relationships of the pongids with each other.

MIOCENE PRIMATE FOSSILS—25–12 mya. The environments of the Miocene Epoch in North America continued the changes that began in Oligocene time. The great mountain ranges of the western part of the continent continued to rise along a Montana-New Mexico axis and further restricted the spread of the moisture-laden winds from the Pacific Ocean to the continent interior. As these winds lost their moisture in crossing the ranges, the continental interior dried out and gave way to the Miocene grasslands that became the Great Plains. At the same time, the great forests of Cretaceous times shifted southward and with these shifts in environments came a great change in the animals that inhabited the region, particularly the grasslands. Camels, horses, rhinos, bears, foxes, rodents, peccaries and other types spread through these grassy plains of North America. But not a single primate fossil has been found and presumably these arboreal denizens had moved southward with the forests.

The Eurasian forests of the time were massive, extending from Spain to China and southward beyond the Mediterranean into Africa. The diversity of Miocene animals that lived in these great forests, in contrast to those of North America, was great and included many types recognizable as related to modern forms. Cats, dogs, pigs and tapirs diversified throughout the Old World and Eurasia had bovids, deer, giraffids and hyenas. A great land bridge between Asia and North America developed across the Bering Straits when sea level dropped to expose a very broad continental shelf. Many species crossed from one continent to the other increasing the diversity on both landmasses. The African mammoths, for example, spread north and east to Europe, Asia and America, and the American horses and camels spread westward to the Old World.

A few fossils of New World monkeys appeared in South America at this time, possibly marking the start of that group. Some primate fossils in Africa are distinctly Old World monkeys and they record the beginning of that group. The most important changes in primate faunas, however, were taking place in East Africa. There, the descendants of the late Oligocene Egyptian ancestral genus *Aegyptopithecus* are found in Miocene and younger strata where they had evolved into the specialized types,

the hominoids, ultimately changing the face of Earth. Though monkeys were present in the vast forests of Europe during Miocene time, they could not overshadow the early apes that had begun to appear in these forests. These apes had apparently spread from an origin in Africa to southern Europe, where they were first identified.

In France in 1837 Lartet found the third primate fossil ever discovered, an ape mandible complete with teeth from Miocene strata about 16 million years in age. Named *Pliopithecus*, it has dentition similar to that of living gibbons and other features that Simons believes relate it to the gibbons, either by direct or near-direct descent. Other workers, however, do not agree with such a relationship. A similar type of ape, *Limnopithecus*, was found in Kenya nearly a century later in 1933. These two genera demonstrate a close relationship, even though found on two different continents, and it has been suggested the two should be combined into one genus. These types, acceptably termed apes, were likely descendants of *Aegyptopithecus* along with the stock that produced the great apes and humans.

Toward the end of the Miocene Epoch, the vast forests of the Northern Hemisphere began to shrink as the climate continued a cooling and drying trend. Forests disappeared in most places and replacement grasslands spread throughout the Old World. Savanna faunas flourished in these vast grassy plains, particularly herd herbivores. The dryopithecine apes of East Africa that were thriving in the Miocene rain forests, spread into southern Europe, across Asia and into China. When the forests declined, the dryopithecines also began to suffer losses in keeping with the changing environments. New versions of these apes began to develop as natural selection favored certain innovative models that could tolerate these changed conditions. One new stock of these apes, known as *Ramapithecus*, appeared during late Miocene time in Africa and spread through the Middle East into Pakistan. The genus *Sivapithecus* also lived in these Asian areas where it was abundant and represents the best-known middle-late Miocene hominoid.

Given these changed circumstances, the interval from the latter part of the Miocene Epoch to the early part of the Pliocene was a time of great importance in primate evolution. Important evolutionary stocks that had appeared during the Miocene evolved into a variety of forms in Pliocene time. These new forms eventually gave rise to the great apes and, more importantly, the hominid line that eventually became humans. But the division between Miocene and Pliocene time is not easily determined in the strata. Exact age determinations are difficult to obtain and the picture of anthropoid evolution and radiations near this boundary are somewhat indeterminate.

PLIOCENE PRIMATE FOSSILS—11–3 mya. The transition from Miocene to Pleistocene time is difficult to pinpoint in terrestrial strata, and some of the important primate-fossil-bearing beds are inconclusively assigned to late Miocene-early Pliocene strata. Clearly, the dryopithecine apes appeared in Miocene time and gave rise to significant descendant stocks in Pliocene time; however, establishing the precise dates for these events is difficult in the absence of radioactive materials associated with the fossils. The Pliocene record in Africa is generally poor, except for the latest Pliocene time from about 6 or 7 to 3 million years ago. Fortunately, strata of this particular interval is well developed in the Lake Turkana area of the East African rift valley and in the Omo district of Ethiopia, both areas that have yielded important fossils that have helped clarify human ancestry.

The most important of these Pliocene primate fossil finds are the australopithecines, the first hominids and the genus believed to have given rise to early forms of our own genus *Homo*. This group was first identified in South Africa in 1924 with the discovery of *Australopithecus africanus*, a small walking biped. Dart, who discovered and named this species, insisted that it was a direct ancestor of mankind because of its bipedalism, essentially human teeth, and the relatively large size of its brain. Another species of more robust nature, *A. robustus*, was also found in South Africa. Different workers interpreted these australopithecines as hominids or hominoids amidst much controversy. This situation continued until 1959 with proponents for the hominid nature of australopithecines pitted against opponents. Then, Louis Leakey found another advanced primate, *Zinjanthropus boisei*, associated with primitive stone tools at Olduvai Gorge, Tanzania.

Though initially believed by Leakey to be a new, perhaps toolmaking genus, *Z. boisei* has been reassigned to *Australopithecus*. Today, however, it generates disagreements about its relationships and position because some workers believe it to be a northern specimen of *A. robustus* while others consider it to be a separate and third species of *Australopithecus*. This specimen is, nonetheless, of immense importance to human evolution in another

way because it was recovered from strata interbedded with volcanic materials. These volcanics permitted radioactive age-dating by analysis of their potassium-argon content. Repeated analyses by different workers produced age dates of about 1.5 million years, more than double the predicted and previously accepted age of mankind. *A. boisei*, thus, ushered in a new age in human evolution.

Leakey had found exceedingly primitive worked stone tools associated with these specimens recovered at Olduvai; they are now known as tool of the Oldowan culture. It was initially speculated that *A. boisei*, was the creator of these tools. However, this species has a very large and powerful jaw, and these aspects strongly indicate a species that dined on very coarse plant materials that required a powerful chewing apparatus. These structures and abilities tended to argue against a toolmaking ability. Fortunately, a second type of hominid was found in Bed I at Olduvai in 1960. This second form consisted of fragments of various body bones that appear to be more modern than those of the australopithecines. On these grounds, Leakey identified this specimen, too, as a hominid but considered it the earliest human. He therefore named it *Homo habilis*, the toolmaker.

THE DRYOPITHECINES

Beginning with some fragments found by Fontan in 1856, a great many discoveries of primates fossil from various sites throughout the Old World have provided an involved picture of a complex and important group of advanced hominoids that arose during the late Miocene-early Pliocene interval. Collectively known as **dryopithecines** after the first genus, these hominoids have anatomical structures and features that allow them to be considered ancestors to the great apes and the hominid lineage that became humans. The dryopithecines were first found in Europe, then in Asia and finally in Africa. This restricted distribution supports the view that, although they were a widespread group living in a variety of different environments in the Old World, they could not spread into the New World because no connecting land bridges were in existence during the time when they were evolving. As these advanced primates spread across Africa, Asia and Europe, they responded to the slightly different environmental pressures by adapting in slightly different ways as they became new species. In time, these new species were also able to gradually evolve into additional new populations under the varied environmental conditions, and these

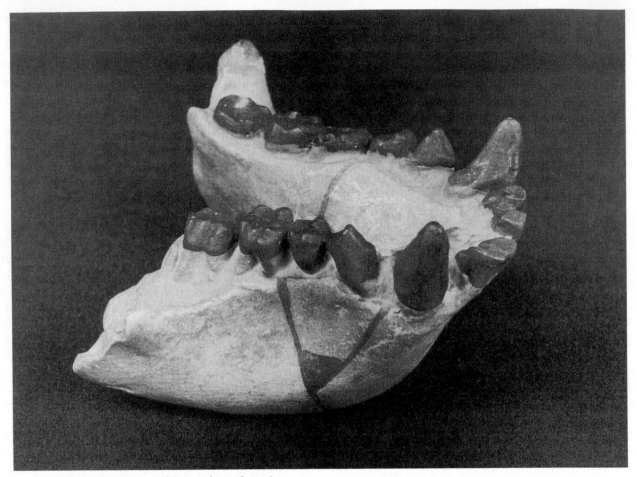

Figure 5.8 Reconstructed jaw of *Dryopithecus fontani*.

newly evolved forms gave rise to the great apes and eventually to humans.

The first find (Fig. 5.8) was from France in late Miocene strata dated at 14 million years ago. It was named *Dryopithecus fontani* after its discoverer by Lartet. The name *Dryopithecus* refers to the oak leaves found in other sites with related faunas and indicates that Lartet believed the animal was a forest creature. The original fossil consisted of jaw parts with some teeth from a chimpanzee-sized animal; a partial limb bone was found subsequently at the original site. Lartet considered this creature to be an early ape based on its features. Gaudry then studied the remains and noted some features which he felt were more hominid and perhaps less apelike. After examining a later, more complete mandible, however, he decided (1890) that the fossil was not a hominid and was also more primitive than modern apes. The natural conclusion was that *Dryopithecus* was perhaps ancestral to these hominoids or was itself an ancestral hominoid.

From this inconclusive beginning based on three dryopithecine jaws found in Europe, discoveries of other related fossils have expanded the size of this crucial group into a large collection of genera and species that show differences in size and habitat. However, these conclusions were not established for many years after the original discovery and only after subsequent workers made the additional finds in widely separated areas. The relationships between the different fossil species were gradually accepted as the experts made discoveries, identifications and revisions of the specimens found in Europe, Africa, the

Middle East, India and China. Today the group is divided generally into European, Asian and African forms; one type, *Ramapithecus punjabicus*, is known in both Asia and Africa.

THE COMPLEX OF DRYOPITHECINE NAMES

The second species of the group was discovered in the late nineteenth century when Lydekker named *Paleopithecus sivalensis* for a fossil primate from Miocene-Pliocene strata of the Siwalik Hills of Pakistan. However, he did not realize that this generic name was already in use for a reptile and was, therefore, invalid for this Asian animal. This primate was later reassigned to *Dryopithecus* as *D. sivalensis*. Pilgrim subsequently (1915) named two new species of hominoids, also from the Siwalik Hills, as *Dryopithecus punjabicus* and *Sivapithecus indicus* (Siva's ape of India) in the mistaken belief that they were related to the European genus. Then in 1934 Lewis erected a new genus and species, *Ramapithecus brevirostris* (Rama's ape), based on a partial jaw from the same productive Siwalik Hills; later finds of partial materials were also assigned to this genus.

By 1965, using much more material from several different genera and species, Simons and Pilbeam decided that the Asian hominoid forms *Paleopithecus* and *Sivapithecus* could not be separated from the European *Dryopithecus*. Accordingly, they reassigned both of them to that genus. Thus, *P. sivalensis* became *D. sivalensis* and *S. indicus* became *D. indicus*. Pilgrim's Indian species *D. punjabicus* was, however, found to be incorrectly identified and was reassigned to *Ramapithecus* as *R. punjabicus*. In this fashion, the Asian species were reduced to three, *D. sivalensis*, *D. indicus* and *R. punjabicus*. To further complicate the matter of the Asian hominoids, Pilbeam (1984, p. 4) apparently accepted *Sivapithecus* as a valid genus based on a number of finds he made in Pakistan in 1973. Using additional fossil material found in the Siwalik series, he accepted that *Sivapithecus* and *Ramapithecus* are sufficiently indistinguishable from each other and that *Ramapithecus* would perhaps be better included within *Sivapithecus*. Investigations of these and additional finds also suggest that *Sivapithecus* is similar in many respects to orangutans and many workers now consider it to have been ancestral to these red apes or a member of the group from which they descended. Accordingly, *Ramapithecus* has been shifted by most workers into *Sivapithecus* and

from the possible position as a hominid near the line to humans to the hominoid stock that led to orangutans.

The African dryopithecine fossils are concentrated in early Miocene strata of East Africa, where they were first found in the 1920s. Hopwood named this first African species *Proconsul africanus* in 1933 after a popular zoo chimpanzee in Britain named Consul. This species is somewhat smaller than a chimpanzee, which it was considered to resemble. Most workers now believe *Proconsul* to be another type of dryopithecine, considerably expanding the concept that includes European, Asian and African species. Two additional species from this region of Africa have been described, *P. major*, about the size of a female gorilla, and *P. nyanzae*, about the size of a chimpanzee. These finds demonstrate that the dryopithecines flourished in this region about 20 million years ago.

In 1962, Louis Leakey erected a new genus, *Kenyapithecus wickeri*, for a fossil maxilla from East Africa. But detailed investigations (Simons, 1972, p. 269) since have determined that it, too, is a dryopithecine and it has been shifted into *Ramapithecus* as a synonym of *R. brevirostris*. A further complication arose later when *R. brevirostris* was discovered to lack sufficient distinction to separate it from *R. punjabicus*. *R. brevirostris*, too, has been reassigned and it is now considered another name for *R. punjabicus*.

The reasons that these dryopithecines were initially identified by other names was because the paucity of material did not allow for clear establishment of the concepts of the different taxa and because the available fragmentary material was inadequate for comparisons at those times. Today, however, the situation is much better understood and clarified because of discoveries of these fossils in both Miocene and Pliocene strata from Europe, Africa, the Middle East and Asia. These additional finds and new information necessitated the revisions that have consolidated the dryopithecines into three genera, *Dryopithecus*, *Ramapithecus* and *Sivapithecus*. Workers examining older literature must be aware of these changes to understand exactly which species are being discussed and how they relate evolutionarily.

Dryopithecus today consists of six species; *africanus*, *fontani*, *laietanus*, *major*, *nyanzae* and *sivalensis*. Together, they exhibit characters that are also found in gorillas and chimpanzees and a few that are recognizable in humans. The European and African species of

Dryopithecus seems the likely ancestor of gorillas and chimpanzees and, perhaps, humans. *Ramapithecus* and *Sivapithecus* each has a single species, *R. punjabicus* and *S. indicus*. These two genera appear to be the ancestors of orangutans or at least of that lineage. They are structurally nearer the orangutans than they are to the other great apes and the branch to humans. Pilbeam (1984, p. 4) says

> Examinations of DNA and several proteins of
> living humans and apes consistently shows
> that we are genetically closer to the African
> apes than to the orangutan.

Today, workers view these dryopithecines as a group of important Old World hominoids that had spread widely into a variety of ecological niches in Africa, Europe and Asia. They evolved by adapting to these environmental niches and by developing strategies and behaviors slightly different from their ancestors to cope with circumstances in these niches. Their diversity and distributions reflect the great successes they enjoyed in this evolution. In making these changes, they shifted in directions that led them, ultimately, to their direct descendants, the advanced hominoids, the living apes that we know today. Surprisingly, few workers have objected to the placement of the dryopithecines near or at the base of man's family tree, an indication of the clear consensus that this position is the most appropriate interpretation of the available fossil material. The dates and ages for the family tree of these dryopithecines are appropriate and no other fossil species has been found that exhibits the characters and traits that would allow an alternative pathway to humans. Without significant new fossil discoveries that could support a different hypothesis for descent to humans (and create a major upheaval in human origins theory), these dryopithecines are the only logical choice today for the ancestral stock of higher primates that could have led to humans.

The dryopithecines, are, therefore, the crucial primates in the evolution to humans. They evolved from anthropoid stock which arose as far back as the Oligocene as the small primate *Aegyptopithecus* found in the Fayum. This type apparently gave rise to some intermediate genera, probably *Pliopithecus* and *Limnopithecus* from Africa and Europe, which can be considered the first apes. These types are so similar that some workers have considered them to be a single, widespread genus rather than two separate ones. A second wave of evolution from this stock likely gave rise to the dryopithecines and they spread throughout the Old World. On the different continents this stock underwent adaptations and became diverse and specialized. The Asian stock became extremely conservative and gave rise to a very large form, *Gigantopithecus*, which became extinct and left no direct descendants. The only living descendants of the Asian stock are the orangutans, which differ from the other living apes by their distinctive solitary behavior. The African stock, meanwhile, gave rise to two lines of descendants, one conservative and the other innovative. The conservative stock gave rise to the great apes of Africa, gorillas and chimpanzees, while the innovative stock gave rise to improved types that became the australopithecines. The dryopithecine stock of Europe did not advance beyond the dryopithecine stage or provide local evidence of any descendants. Perhaps climatic conditions in Europe intervened and hampered the development of innovative models; perhaps conditions prevented it. The facts are simply that no higher descendants of the dryopithecines are known in Europe. Accordingly, we must view Africa as the birthplace of the human line, because there is no evidence to support any other site.

Only one other primate of note appeared in Pliocene strata of the old world. This was the very large species named *Gigantopithecus blacki* by von Koenigswald in 1935 for an immense fossil tooth of human appearance that he had purchased in a Hong Kong drugstore. Such teeth were not entirely unknown to drugstores or apothecaries where they had been used for medicinal purposes for many years. The tooth has been identified as that of a relatively recent ape of large size. Additional fossils, including four massive jawbones found in southwest China and northern India, have led some workers to postulate that this was the largest of all primates. Simons and Ettel state (1970, p. 77) that it

> . . . successfully adapted to life in the savannas
> and forest fringes of Asia at least five million
> years ago and perhaps as long ago as nine
> million years.

The size of the jawbones and more than one thousand known teeth suggest a very large animal (Ciochon, et al., 1990) that ate great quantities of plant materials to sustain its size. No other bones are known. The Indian jawbone is much older than the original species, *G. blacki*, from China and probably represents an ancestral species, *G. bilaspurensis*. From all the information available about this herbivorous giant primate, it has been interpreted as a very conservative side branch of the pongid line. *G. bilaspurensis*, the Indian species, may have descended from *D. indicus*, the large, gorilla-sized dryopithecine found in both West Pakistan and India.

THE SPREAD OF THE DRYOPITHECINES

All of the hominids, with the exception of the single living subspecies, *Homo sapiens sapiens*, are now extinct, a fact that provides much room for speculation and major differences of opinion regarding their origins. The evolutionary line to modern humans began during Miocene time when the dryopithecines appeared. They were cosmopolitan apes typified by the first hominoid fossil discovered, a jaw found in France in 1856 and named *Dryopithecus fontani*. These early apes, the dryopithecines (see Fig. 5.1) of the Miocene-Pliocene Epochs, had spread through the forests of Africa, Europe and Asia through the land connections that had developed between these continents during early Miocene time, about 17 million years ago. Although more than twenty-five generic names have been proposed for these dryopithecine apes, the separate species can be assigned today to only two or three genera. These types arose as they became adapted for survival in different habitats with varying environmental conditions. These adaptations were the structures, features, traits and behaviors that made them so special and distinct. And these adaptations enabled them ultimately to give rise to the other hominoids and the hominids, including the extinct types and those living today, the gorillas, chimpanzees, orangutans, gibbons and, of course, modern humans.

The evolution of the dryopithecine line of apes as a side branch from the main anthropoid stock, probably Old World monkeys, took place during or just after the Middle Oligocene Epoch, some 30 to 35 million years ago. The line to New World monkeys had already separated from the Old World monkeys by then and probably did not contribute to the hominoid line. At least one genus of Oligocene Old World anthropoids, however, can be identified as very close to the Miocene hominoids, perhaps even ancestral to them. This form, *Aegyptopithecus zeuxis* (see Fig. 5.7), is a small, monkeylike species found in 1906 but not described until 1963 (Simons, 1965). It is a small arboreal primate with apelike teeth and a prominent sagittal crest similar to that found in the dryopithecines. Because of these and other important features, Simons believes that it represents or gave rise to the branch that led to the dryo-

pithecines and eventually to the apes and humans. It is, according to Simons, our most revered ancestor.

EUROPEAN DRYOPITHECINES. The dryopithecines are founded on the original species, *Dryopithecus fontani,* that was found in Europe. This specimen and one other, both from France, provided the information for the original definition of the genus, although several other specimens have been found in Germany, Austria and Czechoslovakia. They cannot be distinguished, however, from *D. fontani* because this fragmentary material consists mainly of teeth, one lower jaw and three limb bones. Accordingly, these additional specimens are assigned to *D. fontani*. As an added complication, the European specimens are similar to the Asian dryopithecines, especially *D. sivalensis*.

AFRICAN DRYOPITHECINES. More than one thousand specimens of various Miocene hominoids ranging from 17 to 22 million years old have been found in East Africa (Pilbeam, 1984, p. 91). These are relatively well understood because of the interest they have generated and the numerous data published about them. At least six species of these hominoids, from gorilla to gibbon in size, are believed to have inhabited the forest and open woodland environments in the Rift Zone of Kenya and Uganda, though savannas were also present in that region at that time. These species were initially assigned to the genera *Dryopithecus*, *Proconsul*, *Kenyapithecus* and *Sivapithecus*, but these initial taxonomic assignments were based on fragmentary and sparse material that only confused their relationships. Extensive reexaminations of these older specimens and new fossil finds have necessitated revisions. Both *Proconsul* and *Kenyapithecus* have been shown to be essentially indistinguishable from *Dryopithecus* and have been reassigned (Simons and Pilbeam, 1965) to it. The result of these revisions has been to reduce the dryopithecine genera to two, *Dryopithecus* and *Sivapithecus*.

The geologically oldest of these African dryopithecine species is *Dryopithecus* (*Proconsul*) *africanus*, a baboon-sized arboreal, fruit-eating ape found in 1933 by Hopwood. Two other species, *D. major* and *D. nyanzae*, were found in 1950. These species likely arose by adaptation to life in different environments of East Africa. Pilbeam (1972, p. 44) and Simons (1972, p. 252) believe that *D. major* probably was the ancestor of gorillas, though they did not discount it giving rise to chimpanzees. They based their interpretations on its

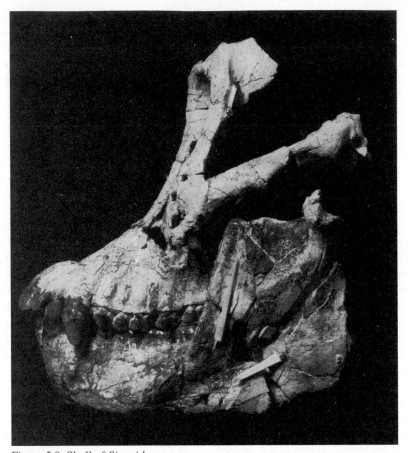

Figure 5.9 Skull of *Sivapithecus*.

powerful dentition and palate, a foot bone, and some other features. All these characters, however, strongly suggest gorilla features rather than chimpanzee. *D. africanus* has the size and some other features that suggest it may have given rise to the chimpanzees.

ASIAN DRYOPITHECINES. In Asia, several additional dryopithecines were found, including specimens initially named *Ramapithecus*, *Sivapithecus*, *Bramapithecus*, *Sugrivapithecus*, *Gigantopithecus* and others. Of these names, only two, *Gigantopithecus* and *Sivapithecus*, survive as undisputed genera. *Ramapithecus* is a name that has been commonly used but several workers believe it to be another example of *Sivapithecus*.

During Middle Miocene time, these Asian dryopithecines apparently split off from the African and European types and began to evolve into more specialized forms. Pilbeam's field research in Pakistan provided a wealth of fossil hominoid materials and allowed reconstructions of the local late Miocene environments. Nonprimate fossil bones found there indicate the region was a combination of woodland, bush and grassland and suggest that these hominoids were less arboreal than their African relatives.

In 1973 Pilbeam found some *Sivapithecus* limb bones and a partial skull (Fig. 5.9) with a face very similar to that of an orangutan. The teeth are especially distinctive and important because they have thick enamel on the chewing surfaces. Of the modern apes, only the orangutan has similar thick enamel development on its teeth. There is no compelling evidence to include *Sivapithecus* in the main line of evolution to hominids and comparison with the oldest hominid, *Australopithecus*, does not reveal any important similarities. Physical evidence does, however, strongly support a branched lineage with *Sivapithecus* being the ancestral stock for orangutans.

The association of *Sivapithecus* with other fossil animals suggests that these large primates lived in open grassland and bush-woodland in contrast to the African dryopithecines, which were forest dwellers. The Asian species, accordingly, had access to different available foodstuffs, primarily harsher and tougher plant materials than the leaves and fruits found in the African forests. *Sivapithecus* adapted to these new and different foods by developing a much more powerful jaw with stronger teeth similar to those of modern orangutans. In these more open environments, therefore, *Sivapithecus* became a more specialized hominoid and diverged from the earlier African dryopithecines.

An African specimen of *Sivapithecus* was also found in 1983 at Buluk, Kenya. These remains are dated to at least 17 million years ago, making it a contemporary of *Dryopithecus* (*Proconsul*) found in the same region. The antiquity of this African *Sivapithecus* demonstrates that the orangutan stock or lineage had split from the hominoid ancestral branch much earlier than previously believed, perhaps even in Africa. This separation also indicates that the main hominoid stock had even more time to become adapted to different environments and give rise to the radical new types that became the hominids.

GIGANTOPITHECUS

The largest of all the primates, living and fossil, is believed to be the extinct Asian dryopithecine, *Gigantopithecus*, known from more than 1,000 teeth and several jawbones from India and China. The teeth and jawbones of this form are immense and indicate an herbivore of great size, much larger than a modern gorilla. It must have consumed great quantities of vegetation for sustenance and the worn condition of the teeth and jaws reflect this harsh diet. Though these teeth and four jaw-

bones are the only known fossil materials of this genus, the wear on these teeth is distinctive and has allowed interpretation of its diet. This in turn has enabled some key interpretations of the evolutionary steps that led to humans.

The wear on the *Gigantopithecus* teeth show patterns similar to those of living ground-dwelling baboons (*Theropithecus*) studied by Jolly (1970) in East Africa. And these similarities suggest that both primate types evolved in the same general way. The tooth wear observed by Jolly in the baboons resulted from their diet that consisted primarily of large quantities of various tough foods found on the savannas. These foods were grass seeds, roots, rhizomes, stems, nuts, etc., and are typical of a **graminivorous** (seed feeding) behavior. Many of these foods are commonly picked directly from the ground or dug from it; they are generally covered with dirt that contains hard mineral grains. All these foods are normally tough and require powerful and constant chewing by the cheek teeth. The natural effect is the production of distinctive wear patterns on the teeth, especially if abrasive mineral grains covered the foods. Jolly named this wear the "**T-complex pattern**" for the genus of baboons being studied, *Theropithecus*.

Jolly believes that baboons originally lived in forests where they dined primarily on fruits and leaves (a **frugivorous** diet). They then moved from these forests into the open grassland savannas and there became adapted to the more varied foodstuffs. Their teeth, however, were still adapted to more restrictive and softer forest diets of fruits and leaves like the other primates and were generally unsuitable for the tougher foods in the savannas. Accordingly, exploiting the new foods in the savannas caused severe wear on their premolars and molars and they underwent a variety of dental modifications to accommodate this tougher fare. These changes resulted in the development of the T-complex wear pattern recognized by Jolly. This different wear pattern has, thus, been interpreted as due to the gramnivorous mode of feeding in the baboons that contrasts to the different wear produced in gorillas and chimpanzees by their frugivorous diet of soft leaves and fruits.

Because *Gigantopithecus* exhibits similar wear on its teeth, Jolly believed it followed a similar, graminivorous food gathering behavior. And such feeding practices would have been expected in *Gigantopithecus* if that primate had moved from forest environment habitats with abundant leaves and fruits to savanna grassland habitats with different and tougher foods. In the savannas, *Gigantopithecus* would logically have developed a T-complex wear pattern on its teeth.

This pattern of tooth wear is important in the higher primates for another reason. The oldest hominid, *Australopithecus*, has similar wear on its cheek teeth, especially the species *robustus*. If Jolly's analysis and interpretations are correct for the baboons, this T-complex wear pattern may indicate that the australopithecines or their immediate ancestors followed a similar ecological niche pattern by moving out of forests and onto the savannas. There, they, too, would have utilized a graminivorous feeding behavior as demonstrated by their tooth wear.

Ciochon, et al. (1990) postulate another mechanism to account for the wear on *Gigantopithecus* teeth. Using an SEM they have found microscopic pieces of plant-produced silica (phytoliths) embedded in the enamel surface of many *Gigantopithecus* teeth, including a scratch mark ending in one of these bits. These phytolithes are the products of ordinary metabolic processes as plants absorb chemicals through their roots. The authors report that these silica bits take the shape of plant cells in which they solidified, and the phytoliths on the *Gigantopithecus* teeth resemble the cells of several different kinds of plants. This view, of course, weakens the argument for the T-Complex wear pattern on australopithecine teeth and its environmental implications, although it does not negate it.

Certainly the australopithecines as we know them now were not newcomers to the savannas but probably had lived there for some time as demonstrated by their well-developed walking feet and legs. Such structures without the opposable large toes are obviously poorly suited to arboreal lifestyle and did not evolve in slow fashion. Even an open forest lifestyle would have created difficulties for them if they had to return to trees quickly to flee predators. Instead, these legs and feet strongly support an open environment habitat in which bipedal behavior is advantageous over the quadrupedal or knuckle-walking practices by apes with their opposable thumbs and toes. That the early australopithecines were bipedal is further supported by the discoveries of *Australopithecus* footprints at Laetoli, East Africa by Mary Leakey in 1975 (Fig. 5.10).

Figure 5.10 Mary Leakey examining australopithecine footprints at Laetoli.

Bipedalism strongly supports the view that the first hominids had moved from the forests to at least the forest-savannas margins where they utilized this important behavior and fed on the available foods. If the T-complex tooth-wear pattern is valid for them, it also supports this view. The evidence of bipedalism argues quite forcefully that the early hominids did not evolve from the hominoids in several intermediate stages but rather evolved rapidly under the selective pressures of the new environments.

SUMMARY

It is generally accepted today by virtually all scientists that humans evolved from primates. This is a large group of primarily arboreal creatures that have many attributes that strongly support our evolutionary relationships. The most primitive of the primates are the Prosimians, a varied group that arose more than 65 million years ago. Their arboreal lifestyles led to the development of structural, physiologic and behavioral traits, features and characteristics that eminently suited them for the demanding and difficult habitats in trees.

The Prosimians gave rise to the Anthropoids, probably in the great forests of Africa during early Eocene time. Within a short time, these new Anthropoids evolved into both Old World monkeys and New World monkeys and continued to diversify into a variety of types. Ultimately they produced the living species that are present in great numbers in some places in the tropics. One African species they gave rise to was *Aegyptopithecus zeuxis*, a small form found in Oligocene strata in the Fayum of Egypt that may have given rise to the more advanced apes, the dryopithecines.

Abundant fossil evidence indicates that the earliest apes, the dryopithecines, became a cosmopolitan group of apes that spread across the Old World where they inhabited different environments. There they experienced a variety of conditions that exerted different selective pressures and they adapted to these different circumstances. Under the pressures of natural selection they evolved into a variety of forms; more than 1,000 specimens from 17 to 22 million years ago have been found in East Africa. Some of these species left descendants that were conservative in nature; they eventually became the great apes. Two or three of these species are potential candidates for gorillas and chimpanzees. One stock, *Sivapithecus*, became the probable ancestor of orangutans, perhaps 15 million years ago. Another conservative stock continued to evolve and split about 9 to 10 million years ago into the gorilla and chimpanzee lineages. Then finally, about 5 to 8 million years ago, a more daring group, the hominids, separated from the chimpanzee line.

There is a gap of several million years between the latest possible dryopithecine ancestor and the earliest type of humans, the australopithecines. No fossils have been found in this interval to clarify the specific details of this evolution, but the threads of continuity are present in so many characters of the dryopithecines and australopithecines that any other conclusions are illogical. Given the nature of dealing with events of the past and determining the likely directions for descent, we are confident of our ability to make positive statements about these fossils and their descendants. This, then, is the direction that nonhuman primates, the hominoids, took in becoming hominids.

STUDY QUESTIONS

1. Describe the evolutionary path that primates took from insectivore ancestors to humans.

2. What Oligocene primate could have given rise to the dryopithecines?

3. What is the evolutionary significance of human and dryopithecine dental formulae?

4. Why are the dryopithecines considered a cosmopolitan stock?

5. Why was there so much confusion with dryopithecine names?

6. Identify the dryopithecine species that probably served as the ancestors of the living great apes.

7. What dryopithecine stock probably gave rise to *Gigantopithecus*?

8. Describe the T-complex pattern of tooth wear. What is its significance in interpreting environments?

9. Identify the discoverers or the principal workers associated with the following fossils:

    a. *Aegyptopithecus zeuxis*

    b. *Gigantopithecus blacki*

    c. *Proconsul*

    d. *Dryopithecus*

    e. *Kenyapithecus wickeri*

10. Differentiate between gramnivorous and frugivorous feeding behaviors.

11. What evidence places *Sivapithecus* outside the main evolutionary line from apes to humans?

12. Distinguish between hominoids and hominids.

6 The Rise of Hominids

INTRODUCTION

The various attributes of the living apes and humans, their structural characters and features, physiological systems, biochemical makeup and behaviors, all strongly support a close relationship between them. Further, both of these animal stocks are linked to earlier fossil primates on the basis of skeletal structures, although full comparisons of the nonskeletal attributes are not yet possible. The relationships between the living and fossil hominids and the hominoids are best explained by an evolutionary descent model from a common ancestor, a primate that lived perhaps 25 to 30 million years ago. In this model, initiated by Darwin's work, the nonhuman hominoids continued to practice the conservative lifestyle of their ancestors and remained essentially the arboreal jungle and woodland apes that we know today. The hominid stock, however, diverged from this more conservative hominoid lineage and evolved into humans, rational creatures that expanded into virtually all of Earth's environments. This human stock, *Homo sapiens*, has developed complex cultures and great societies and has become the predominant species on Earth. The separation of hominids from hominoids took place perhaps 5 to 8 million years ago without leaving much trace, though the fossil evidence supports a likely path of descent.

Humans evolved from primate ancestors by means of significant changes in skeletal anatomy, physiology, behavior and abilities. The cause of these changes is not known for certain but it is highly likely that these animals were strongly influenced by their preferences for certain habitats. Our ancestors, therefore, were originally arboreal primates that probably descended from the trees because they found a better, more plentiful, or more suitable supply of food on the ground. Being typical arboreal primates, however, with structures, features and behaviors eminently adapted for life in trees, these ancestors were not well suited for life on the ground. Despite the inherent difficulties for arboreal species in terrestrial environments, some of this ancestral stock, at least, found this new terrestrial lifestyle so appealing that they elected to leave the trees for a permanent life on the ground.

Arboreal primates today feed largely on the plant materials commonly and abundantly found in the trees, mostly fruits and leaves, like the diet of modern chimpanzees. The foodstuffs available on the ground in woodlands and on the savannas, however, are much different. Terrestrial plants produce less fruit, but there are abundant leaves and a much more varied supply of other foods, especially nuts, berries, seeds and roots. Any shift from arboreal to terrestrial lifestyle had to take into account some impetus that dictated or encouraged this change. It is possible that they made this move because of foods. Purely arboreal primates were probably restricted in their numbers because many of the foods upon which they fed were seasonal foods that could only support limited populations. During other seasons these arboreal primates had to rely on other plant materials, perhaps less appealing foods. The great variety of terrestrial foods, however, could provide a near continuous food supply during an entire year. Nuts, berries, seeds, and roots, for example, could be had at almost any time, as well as fruits and leaves. Such more varied and available foods would have attracted some arboreal primates to descend from the trees during the different ripening seasons. In time, their descent for seasonal food gathering could have become a routine pattern. Obviously, there must have been some advantages to be gained by descending to the ground and selecting a radical new lifestyle. Whether this advantage was in the variety and abundance of terrestrial foods or some other incentive, it encouraged them to develop a terrestrial lifestyle. And this new lifestyle did not prevent them from climbing back into trees for particular foods.

These newly terrestrial primates probably were much like modern baboons in behavior, though more like modern chimpanzees in anatomy. At first they would have retained all the original arboreal hominoid features, structures and behaviors, but some of these were not particularly useful in the new circumstances. As they spent more time in this new environment they began to develop more specialized adaptations that were better suited for a terrestrial lifestyle. Their grasping feet, undesirable for life on the ground, began to develop into walking feet with loss of opposability in their big toe. Walking feet were much better suited for terrestrial life and offered increased fitness and greater advantages than grasping feet. Modern baboons, for example, have a flattened foot more similar to the human walking foot than any other primate. This baboon foot probably evolved in the same fashion as our own human foot, because of the largely terrestrial behavior of baboons. Our own ancestors developed their walking feet some 5 million years ago, probably in conjunction with the adoption of an erect bipedal behavior. Baboons, however, never made the total shift from the monkey pattern as did the humans but retained their quadrupedal gait. Hominid upright bipedal locomotion is not an easy mode to adopt. It demands balance and coordination, abilities that made other demands on the entire body. The adoption of such a terrestrial lifestyle led to skeletal and muscular changes in hominid feet, legs, pelvis and spine and appropriate neurological areas in the brain.

These were not the only anatomical changes that took place in the early terrestrial hominid ancestors. There is evidence in the wear of hominid fossil teeth to support the view that terrestrial foodstuffs gradually became their normal diets. These diets included leaves, fruits, seeds and nuts, like those of modern chimpanzees and baboons. In addition, they probably collected small insects, grubs, worms, eggs, and even small animals, also like chimpanzees and baboons. This change in diet began to be reflected in sizes and shapes of their teeth, jaws, jaw muscles and skulls. With harsher foods to crack open and with greater demand for chewing, their cheek teeth became larger. Simultaneously, and for the same reasons, their front teeth became smaller. The jaws also started to become smaller and this was accompanied by reduction of the chewing muscles. Sagittal crests to which the large chewing muscles were attached became superfluous and disappeared. Their hands became much more effective instruments as they were used delicately to pick small seeds and other items. With the reduction in size of the jaws, greater use of the hands, and some other changes, the prognathous condition of the face began to recede and their faces became more flattened. As their faces flattened, they also began to develop noses and chins. Slowly but surely these ancestral primates were becoming more human.

The changes in environment and lifestyle not only initiated changes in muscular and skeletal anatomy and physiology but demanded other changes in the body as well. Survival on the ground with new structures and behaviors placed greater-than-ever demands on their brains. As the pressures of the environment favored individuals with better brains, these began to increase in size and, undoubtedly, in quality. In particular, the areas devoted to memory and motor control increased significantly in both size and quality. This trend is demonstrated by increases in brain volume in the fossil skulls from about 25 million years ago to those of relatively recent time. Though increases in brain size can be calculated directly, changes in brain quality must be inferred from the increased behaviors, abilities and skills of the early hominids. These are poorly known from the stock that was making the transition but are well seen in the early hominids.

Life in the terrestrial environments for these early hominids produced changes in their behaviors that accompanied their physical changes. Arboreal behaviors are not well suited for life on the ground and, though these primates could have survived on the ground, they would have been clearly at a disadvantage. This early stock that came down to the ground undoubtedly retained the social behaviors of their arboreal ancestors that were not compromised by the new terrestrial lifestyles; there was no need to change something that was not a problem. Some of these animals changed or modified some behavior patterns as would be expected of any species that moved into a new set of circumstances. Probably these individuals adopted behaviors similar to those of modern baboons, a compromise between terrestrial and arboreal lifestyles. Living apes are well known to employ a variety of complex social behaviors that successfully regulate their lives. These social behaviors likely originated in the early arboreal primates and have been transmitted to their descendants, the living apes, either genetically or by training. As such, the behaviors of living apes are probably good models of those employed by their ancestors, who are also our ancestors. Many of these social behaviors would have been quite useful to terrestrial primates and there is every reason to believe that these ancestral hominids used any and all behaviors that conferred greater fitness and survival benefits. Many of these are the same social behaviors recognized in humans today and they support, therefore, an evolutionary relationship with living apes and demonstrate a connection by inheritance from our primate ancestors.

THE EVOLUTIONARY PATH FROM DRYOPITHECINES TO HUMANS

The evolutionary path from the ancestral dryopithecines to humans is not well defined. It is unknown how many different species, if any, intervened between dryo-pithecines and humans, though the present evidence suggests few stages beyond those already identified. Fortunately, we can trace the lineage through features common to all these species. Tooth wear and bipedalism are just two lines of evidence that are useful. Another way to reconstruct this evolutionary pathway is to reverse the direction of progression and go from living humans to the ancestral types. If we can identify the same or similar structures, attributes, features and behaviors in the fossil specimens that are found in the living types, we can make a reasonable case for the evolutionary pathway.

On the basis of several lines of this evidence, it seems clear that the structures, attributes, features and behaviors that we recognize today in modern humans have some direct parallels in living apes and monkeys. The chances of these being developed independently in us and our primate relatives are simply too great to be coincidental. A more reasonable explanation is that these parallels represent aspects that both types inherited by evolution from a common ancestor. If so, some of these aspects, however modified with time, should be recognizable to some degree in the Miocene dryopithecine stock that we identify as our ancestors. We are, in fact, able to do precisely that; to identify many physical features, behaviors and attributes in these animals that we have already recognized in ourselves. Predictably, therefore, any intermediate evolutionary stages should possess these same aspects, though perhaps in some intermediate degree or fashion. Who then were the animals that descended from the dryopithecines? What were they like? And why did they eventually become hominids?

THE AUSTRALOPITHECINES

Hominid evolution to modern humans takes into account a lengthy and intricate pathway leading from the primates, through the dryopithecines and apes and finally into the hominids. The oldest hominids are the australo-pithecines, those erect bipeds of late Pliocene time and the same group that generated great dispute when first discovered. Initial fossil finds of these australopithecines were originally assigned to several genera, but subsequent fossil finds and detailed analyses revealed that these many specimens comprise only a single genus, *Australopithecus*. Two distinct types comprise *Australopithecus* and they are assigned to three or four species found in South and East

Africa; the number of species is in dispute because of differing opinions regarding the nature of each type. Numerous specimens of these australopithecines have been found during the past seventy years, particularly the past thirty years, and permit excellent reconstructions. Subtle changes in their anatomy are well known today and demonstrate their range of variability. Further, these specimens offer excellent evidence of progressive evolution leading to humans. At the same time the chronological distributions of the australopithecines match their evolutionary stages of development. As a result, the australopithecines are now firmly established as the direct ancestor of the genus *Homo* with its several species, including modern *H. sapiens*.

The ancestors of these australopithecines, the dryopithecines, left two sets of evolutionary tracks that pointed in prominent but different directions. One led to the living apes and the other to humans.

THE HOMINIDS APPEAR

The most obvious attribute that differentiates humans from the apes is **bipedalism**. This is our unique ability to stand erect, walk and run on two legs. Bipedal locomotion involves the coordination of various muscular activities that operate through straightened knees in our skeletal structure. It has become a fundamental character of our species and has been credited by some workers as being the primary driving force that enabled our ancestors to become humans. It is undoubtedly related to and probably derived from the VCL posture utilized by apes and most other nonhuman primates and characterizes them. Straus (1962) stated,

> . . . that primates early developed the mechanisms permitting maintenance of the trunk in the upright position . . . Indeed, this tendency toward truncal erectness can be regarded as an essentially basic primate character.

Bipedalism has, however, clearly progressed much further in humans than in the other primates and is possible only because of further significant structural modifications to hominid skeletal, muscle, neural and other systems.

Humans are the only primates that habitually walk in an upright, bipedal posture. Birds and some other animals are bipedal, but none of the higher animals uses it exclusively or even regularly. We are totally terrestrial and can stand erect and move quickly with ease on only two legs. Further,

Figure 6.1 Gorilla knuckle walking.

we do this in relaxed fashion by constantly and effortlessly shifting our balance to maintain equilibrium. Other terrestrial primates, the baboons and gorillas, for example, can also walk bipedally; but they can do so only in limited and awkward fashion using a bent-knee style and for short distances. They are simply not designed for bipedalism. Most primates normally use both hands for support in a **quadrupedal** (four-limbed) gait or one hand for a three-limbed, **knuckle-walking** gait (Fig. 6.1), the style favored by the great apes when on the ground.

The origin of bipedal behavior has been attributed by different workers to a variety of possible causes. Many workers believe it developed in response to certain demands established by the ancestral species and is likely related to the terrestrial behavior of chimpanzees and gorillas. Napier (1967, p. 64)) doubted that bipedal behavior could have originated on the African savannas because of the hazards of life there. Instead, he believes that this new behavior and the origin of humans took place in the savanna-woodlands, neither forest nor grassland, but a

blend of these environments that offered them the advantages of

> . . . trees to provide forest foods and ready escape from predators. At the same time its open grassy spaces are arenas in which new locomotor adaptations can be practiced and new foods can be sampled. In short, the woodland-savanna provides an ideal nursery for evolving hominids, combining the challenge and incentives of the open grassland with much of the security of the forest.

Bipedalism was particularly important to evolving hominids because it totally freed the hands for other purposes, including tool use and the carrying of various items, infants, food and possessions, from one place to another as they walked erect about the countryside. Tool use, thus, was a significant step for some of these primates because it enabled them to perform acts not otherwise possible. This act, so emphatically reported in chimpanzees recently by Goodall, records some beginnings, at the very least, of rational thought. These ancestral hominids did not invent tool use but likely derived it from primate ancestors. The early hominids continued to make tools, like the chimpanzees, by modifying natural materials.

Tool use by some animals is one ability that identifies a level of achievement beyond that of most other animals. Chimpanzees not only use tools but also make them, a significantly higher reflection of their intelligence. In this process the "crafts-chimp" does not randomly perform an unspecified act. Instead, it must establish a use for the tool to be made, a need for this tool, the materials to be used, and then finally make it. Above all, the toolmaker must formulate a concept of the tool to be made by prior acknowledgment of the task to be performed, an inherent aspect of toolmaking in chimpanzees or any other creature. They must be able to project ahead in order to make a tool for some particular use. Chimpanzees have also demonstrated the ability to use tools as weapons, however simple. Though they have a recognized ability to fight with their hands and teeth, they will often use objects as weapons! The intellectual advance from toolmaking and tool use to weapons is not particularly great in humans, but it is in chimpanzees.

If bipedalism offered significant advantages to the early hominids, walking in upright fashion was only one of them. Some experts think that the actual development of terrestrial bipedalism resulted from much more than the direct advantages of upright posture and locomotion. Bipedalism freed the hands for many different functions and toolmaking and tool use may have been of great influence. Whether the adoption of erect, bipedal posture offered significantly greater ability to see an approaching predator, freedom to carry food or manipulate tools, or some other advantage is unknown, and no single one can be identified as the driving force. All of these activities are recognized behaviors in chimpanzees and could easily have been behaviors of their ancestors. Because of our close evolutionary relationship with chimpanzees, these could also have been influential factors in our own ancestor's choice of bipedalism. Perhaps, however, all of these factors collectively offered advantages, slight or great, to our ancestors when they became terrestrial, thus influencing them to adopt bipedal locomotion.

TERRESTRIAL BIPEDALISM MODEL

A likely hypothesis for the origin of hominid bipedalism proposes that it resulted from adaptations to terrestrial life in environments that were changing from forest to woodland. The best evidence indicates that our ancestral stock likely evolved from arboreal apes that inhabited the forests of East Africa. When the climates began to change, some forest areas thinned into savanna-woodland with more open spaces and fewer trees. The grassy open ground offered incentives for life that attracted some of these arboreal apes down from the trees. These individuals probably spend much time on the ground where food is plentiful, where movement across the ground is easy, and where nearby trees can provide refuge. They were able to adapt to new conditions in the savanna-woodland habitats and became like some modern apes.

These originally arboreal hominoid primates adopted new behaviors that were better suited to a terrestrial bipedal behavior that encouraged adaptation to the terrestrial ecological niches. This new, bipedal lifestyle was obviously advantageous to them in these terrestrial environments and, at some point, they became the ancestors of hominids. If they had not become terrestrial and bipedal, they would have stayed essentially arboreal primates like the chimpanzees with all the typical arboreal structural features, particularly the grasping feet. Such features, however, would have been disadvantageous on the ground and the fact is that our earliest ancestors identifiable as hominids had already developed these structural modifications in response to their new behaviors and adaptations. They did not develop almost-walking feet and other structures. Specifically, they became bipedal with walking feet, a clear indication that they had become fully terrestrial. There is no known ancestral hominoid that can be placed between the arboreal primates and the terrestrial hominids; perhaps there never was such an intermediate stage. Those evolutionary changes that differentiate hominids from hominoids can be determined from the structures and behaviors of living chimpanzees. These chimpanzees retained the conservative arboreal structures of their arboreal hominoid ancestors and, probably, many of their behaviors as well. The hominids, however, changed radically!

Modern chimpanzees inhabit both savanna-woodland and thick forest environments and use slightly different behaviors in each. The forests or jungles generally have very closely set trees, and the ground is very thickly vegetated with shrubs, bushes and other plants; there is very little open space. The savanna and woodland environments, however, consist of trees and grassy open ground.

Figure 6.2 Chimpanzee using bipedalism.

Reynolds and Reynolds (1965) studied chimpanzees in thick rain forests with little open ground and found that they used bipedal behavior much less frequently than arboreal behavior. These forest chimpanzees simply did not have much open space on the ground and stayed in the trees rather than struggle through thick underbrush. Kortlandt (1962) reported that chimpanzees studied in a plantation woodland environment exhibited mostly terrestrial behavior in the open ground spaces and some cautious arboreality. They used bipedal behavior about 10 to 15 percent of the time on the ground for a variety of situations, including threat displays and crossing small brooks without getting wet. Goodall (1965) reported that her chimpanzees, in a woodland environment similar to that of Kortlandt's subjects, also used bipedal behavior for various reasons. They stood up to look over the tall grass for danger or to spot other chimpanzees or some unusual activity nearby. They also stood up when their arms were filled with food or other objects (Fig. 6.2), during greeting and courtship, in threat displays, and when the ground was wet from heavy rains. In both research studies the woodland apes also spent much time feeding and nesting in trees where they often used bipedalism to walk along the branches.

Sigmond (1971) suggests that the degree of chimpanzee bipedalism is related to differences in ecological niches. Field studies suggest that the relationship between thickness of the forest cover and the related amount of open ground may be important factors in chimpanzee selection of bipedalism, however imperfect, over knuckle walking. Thicker forests tend to restrict any advantage to be gained by seeking food in the little available open ground space; food is primarily found in the trees. The grassy open ground in woodland areas, however, has greater variety of foods and ease of movement to attract and encourage the chimpanzees to spend more time on the ground, where they can use bipedal locomotion.

These studies suggest that in particular environments chimpanzees will opportunistically use any behaviors that are advantageous, whether the behaviors are arboreal or bipedal or both. Chimpanzees are, after all, not stupid apes; they will use behaviors that offer successes. If this interpretation is valid for chimpanzees, our nearest relatives, then it may also have been an important factor in influencing our ancestors to select bipedalism in the changing African environments that they had begun to inhabit.

Bipedalism probably originated in a very early hominid or advanced hominoid stock that had become largely ground dwelling or had lived in mixed ground-arboreal environments, like modern chimpanzees. This protohominid probably had a social structure similar to modern chimpanzees, had similar behaviors, was curious and intelligent, and probably made or used simple tools on occasion. It was more than an ape but also not quite human. It shared a common ancestry with modern chimpanzees who, instead of advancing, retained their conservative ways and the ape body pattern. This stock changed for one or more reasons, however, and became more inventive and experimental. This stock adopted bipedalism with all the advantages that went with it, including the freeing of hands for new tasks and for tools. Such structural and behavioral modifications may have provided the crucial push that directed these protohominids on the irreversible path to humans. Chimpanzees stayed arboreal with some bipedalism but retained the opposable digits on their feet and all the hominoid behaviors. Our ancestors made the transition from arboreal to terrestrial behavior and, as they adapted, developed the required skeletal structures and brain abilities then or shortly afterward. Our ancestors, therefore, developed walking feet and true bipedalism, probably shortly after separation from the more conservative pongid stock. Our stock became humans; the rest remained apes.

It is unlikely that the transition of hominoids into hominids took place simply because some primates became bipedal. Humans arose because of many factors and adaptations that initiated changes. In particular, social behaviors contributed significantly to hominid evolution. These behaviors were patterns of living that regulated our propensity to live in association with other humans, even other species of animals. Behaviors or behavioral strategies that confer greater fitness on any individuals or groups are likely to be selected and integrated into any species, and social behaviors were important factors in the success of early and later primates. The acceptance and adoption of many such behavioral adaptations from our ancestral stock and their prevalence in our complex societies indicates their value for increasing stability and survival in our species. Several social behaviors, especially dominance,

aggression and territoriality, among others, provide increased group fitness in both humans and apes. They are found in many other species of animals and are common to all higher primate societies. They were also probably crucial to our ancestors when early hominid social groups were first forming. Today, without laws and police, many humans would quickly resort to a dominance-based social structure in which aggression, strength or wealth would determine the leadership. Despite our systems of laws and police, some individuals still operate in such fashion. The commonality of these behaviors in genetically related animals suggests that they are genetically inheritable to some degree.

TOOL-USE MODEL

It is apparent that advantages conferred by early bipedal behavior in the earliest hominids enhanced bipedalism by feedback. Freeing their hands from locomotion was an obvious advantage that had long been developing in primates. Once hand ability had become important in primates, its obvious benefits resulted in it becoming increasingly emphasized. Terrestrial behavior further emphasized this enhanced hand ability as these primates began to develop better hand-eye coordination and better brains to operate these functions. These adaptations also shifted them further away from the pongid line even though they probably behaved initially much like modern chimpanzees, making and using tools of nonresistant materials gathered opportunistically. One hypothesis proposes that manipulation of tools (Fig. 6.3) was a major factor in the appearance of humans. Others have opposed this, arguing that the lack of stone tools in timely association with known bipedal fossils makes such assertion unlikely. However, the earliest tools are not likely to have been prepared stones but unworked stones or wood sticks which are not easily fossilized. Neither of these materials, unworked stones or wood sticks, would have been easily recognized for what they are even if they had been preserved and found.

There was much to be gained by the use of tools among early terrestrial protohominids. Tools reduced dependence on brute strength for survival and well-being. This independence meant that curious and intelligent animals could compete with the stronger ones that were not quite as smart. From this point, dominance and other behaviors in hominids would no longer be dependent primarily on strength and size. Increased use of hands and tools would tend to

Figure 6.3 Chimpanzee going termite "fishing."

have an affect on the physical structures of their bodies and brains, favoring and emphasizing those that were useful and deemphasizing those that were less suitable. As these early hominids stood up to utilize their increasingly versatile hands, their grasping feet would have proved to be handicaps. Eventually this trend resulted in the development of total bipedalism and walking feet, both reinforced by the increased use of hands and tools. These structures and behaviors would have provided specific and clear advantages to ground dwelling primates and they were on the right track to becoming hominids.

The faunas of the savanna-woodland environments where the primates practiced bipedalism had to be of some influence in this tool-use model; hominids did not evolve in a vacuum. The predatory animals and scavengers that roamed the region would have been formidable opponents to any primates trying to survive on the ground. These protohominids were probably only three to four feet tall and weighed fifty to eighty pounds. But how could small primates have survived in environments filled with hungry predators such as lions, leopards, cheetahs, wild dogs, hyenas and jackals? When danger developed, they may have survived simply by fleeing to trees for refuge. The ability to stand erect might even have given them a slight advantage in seeing the approach of dangerous predators. Modern humans are not as swift as most predators and we have little reason to believe that our diminutive ancestor was gifted with any greater ability. They probably could

not outrun any of these predators and even a quick escape into trees may have been only a partial solution. Lions and leopards often climb trees, though the lion is generally too heavy to follow into the higher and more delicate branches. Our ancestral stock was obviously incapable of competing with these predators and scavengers on strength alone. Few modern humans, although larger and stronger, would consider unarmed combat with any of these predators, especially the larger ones. How then could a smaller ancestral primate have survived against such odds?

The answer to survival may have been in their use of tools and group defense. If these terrestrial primates had learned to use branches, sticks and stones as tools, they might have been able to survive on the ground in the savanna-woodland environment against a variety of meat-eaters. Though they probably would have been unable to do any serious damage to large predators with branches, sticks or stones used as weapons, they might have frightened them enough to escape. Chimpanzees are known to use such tools as weapons and can drive leopards away. Perhaps the early hominids did precisely the same. The use of weapons could have given them a slight advantage so that some individuals survived. These survivors were probably those individuals that had developed better tactics to use against the predators, and they passed these abilities on to their offspring.

WEAPONS

Arguments that primates cannot use weapons are no longer tenable. Both tool use and toolmaking activities have been well established in chimpanzees through numerous field studies of the past four decades. Though their toolmaking ability clearly lacks the degree of sophistication we might demand for toolmaking behavior in modern humans, it is the same basic ability. Chimpanzees often use sticks or branches to frighten other chimpanzees or even to ward off leopards. Kortlandt (1962, p. 138) observed a chimp with a club attacking a leopard and said, "Apparently chimpanzees use clubs instinctively." Goodall even filmed an encounter between a stuffed leopard and a band of chimpanzees in which they began to scream and run about the "predator." One of them began to thrash branches around and eventually began to wield them more directly against the leopard. Another threw a rock. And Galdikas filmed an orangutan wielding a stick, perhaps in play, against a dog.

Figure 6.4 Chimpanzee using a tool, throwing a branch.

None of the apes, however, use such weapons with ease or even modest facility. They flail around with sticks or branches and may eventually and luckily direct them somewhere near the objective. As they wield these weapons, they scream with fearful noises and jump about frantically, creating frightful images that can deter any animal not conditioned to such a display of behavior. Goodall (1965, p. 810) reported of her Gombe chimpanzees that

> When frustrated or excited, chimpanzees often perform "charging" displays. They rush along, dragging fallen or broken-off branches, throwing rocks or sticks, leaping up to sway branches, stamping or slapping the ground.

She also reported (1965, p. 812) the remarkable exploits of one chimpanzee named Mike (Fig. 6.4) who became the dominant male in his band by the quite novel discovery of weapon use. Goodall explained

> The year before we set up New Camp, Goliath, J.B., and Leakey were the top-ranking males. Mike, though just as big, ranked low in status. We didn't know why, but he was constantly attacked or threatened by nearly all other males.

When we left the reserve at year's end, Mike was cowed and nervous, flinching at every movement or sound.

On our return we found a different Mike. He was feared by every individual in the community. We shall never be sure, but it seems likely that by leaving empty kerosene cans lying about, we ourselves had helped his rise to power. He had learned to throw and drag these cans along the ground, and they made a tremendous noise.

Mike often walked to the tent while a group of chimpanzees was resting peacefully nearby, selected a can from the verandah, and carried it outside. Suddenly he would begin to rock slightly from side to side, uttering low hoots. As soon as the hooting rose to a crescendo, he was off, hurling his can in front of him. He could keep as many as three cans in play, one after the other.

Chimpanzees as a rule hate loud noise—except for their own screams—and so Mike, with his strange display, frightened the others. We ourselves grew to dislike his behavior and hid all cans. But by that time, if artificial props had indeed raised Mike's rank, he had no more need of them. The other chimpanzees, at his approach, would pant nervously and bow to the ground, acknowledging his dominance.

Mike had discovered or learned how to use some specialized objects as weapons. How long will it be, if ever, before his fellow chimpanzees, if any, learn the advantages of such a technique?

DISCOVERY AND LEARNING

Any techniques that produce enhanced survivability against the difficulties of life in any environment constitute behaviors that can be discovered by any individual. The discovery of a particular new behavior can be extremely important to a population, whether it be in finding food, a better technique of finding shelter, or tool use. Such behaviors can also be learned by others once the technique is discovered and found to be successful. All animals learn

Figure 6.5 Japanese macaque washing food.

survival techniques. Mammals and primates, in particular, are the quickest learners among animals. Not only do they learn, they also teach others, especially the young. New behaviors appear from time to time in all species as they try to survive in hostile environments and the successful behaviors must be taught to the young, otherwise animal species would remain static and never change. If tool use or weapon use is a behavior found to be successful in repelling predators, it can be acquired or learned by others, just as it is in macaques and chimpanzees.

The classic discovery of a new behavior and learning is seen in Japanese macaques studied in the wild by researchers on the small Japanese island of Koshima. There, a female macaque identified as Ito learned new food-handling techniques that provided her with more and easily obtained food. Scientists had placed sweet potatoes on the beach to attract the monkeys to where they could be easily observed. The sweet potatoes became encrusted with sand, however, and they were difficult to eat. Ito discovered that she could wash the sweet potatoes clean of sand in a nearby stream (Fig. 6.5). When they were fed wheat, Ito also discovered how to separate easily wheat grains from wet sand, usually

a tedious task requiring the removal of each grain by hand. She simply dumped the sand-coated wheat into the stream; the sand sank but the wheat floated and was easily skimmed and eaten. Other members of her troop quickly learned both techniques from her and soon most were busily engaged in washing sweet potatoes and separating wheat from sand. Some very enterprising macaque youngsters also discovered that the wheat washing method inevitably resulted in some wheat grains being washed downstream and lost. They learned to wait downstream and recover the grains that floated away from others upstream. Learning had spread quickly through the troop whose members generally accepted these better techniques for cleaning food. The youngsters tended to learn these techniques much quicker than old individuals and many of the very old macaques never seemed to learn the methods. Perhaps there is some truth to the old adage that you can't teach an old dog new tricks. In one interesting development, some young macaques began to wash their sweet potatoes in salt water. Apparently they enjoyed the salty flavored sweet potatoes better than plain ones washed in fresh water.

Figure 6.6 Chimpanzee using a tool (stick) to open a box.

Chimpanzees are much more intelligent than macaques and have a greater curiosity. Their ability to learn from each other is much greater than that of the macaques, and chimpanzees are more than willing to experiment. If macaques can discover and learn new techniques, what can chimpanzees discover? Few things are safe from their innate drive to pry and poke into things (Fig. 6.6). One captive chimpanzee, Washoe, was the subject of early experiments in sign language and provided valuable information about such chimpanzee learning. She lives with several other chimpanzees, including an adopted son, in a study group in Washington. This "son" learned more than forty signs directly from other chimpanzees in a controlled experiment in which humans never signed with him or in his presence. To chimpanzee learning we must now add a capacity to teach newly learned abilities. This must include the possibility of teaching the use of tools and weapons.

Today we acknowledge that the potential for making and using tools is not restricted to humans; it has been identified in another animal species. The fact that this skill has been identified only in chimpanzees strongly supports their status as our nearest relative. If such skills require a minimum level of intellectual capacity and achievement, the chimpanzees may well be near this threshold that many experts believe is somewhere between humans and apes. We might even speculate that if humans had not evolved on Earth these chimpanzees might have been able eventually to develop into simple hominids. However, humans did evolve and are much more efficient at any intellectual task than chimpanzees. These advanced apes cannot hope to compete with humans in any human behavior. Though they are advanced apes, they are doomed to remain apes, perhaps even doomed as animals endangered by ordinary human activities.

COOPERATIVE BEHAVIOR

Richard Leakey and Roger Lewin stress **cooperative behavior** as perhaps the most important factor that led to the successes of the early hominids and the origin of humans. They state (1977, p. 10),

... that humans could not have evolved in the remarkable way in which we undoubtedly have unless our ancestors were strongly cooperative creatures. The key to the transforma-

Figure 6.7 Cooperative behavior in early humans.

tion of a social apelike creature into a cultural animal living in a highly structured and organized society is sharing: the sharing of jobs and the sharing of foods.

Such sharing behavior probably originated in our distant ancestors when they collaborated for mutual benefit (Fig. 6.7). Cooperation in common defense against a strong enemy is clearly an action that affords benefits to a group's members and has been well described and identified in baboons and numerous other species. Chimpanzees occasionally cooperate successfully in hunting for small animals to obtain meat. They share the kill, sometimes with nonparticipating members of their group. In such examples, these cooperative actions significantly multiply the potential abilities of single individuals for the benefit of all participants. Any accidental discovery of cooperative behavior that was successful could easily have been repeated. This would naturally have lead to familiarity and more frequent practice. With increased practice these animals would have increased their cooperative skills until they became a natural and normal behavior for many situations, not just for hunting and defense.

ALTRUISM

A behavior related to cooperation, **altruism**, is common in humans and has now been recognized in some primate groups and other animals. Altruism is any behavior that is good for another individual or the entire group, even though it may also result in bad, harmful or even fatal effects on the altruistic individual. It differs from cooperation in that the altruistic individuals may not gain personally from their actions. Baboons and chimpanzees, for example, practice altruism when they unhesitatingly attack predators that jeopardize their group. They act altruistically, even though the act may result in their own death. This sacrificial behavior represents an apparent paradox in which an individual's altruistic behavior is somehow beneficial to it, even though it dies. It is a beneficial action only if the individual's altruism results in its ideals, offspring, relatives, group or territory being preserved. Individuals that sacrifice their own well-being for blood relatives are protecting their own genetic materials present in their kin. A sacrifice for territory may have the effect of preserving their genetic materials by reserving the food-gathering region. This is true in baboons, chimpanzees, bears, humans and countless other animals. Dawkins (1976), however, argues that although altruism is quite common, it is a selfish behavior, not altruistic. Selfish individuals, however, tend to be considered misfits in societies, whether elephant, horse or human. Their disadvantageous emotions and actions tend to promote disunity in a society to its detriment. In contrast, an altruistic individual is much more welcomed in society because most perceive the benefits and desire them. Altruistic behaviors, emotions and actions tend to promote group survival and well-being, even though a specific individual suffers in performing the altruistic act. Because such acts commonly have survival value, some workers (Fox and Tiger, 1971; Wilson, 1975) have argued that many such behaviors are immensely valuable to a species. They maintain that individuals act altruistically because there is some benefit to be gained by them from such acts.

Any pattern that produces increased fitness in an individual, whether it be behavior, structures, features, traits, etc., is desirable and any species that can pass this fitness on to offspring will be better at adapting than one that does not. Successful behaviors that lead to greater fitness in a species must either be learned or inherited. Learning, however, cannot account for many obviously instinctive acts and some desirable behaviors are, therefore, believed to be genetically based. In the simplest terms, these acts are likely transmitted through adaptive or suitable gene combinations. They are good or suitable simply because they result in successful reproduction and offspring that are similarly successful. Wilson (1975, p. 3) points out that

> As more complex social behavior by the organism is added to the genes' techniques for replicating themselves, altruism becomes increasingly prevalent. . .

Any gene combination that tends to confer greater fitness by altruism is desirable and will be affected by natural selection to increase in a population. In contrast, many gene combinations produce unsuccessful behaviors that result in reduced fitness. Evolution is blind as to which genes are good or bad. Those behaviors that are better for a population in a given set of circumstances will tend to be retained, and the genes that produce such behaviors will increase. In this fashion, natural selection constantly improves a species. Cooperation and altruism are two such behaviors that confer increased fitness on a species and can become genetically selected, or instinctive, behaviors.

SUMMARY

The idea that humans can inherit the capacity for various behaviors in addition to physical anatomy is based on continuing studies of the pongids. Such a concept signals a radical departure from the long-held belief that humans are unique and separate in the animal world. Initially this view, which is based primarily on philosophical and religious beliefs, held that humans are creative creatures different from the others. In the western world, this was derived primarily from the Old Testament book of Genesis in which humans were given dominion over the animals and Earth. Humans were, therefore, not considered as animals like the other creatures but occupied a special part in the biological world. In time this view changed and humans were accepted as animals like the others but still separate and distinct because of uniquely human traits and behaviors, language and speech, tool use, toolmaking, societies, murder, thought, self-acknowledgment and others. The relatively recent recognition in other primates of many of these traits and behaviors has demanded a reassessment of the traditional separation of humans from other animals and the idea of our own uniqueness.

The obvious anatomical character that can be considered unique to humans is bipedalism. It is particularly useful in separating fossils of early humans from ape fossils. Several different evidences of bipedalism can be found in these hominid fossils, a knee joint, a foot, a preserved footprint, or some other direct evidence. Another essential character of humans is our high degree of thinking ability and the social structure and culture that resulted from it. However, it is rather difficult to determine thinking ability in a particular fossil without some incontrovertible evidence.

Toolmaking can provide reasonable evidence of thought because such an activity requires planning and conscious effort to produce the implement. Thus, toolmaking in chimpanzees can be considered a significant step toward thought like that of humans, even though it is only a small step. The making of stone tools, however, is ample evidence at this time of human activity; only humans are known to chip stones to make tools. The making of tools from woody materials, as practiced by chimpanzees, is evidence of some thinking and planning but these soft materials are rarely preserved and may not be recognized if found.

Though the recent work on biochemical analyses of the anthropoids and the intriguing sociobiology studies of behaviors identify our relationships with the other primates, such evidences are not easily determined from the fossil record. Skeletal remains and worked stone tools remain the best criteria for identifying fossils as either human or ape.

The separation of the human line from the ancestral stock of apes has been reasonably determined from paleontological evidence based on materials collected at field sites in Africa, Europe and Asia, from biochemical analyses of living primates, and also on estimates of rate of molecular mutations. The paleontological evidence is relatively exact and has been accompanied by excellent absolute age dates obtained by radioactive mineral age dating techniques. Repeated tests of such samples provide similar results and tests made at different sites show little variation. Thus, we have great confidence in radioactive age tests and the dates they provide.

The similarity of behaviors of humans and some living apes supports this evolutionary picture. As social animals, we are the inheritors of behaviors that originated in our ape ancestors. With our physical structures, superior brain, and these behaviors we advanced far beyond the ape stage and developed specialized techniques and abilities for survival and for much more. A critical departure for the earliest hominids was the acceptance of bipedal behavior and all that went with it.

Bipedal animals are a very insignificant minority out of all the millions of species that have lived, and we are the only bipedal mammals. Why did most other animals fail to evolve bipedalism? Why no other mammals? There had to be some benefits to be gained by these conservative hominoids in becoming terrestrial and bipedal. Certainly the earliest known hominids were too small and weak to offer significant defense against predators, they were not swift enough to flee; and with walking feet they could no longer adequately climb trees. Yet they survived.

Universal bipedalism among hominids was undoubtedly derived from the largely terrestrial behavior of our ancestors, which also freed their hands to exploit great capabilities and potential. The greatly enhanced usage of hands for manipulating materials of their environments offered them significant advantages. The picking and carrying of foodstuffs was the most obvious benefit, and the use of sticks as tools for digging was of great value. And once tools began to be used, they became important. Foods previously available only with difficulty were more accessible. No other animals rely on anything beyond their own fang and claw for gathering their food, providing shelter, defense and more.

The acceptance of and increased dependence on tools by the early hominid ancestors led to the associated improvements in their brains that turned them from another type of ape into thinking creatures, humans, that eventually came to become the most significant and powerful species on Earth. The trend that began some 5 to 8 million years ago, when hominids split away from the chimpanzee stock or its ancestor, has not yet ended. I sit preparing this manuscript on a modern tool, a computer, and have found another use for one of the stone tools made by early hominids; it is an Oldowan chopper, now used as a paperweight.

We have now identified many lines of evidence that demonstrate that humans are inextricably linked by evolution to the great apes, most closely with the chimpanzee. When Darwin's idea of this relationship between apes and humans was published and became known, a popular cartoon depicted a Victorian matron commenting to a friend,

> I hope that it is not true that we have descended from apes, but if it is, let us hope that it does not become commonly known.

I hasten to add, probably to her regret, that it is now well known.

STUDY QUESTIONS

1. Identify environments where bipedalism probably arose in the early hominids.
2. What probably drew the developing hominid stock to a terrestrial lifestyle?
3. What advantages did bipedal behavior offer to the early hominids?
4. Why didn't nonhuman primates adopt bipedalism?
5. List and describe factors that might allow a hominoid or a hominid to become the dominant individual in its society.
6. Explain how the use of a tool by the chimpanzee Mike to achieve dominance in his troop relates to or is different from tool use in humans.
7. Describe altruism in hominoids. Cite an example.
8. Apes are capable of learning many new behaviors. Cite an example.
9. Postulate a nonevolutionary scientific hypothesis for the origin of humans.
10. Explain why thinking and learning are so important in the evolution of humans from apes.

7 The Australopithecines

INTRODUCTION

The search for the ancestors of humans began even before the publication of Darwin's second great book on evolution, *The Descent of Man and Selection in Relation to Sex* (1871). In his earlier book, *On the Origin of Species*, Darwin sought to avoid the issue of human evolution though he had implied that mankind must have been involved in evolution. He believed (1872, p. 373) that eventually "Light would be thrown on the origin of man and his history." But with the second book, Darwin had established (Simpson, 1966, p. 472) two essential and fundamental points with regard to the evolution of humans: 1) that humans had evolved from earlier species by the process of evolution through natural selection, and 2) that humans had evolved from the primates, either through apes or monkeys.

Today many experts believe that Darwin tried to exclude mankind from his evolutionary process because he anticipated the negative impact such an idea would have had on Victorian British society in the mid-nineteenth century. He was exceptionally sensitive to the religious beliefs of the public and sought to avoid any views that might damage their faith. Slowly, however, the idea of human evolution forced itself into the scene as more and more of his British confidants and associates supported his views. Many of these were the leading scientific minds of Europe with widespread influence, and their advice to Darwin was of considerable value. Eventually, his friends urged that he publish his views about the evolution of humans before someone else preempted these ideas. In time, even his family began to acknowledge the connection between evolution and humans and supported the publication of his views. When he finally published *Descent of Man* it was in a milieu in which the opponents of evolution had been organized in opposition to his ideas for a number of years. But for all the difficulties raised by clerics, scientists and ordinary citizens, Darwin found that he had a great many supporters at home and abroad.

Darwin's *Descent of Man* (1871) may be considered the formal beginning of the greatest search humans have ever known, the attempt to find our ancestors and our place in the world. Huxley had already (1863) established man's place in nature as a member of the animal world and raised

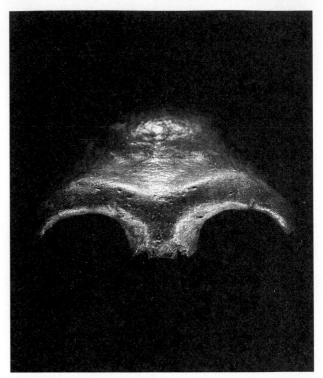

Figure 7.1 Neandertal skullcap.

many questions about our present status, such as how we had evolved, whence our race had come, about the limits of our power over nature and nature's power over us, and to what goals humans ultimately strive. Huxley considered the sum of mankind's anatomical structures and features in an effort to determine what, if any, relationships humans had with the other animals. He affirmed that humans were biological creatures, clearly a mammalian vertebrate and specifically a primate. He also noted that among the primates can be found an almost continuous series of mammalian brains with only one structural break, not between hominids and apes as might be expected, but between the prosimians and the simians. It was clear Huxley recognized the brains of the monkeys and apes as nearer hominids than the prosimians and equally clear that mankind's place in the primates was very close to the apes and monkeys. Accordingly, he postulated a descent of humans from these anthropoids.

At this time, so shortly after publication of Darwin's *Origin of Species*, only two very old fossils of humanoids were known. The key fossil was a skullcap (Fig. 7.1) and some bones that had been found in the Neander Valley in

Figure 7.2 Gibraltar skull.

Germany in 1856 and which ultimately gave rise to the concept of the Neandertals. The second was a skull (Fig. 7.2) of uncertain nature that had been found in 1848 at Gibraltar but which remained unclassified for many years. Though both were of unknown affinity at first, subsequent fossil discoveries demonstrated that the Neander Valley fossils, Neandertals, were a valid type of ancient humans long gone from Earth. The Gibraltar skull was thereupon identified as a Neandertal.

Hermann Schaaffhausen, who had first described the original Neander Valley remains, and Huxley both considered these Neandertals to be essentially human. Huxley insisted that these remains could not be an intermediate stage between humans and apes. He believed that the Neandertal skull was a more pithecoid end member of a series that included the known variety of humans, but was, nonetheless, a human. Although Huxley was clear in his own mind of man's place in nature and how it had occurred, Darwin for a long time hesitated to ascribe the same evolutionary origin to humans that he had ascribed to other species. Finally, in the face of growing interest and evi-

dence, Darwin accepted the arguments and published *The Descent of Man*, his sequel to *Origins of Species*.

THE SOUTHERN APE-MEN

Until 1925 the prevailing concept of ancient humans was based on scattered incomplete fossils of mostly indeterminate taxonomic identity and poorly defined relationships. At best the concept was a confused collection of conflicting ideas about widely separated fossils that were restricted to three types, Dubois' *Pithecanthropus erectus*, the Neandertal man, and Cro-Magnon man. In 1925, however, **Dr. Raymond Dart** announced the discovery of another important fossil and changed forever the way in which scientists and others looked at mankind's history and our evolutionary path. Dart was an anatomist who had studied under two great biologists and trained in Australia, England and the United States. In 1922 he became an anatomy professor in the Medical School of the University of Witwatersrand in Johannesburg, South Africa. There he was fortuitously placed to render critical judgments on the major fossil discovery soon to be made.

For a number of years, fossil baboon and ape skulls had been found in South African caves. These finds were significant because none of these animals were previously known as fossils south of the Fayum in Egypt, and the living apes were unknown south of the Lake Kivu region of Zaire, about two thousand miles north. In 1924 a student of Dart's, Miss Josephine Salmons, had noticed one such fossil baboon skull in the home of the owner of a limestone company and, knowing Dart's interest in skulls and brains, brought the information to him. Dart inquired after these cave fossils and soon received some limestone breccia blocks from the quarries at Taungs near Kimberly, where the baboon skull had been found. Dart found in one of these fossil-bearing blocks a natural endocranial cast of the inside of a braincase, a remarkable discovery (Fig. 7.3). Within a short time he found in another block the skull from which the braincast had been formed. The skull itself consisted of the entire face, jaws and teeth, and braincast of a juvenile animal about five or six years old. Its age was easily determined by examining the nature of the distinctive bone sutures and the teeth, which were milk teeth, the first set. The braincast resembled brains of both chimpanzees and gorillas, but, in Dart's opinion, the animal was neither a chimpanzee, gorilla, or baboon because its brain was too large for an adult of these types, especially consid-

Figure 7.3 Raymond Dart with Taungs skull.

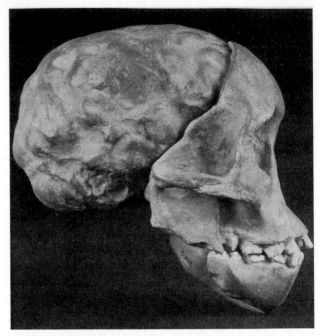

Figure 7.4 Skull and brain cast of Taungs child.

ering the skull was of a juvenile that had not yet reached full adult size. If not a monkey or ape, then what was it?

The skull's face (Fig. 7.4) was rather flattened and vertical without a protruding or prognathus jaw condition like that of an ape or baboon; it also lacked the prominent brow ridge and sagittal crest of an ape. The jaw was small and the teeth were neither simian nor pongid in appearance; the molars were generally human though slightly larger. A very important feature of the skull, the foramen magnum, was situated directly underneath, not at the lower rear as in apes. This position indicated that the animal held its head erect, something that made sense only if it had an upright posture. But Dart knew that upright posture is totally non-ape in character based on both fossils and living specimens. If the creature was neither simian nor pongid, neither monkey nor ape, could it possibly be human? The brain was believed to be too small to be human although it was too large for an ape. It was obviously different from any

known type of human and too advanced to be an ape, but just how different was it from living humans? Dart eventually concluded that the **Taungs child**, as it became known, was of

> . . . importance because it exhibits an extinct range of apes intermediate between living an-thropoids and man.

It was to become a new species, too primitive to be human and too advanced to be an ape but intermediate in relationship to both. Dart named it *Australopithecus africanus* (southern ape of Africa) and submitted the infor-mation for publication in the prominent English scientific journal *Nature* (Feb. 7, 1925) within a few weeks after discovery and analysis. This haste to publish, however, caused years of grief to Dart.

The discovery of this new "**missing link**" (the earlier "missing link," **Java man** or *Pithecanthropus erectus*, had been found in 1890 in Java) created almost as much furor as Darwin's *Origin of Species*. Few experts from Europe and America agreed with Dart's interpretations, possibly because Dart was neither a paleontologist nor an anthro-pologist. Besides, he had not consulted with the European experts in his rush to publish the information. They also believed its brain was too small to have been a near human because of their preconceived idea that such a missing link would have needed a much larger brain. His two prominent teachers, Sir Arthur Keith and Sir Grafton Smith, both leading experts denied Dart's claims of importance by insisting that the skull was simply that of a new or aberrant ape similar to a chimpanzee.

Dart proposed later (1926) that humans had passed through three stages in developing from the arboreal primate ancestry to the modern form. The first two stages were essentially theoretical and based on the belief that humans had evolved from apes. Stage one was a semiarboreal or semiterrestrial phase as exemplified by living apes that lived largely on the ground but had not yet lost their arboreal structures and abilities. This stage was still undiscovered. The second stage was also theoretical and was an "entirely terrestrial man-ape phase" intermediate between terrestrial apes and terrestrial humans. Many experts had insisted for years that Dart's find was not in the direct lineage to humans but was at best, perhaps, a distant cousin. However, later discoveries of numerous specimens of *Australopithecus africanus* demonstrated its reality as this "theoretical" second phase. Modern humans are the living reality of the third phase that demonstrated the full status of a terrestrial human with all the characteristics and behaviors of living people.

Dart believed that these three phases had evolved under the influence of some unverified forces, probably the likely result of changes in behaviors from arboreal to terrestrial habitats. The forces that caused these shifts in behaviors may have been related to the climatic changes that were assumed to have taken place in the Old World beginning in late Miocene time. A great drying up was believed to have begun about 7.5 million years ago as seen in the fossils and the sedimentary rocks. This period of desiccation was so severe that the Mediterranean Sea had dried up several times about 5.5 million years ago, after its connection with the Atlantic Ocean had been blocked by mountain building from Morocco to Spain. Precipitation was reduced at this time all across Africa and this drying-out caused the great forests to decline in size and expand into savanna-woodland habitats. The arboreal primates that inhabited these forests would have been seriously affected by such changes when their habitats suffered shrinkage, much like modern apes are suffering today in some areas. Any ancient arboreal primates

that possessed the potential ability to live as terrestrial animals, however poorly, could have survived in the savanna-woodlands when their forests suffered declines and retreated. Naturally, these primates would have had to learn to eat new foods and develop new behaviors because they were forced to survive in changed environments with different available foodstuffs. Fortunately, such changes in the habitats were very gradual, over thousands of years and there was ample time for favorable adaptations to appear in the primates and take precedence over unfavorable ones.

An indication of how serious any change in environment could be for a species is seen in modern gorillas. These highly specialized animals are finding it increasingly difficult to survive in their forest habitats that are being cut down and replaced by agriculture activities. At the same time, they are being invaded by hordes of camera-carrying tourists seeking glimpses of them before they disappear completely from the wild. If modern gorillas, which are a very conservative species, find survival difficult today in the face of human expansions and inroads, then late Miocene hominoids probably suffered similar serious survival stresses when the climate began to dry up and the forests responded by shrinking. How animals respond to stresses of the environment is a critical aspect of their survival and relates directly to evolution and natural selection; these responses are a direct consequence of their genetic makeup.

Shrinking habitat space and resources tends to create greater competitions between and within inhabitant species for the reduced territory and resources. Increased competition produces stresses felt by the animals; some can handle the increased stresses, others cannot. Those species or individuals that can handle the stresses may be flexible enough to develop new behaviors (e.g., eat new foods) or modify existing ones (e.g., become more aggressive in defending space) for the new or different habitat conditions. Most of the Miocene hominoids probably already possessed traits or behaviors that contributed to their success in these changed habitats. These species tended to survive although the new environments exerted pressures that caused them to change their ways of life and become very different in time, ultimately to the point where they evolved into new species.

Animals that are highly specialized and adapted for a particular environment suffer greatly when their environment changes; they are highly dependent on specific resources and cannot easily adapt to new resources when the conditions change. They are said to be genetically inflexible and, as a result, they are likely to become extinct under new conditions in new environments. The gorillas fit into this pattern; they are highly specialized animals that have difficulties adapting to new or changed conditions, and they are in grave danger of becoming extinct in the wild. If the African continent was experiencing conditions of lessened precipitation during Miocene time, especially in the forests inhabited by the dryopithecines, then those hominoid inhabitants probably experienced stresses similar to those facing the modern gorilla. Some of these hominoids survived by adapting, but others, unable to adapt, died out.

Isaac disagrees with those who invoked major climatic changes to explain the impetus for the evolution of mankind. He said (1976, p. 507)

> . . . it has been customary to invoke as causes, more or less drastic environmental changes such as a Pliocene drought (Ardrey, 1961; and many others) or dramatic dietary specializations such as hunting or seed eating (Jolly, 1970). However, . . . there has probably not been any great environmental trauma in the late Tertiary of Africa. The continent seems to have supported throughout this time a fluctuating mosaic of forest, woodland, savanna, grasslands and steppe. The most dramatic faunal change in Africa appears to have occurred 4–5 million years ago and to relate more to intercontinental connections than to climatic change (Maglio, in press).

Whatever the circumstances that impelled some of the Miocene hominoids to shift to different environments, the facts are that some of these primate ancestors of hominids made the transition from forest habitats to those of the more open savannas and woodlands.

THE ROBUST APE-MEN

A second individual of determined character entered the search for ancient humans in South Africa in 1934. **Dr. Robert Broom**, a doctor of medicine, retained a lifelong love of nature and an interest in the origin of mammals. He practiced medicine initially in Australia where he published numerous scientific papers on paleontology, particularly of extinct marsupials from cave deposits. He soon learned of some fossils of mammal-like reptiles that had been found in South Africa and moved there in 1897 to find them. Shortly thereafter, with his background in paleontology and record of scientific publications, he gave up medical practice and began to work full time in paleontology and as a professor of geology and zoology. In seven short years he published nearly one hundred scientific papers, but financial problems forced him back into medicine. His research, however, continued and he became expert on South African pelycosaurs, an important group of reptiles found in the Karoo beds. Before long he made the important discovery that mammals had descended from these pelycosaurs and effectively established the relationship between these two major groups. He clearly was not an amateur paleontologist and commanded great respect in the paleontological community.

Upon hearing of Dart's discovery of *Australopithecus*, Broom immediately rushed to see this specimen of an ape-man and was converted to this radical new view of early humans. Dart and Broom interpreted the Taungs specimen as evidence that *Australopithecus* had lived in caves, and both thought that they might have lived perhaps two million years ago. Bones of numerous animals in the caves led Broom to conclude that *Australopithecus* had probably caught the animals in ambush at waterholes and carried the remains back to the caves for food. Some fossils from the caves showed evidence of trauma to the head and many baboon skulls showed fracture damage that could have been caused by blows. But such ideas were for the future, and more hominid fossils had to first be found.

In 1934 at the age of sixty-eight Broom was rewarded with a position as curator in the Transvaal Museum and in 1936 he began searching for fossils in the limestone caves of the region . He learned of baboon skulls found in a cave at **Sterkfontein**, near Johannesburg and requested that the quarry manager inform him of any future specimens that might be of value in his research. Fortunately the manager, G. W. Barlow, had worked at Taungs and knew about such fossils. Within three days he handed Broom the fossil skulls of three baboons and a saber-toothed tiger. Within another week Broom received the major part of a fossil braincast that he believed came from an ape-man. Digging through the blasted limestone fragments for hours, Broom eventually found most of the skull from which the cast had come. He was elated to discover it was a new type of ape-man and named it *Australopithecus transvaalensis*

Figure 7.5 Skull of *Australopithecus robustus*.

(a)

(b)

Figure 7.6 Reconstructions of (a) *Australopithecus robustus* and (b) *A. africanus.*

(southern ape of Transvaal). Later he decided it was different enough to warrant a separate genus and renamed it *Plesianthropus* (near man), though it was later reassigned back to *Australopithecus*. Still later it was recognized as simply another specimen of the original australopithecine, *A. africanus*, its present status.

Through 1936, 1937 and 1938, Broom combed the limestone quarries of the area around Sterkfontein seeking more fossils and finding skull fragments, teeth and some limb bones. In 1938 he received a fine palate with a single molar in place and learned it had been found by a schoolboy at **Kromdraai**, only two miles from the Sterkfontein cave. From the discovery site he recovered more pieces of the skull until he had most of the left side of the skull and the right side of the lower jaw with many teeth. But this skull was different from the Taungs type (Fig. 7.5). It had a flatter face, much larger and more powerful jaw, and larger teeth. In these respects it had some similarities with apes. Broom named this specimen *Paranthropus robustus* (robust near-man) and it has been accepted as the second type of early hominids from South Africa. It effectively demonstrated that there were two different but related South African australopithecines (Fig. 7.6), one a large **robust** type, and the other, a smaller and more delicate form commonly known as the **gracile** type. Broom's second species, *Paranthropus,* has since been reassigned and renamed *Australopithecus robustus*.

World War II interrupted much scientific work in South Africa and elsewhere, but afterward Broom resumed his explorations of the cave breccias at Sterkfontein and Kromdraai. Using dynamite he blasted the rock apart to expose the fossils and had great success; such blasting fortunately is not too damaging to the fossils and they can

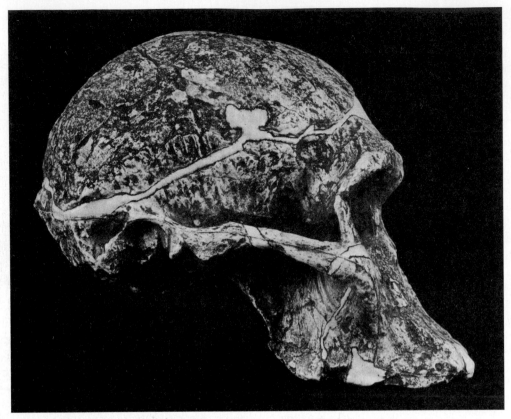

Figure 7.7 Skull of *Australopithecus (Plesianthropus transvaalensis) africanus,* Sterkfontein V.

be pieced together. In April, 1948 he found an almost whole, nearly perfect skull of an adult female of his earlier genus *Plesianthropus* (Fig. 7.7) This specimen, known as "Mrs. Ples," has a brain size of about 480 cc and is now assigned to *A. africanus.* Broom then found an almost perfect male jaw and noted that

> . . . though the canine tooth is larger than in man it has been ground down in line with the other teeth exactly as in man.

This was followed by the discovery of several other post-cranial bones including a nearly perfect pelvis with features that substantiated his belief that these creatures walked upright. He said,

> This structure, human in all essentials, proves that the ape-men walked on their hind legs.

In 1948, spry and able at the age of 81, Broom found a lower jaw at **Swartkrans**, a new site within a mile of the Sterkfontein cave. He found another lower jaw of a larger type which he named *Paranthropus crassidens* (now assigned to *A. robustus*). Working there between 1948 and 1952 with **J. T. Robinson**, Broom also found a nearly complete skull, several fragmentary skulls and upper jaws, some lower jaws, some limb bones and more than one hundred individual teeth and other fragments of this large type. By the 1950s, the robust type of australopithecine, *A. robustus*, was very well known and characterized from these numerous fossils. By this time also, evidence that these australopithecines were separate and distinct from the apes was so unequivocal that no one questioned the australopithecines' status as more advanced anthropoids than the apes and very close evolutionarily to humans.

THE GREAT MAKAPANSGAT BONE CONTROVERSY

About 150 miles northeast of Sterkfontein is the cave of **Makapansgat**, a national monument where an historic battle took place between Boers and Bantu warriors. In 1936 one of Dart's students, **Phillip Tobias**, found a baboon skull in the ash of this cave. Dart was unable to explore the cave then and returned to teaching during World War II. In 1947 he finally raised enough funds and was able to explore for fossils in the cave floor. There he found numerous bones of gracile australopithecines, including a mostly complete skull, parts of other skulls and mandibles, teeth, broken pieces of pelvis and limb bones. The presence of carbon in the floor sediment suggested to Dart that this type of australopithecine had used fire, and to indicate this he named the specimen *A. prometheus*. This idea of fire use is now discounted, and the "new species" is recognized as another specimen of *A. africanus*.

Dart found something else in these cave deposits that led him to draw remarkable conclusions about *Australopithecus* that even today are being debated. The limestones from Makapansgat cave and other caves in the vicinity yielded some 150,000 fossils of numerous types of animals over a period of twelve years. Dart concluded that the bones represented food and meat that had been brought into the cave by these australopithecines. Also found were some fossil skulls of australopithecines and baboons. The baboon skulls were remarkable, if only because a number of them showed trauma that consisted of fracture depressions with a fine network of cracks radiating outward from the center; some also showed a raised ridge across the depression. Dart rejected the contention by some that these animals were struck by rocks falling from the cave ceiling. He believed that the evidence was enough to demonstrate deliberate blows to these animals with some instrument, though he was uncertain who caused the blows or when or why they were struck.

One australopithecine fossil had also sustained serious damage to the jaw. This was a twelve-year-old individual, according to Dart (1956, p. 325–6), who

> ... had been killed by a violent blow delivered with calculated accuracy on the point of the chin, either by a smashing fist or a club. The bludgeon blow was so vicious that it had shattered the jaw on both sides of the face and knocked out all the front teeth.

Dart argued that these fossils indicated the use of weapons. He believed that the damaged baboon skulls represented individuals that had been killed by australopithecines using bones as weapons. And further, these australopithecine hunters also used these weapons on each other. Most workers, of course, disagreed with this idea. **Wilfrid Le Gros Clark** the great comparative primate anatomist, suggested that a statistical analysis of these bones might supply some answers to settle the mystery. Dart thereupon collected for analysis fifty-eight baboon skulls from Makapansgat, Taungs and Sterkfontein, sites spread over 150 miles. Sixteen of these were too fragmentary to be of use in trauma analysis but forty-two others showed some damage to the skull or the jaw. Six skulls (14 percent) had damage in the rear and twenty-seven (64 percent) had damage to the front. Some of the australopithecine fossils also showed damage, including one jaw that had been shattered at the front by a powerful blow.

Dart refused to believe that such a high percentage of baboon skulls could have been accidentally damaged by rocks falling from the cave ceiling. He thereupon enlisted Dr. R. H. Mackintosh, a professor of forensic medicine and a medical-legal expert, to examine the fossil skulls and venture an opinion. Mackintosh stated (Moore, 1963, p. 312) "I have seen people hanged on evidence similar to that..." Dart thereupon produced a fossil antelope humerus from the caves and offered it as the blunt instrument that caused the damage to the skulls. Its double-knobbed distal end fitted perfectly into the double depressions in the skulls. Because of their great numbers these humeri were his choice for the weapon. He later wrote (1949) that the damage to Specimen One from Taungs was from

> A powerful downward, forward, and inward blow, delivered from the rear upon the right parietal bone by a double-headed object.

He also said of Specimen Six from Taungs,

> The V-shaped island of bone left standing above the obvious depression of the cranium shows that the implement used to smash it was double-headed.

Robert Ardrey took up this argument with great vehemence in his initial book (*African Genesis*, 1961) on the origin of humans. He quickly noted that all the bones recovered from the caves represented an abnormal distribution. Ardrey had been a professional statistician early in his career and claimed the ability to recognize such an abnormal distribution. He reported (1961, p. 195) that Dart and his students had collected 7,159 fossils from twenty tons of fossiliferous limestone from the Makapansgat dumps. Of these, 4,560 bones were sorted, identified to genus and described; they represented at least 433 different individuals. A total of 518 bones were antelope humeri, a disproportionate share (11 percent of the total) if this collection was a normal distribution. Only 101 bones (2 percent) were antelope femurs, another abnormal distribution. Of all the antelope bones, 30 percent were of a medium-sized animal about the size of a waterbuck. But these same medium-sized antelopes provided for 60 percent of the antelope humeri, again, a disproportionate share for a normal distribution. Everything about the medium-sized antelope humeri that Dart claimed were weapons indicated an abnormal distribution.

Ardrey also asked, if these bones were the blunt instruments responsible for the death of the baboons, could their distribution in the caves be explained as accidental? Three parts of the medium antelope humeri were found, the distal (distant) end with double condyles, the proximal (near) end which had the shoulder, and the middle without either end. All the humeri were broken, none were complete. The two ends were the likely choices for the instrument, but only the distal end of the bone had the double knobs or condyles that fitted the fractures. These distal-end humeri outnumbered the proximal ends by 238 to 7, another abnormal distribution. Ardrey's recognition of these abnormal distributions supported Dart's contention that these bones were not the result of simple accumulations by random animal feedings but that they were somehow selectively accumulated. He, therefore, enthusiastically endorsed Dart's belief that the bones had probably been collected by the australopithecines.

Dart's ideas of the bone use by *Australopithecus africanus* were formulated into a proposed **"Osteodontokeratic" Culture** (culture of bone-tooth-horn) in which these objects were used as tools and weapons. This culture in australopithecines has been largely discounted today by most workers, especially Brain (1981) who cited a variety of modern situations that produce the same general pattern of bone accumulations as Dart found in Makapansgat. In addition, much of the trauma sustained by these fossils' skulls is ascribed to postdepositional effects, though one must wonder about the source of fortuitous damage to baboon skulls that produced so many similar traumas. One must also wonder how bone collections from open ground in modern native villages, cited by Brain, are comparable to bones from ancient caves.

An alternate proposal to Dart's hypothesis has been proposed. Washburn, von Koenigswald and others have ascribed the volume of accumulated fossil bones from the many different caves to hyenas and other carnivores which carried the bones into the caves, presumably for feeding in shelter. This view was originated by Buckland in 1822 who identified such prehistoric bone accumulations in Kirdkale Cave, Yorkshire as resulting from the activities of spotted hyenas. Both Dart and Ardrey, however, argued against hyenas as the source of the Makapansgat cave bones.

Ardrey consulted **A. J. Sutcliffe**, a paleontologist specializing in hyenas, in an effort to clarify the situation. Sutcliffe had studied British Pleistocene caves inhabited and frequented by hyenas (1969) and from which a great quantity of fossil bones were recovered. As Ardrey pointed out (1961, p. 310) the fossil bones examined by Sutcliffe were, however, from British caves and consisted largely of hyena bones, not the remains of many different animals, as is the case with the australopithecine caves of South Africa. The hyena-bearing layer of a Devon cave, for example, yielded the remains of 110 individual hyenas, 40 of them juveniles, yet only 20 other individuals of all species, a ratio of 6.5:1. At Makapansgat the ratio was 17 hyenas to 433 other individuals, a significantly different ratio of 1:26. The bones studied by Buckland from the hyena layer in the Devon cave had more than 1,000 loose hyena teeth and only 100 nonhyena teeth, a ratio of less than 11:1. This compared with 47 hyena teeth that had been collected in South Africa along with 682 teeth of other types for a ratio of 1:15. The comparisons of bones from the two caves not only did not come close to matching, they showed significantly reversed ratios.

Ardrey (p. 308–9) also went to the British Museum to examine fossil hyena bones. The museum had previously obtained a large block of the Makapansgat bone breccia and from it had recovered sixty identifiable bones. Zapfe (1940) had previously done studies of bones that had been fed upon by hyenas in zoos and discovered that the proximal and distal bone ends were gnawed, split, or grooved by tooth marks. He examined the recovered Makapansgat material in the British Museum and found that none of the bones showed any such damage. Further, he noted that a great quantity of crushed bone and splinters is produced by hyenas as they chew, gnaw and crush bones seeking meat and marrow. The Makapansgat breccia, however, produced only a very small quantity of splinters. The final decision came from Sutcliffe who had studied the hyenas of the Devon cave. He stated emphatically that hyenas were not responsible for the Makapansgat bone material in the British Museum!

Sutcliffe later (1972) examined caves and lairs used by hyenas in Kenya and Uganda to identify the bone collections and any effects on them. He made a distinction between the two types of habitations; caves were larger and easily explored by humans, but lairs were of small dimensions that prevented human entry, though they included numerous and lengthy tunnels accessible to hyenas. The caves held relatively few bones and Sutcliffe stated (p. 144), ". . . nowhere could the bones be described as an accumulation." In contrast, the lairs contained many bones of different animals, and the ground outside was littered with great quantities of bone splinters. **Brain** (1981, p. 55–105) cited numerous researchers who examined bone accumulations from caves, lairs and dens and that were attributed to hyenas and other carnivores. Again, the majority of these researchers implied that hyenas and other animals may have been responsible for the bone accumulations in the South African Makapansgat cave. It is by no means clear, however, exactly how and why these bone collections developed. These researchers examined mostly hyena lairs, dens and small "caves" but not large caves such as those in England and at Makapansgat. The cave size may well be a major factor in the radical differences between the bone collections and may indicate different origins. The findings to date are hardly conclusive.

Both Dart and Ardrey were vehement in their arguments for the australopithecines as collectors and users of bones, especially the antelope humeri. Both carried arguments far beyond that reasonably expectable from evidence of the type and degree indicated by the bones. Dart, for example, had insisted that the transition from apes to humans was largely conditioned by their hunting behavior as demonstrated by these bones. He argued (1953, p. 209),

> On this thesis man's predecessors differed from living apes in being confirmed killers: carnivorous creatures, that seized living quarries by violence, battered them to death, tore apart their broken bodies, dismembered them limb from limb, slaking their ravenous thirst with the hot blood of victims greedily devouring livid writhing flesh.

Dart's words conveyed a vision of these australopithecines that could be recognized as typical of the blood and gore genre of movies rather than scientific appraisals of fossil remains. These graphic phrases did not help his case with skeptical colleagues, and they antagonized many others into opposition. After all, do we really want ancestors who were blood-thirsty hunters? However, if these ape–men were hunters, they would surely have used the methods described by Dart that are no different from those used by modern carnivores, lions, tigers, wolverines, and others, including chimpanzees who occasionally hunt and devour prey.

Dart's arguments for bones as weapons used by the australopithecines led to the search for other types of tools, specifically stone ones that would clearly indicate a human origin. In 1956 Brain discovered 129 pebble tools in the australopithecine bone breccia in the Makapansgat cave and within twenty-five feet of the horizon where the australopithecine remains had been found. These were simple pebbles obviously modified at one end by someone into relatively sharp edges. Such edges are easily produced by striking pebbles together and chipping off flakes. But were these stone tools made by australopithecines or others? In 1957 at the Sterkfontein cave, J. T. Robinson and R. J. Mason found more than two hundred flaked pebble tools in a reddish breccia overlying the australopithecine layer. The materials of all these tools were unrelated to the cave limestones; they were made of quartz, chert, and other hard materials that had been carried into the caves. If these tools were not made by the australopithecines, and the evidence was indeed neutral, then who were the crafters? Could they have been made by someone else who lived shortly after these australopithecines, perhaps their descendants? Or perhaps by a contemporary?

EAST AFRICAN AUSTRALOPITHECINES

By 1959 **Louis** and **Mary Leakey** had been seeking fossils in the Pliocene and Pleistocene sedimentary rocks of Olduvai Gorge, Tanzania (Fig. 7.8) for twenty-eight years. They had found thousands of animal bones as well as stone tools. These stone tools, first found in 1931 and named **Oldowan** (Fig. 7.9), were simple worked stones. No traces of any australopithecines, however, except perhaps these stone tools, had ever been found in these one hundred-foot-thick Pliocene-Pleistocene sediments until 1959. Mary, working alone, found the first hominid

Figure 7.8 Olduvai Gorge where *Australopithecus boisei* was originally found.

evidence, a piece of skull and two very large teeth embedded in rock. Louis Leakey, with the consummate skill of a P. T. Barnum, arranged for photography of the site with the specimens still in place to provide a record of the event. For the next twenty days they worked carefully to free the fossils and they eventually recovered more than four hundred fragments of the skull and upper teeth; no lower jaw was found. This find (Fig. 7.10) stirred the imagination of people around the world, perhaps because of the dramatic filming and the publicity accorded it. Certainly the ability to communicate such news around the world almost instantly via television and other media contributed to the appeal of this discovery. And people were also ready for some news of fossil humans. After all, they had heard about these fossils for many years and this was an important new one, according to Leakey.

Leakey named this species *Zinjanthropus boisei* (East African man and to honor Charles Boise, a financial supporter). Though larger and very similar to the South African robust type, *Australopithecus (Paranthropus) robustus*, Leakey's specimen was an even larger robust type. Others challenged the need for a new genus, and Leakey eventually conceded and renamed it

Australopithecus boisei, indicating its relationship to the South African types. His biographer Cole (1975, p. 229) states, however, that he realized the skull was a robust australopithecine from the very first, not the specimen of *Homo* that he had hoped for. Some experts today believe that *A. boisei* is not even a new robust species, but that it is simply another specimen of *A. robustus*, though slightly larger. The differences between the South African and East African robust types may be due simply to environmental differences. Interestingly enough, these two types of robust australopithecines have not been found outside of their original discovery regions and this restrictive distribution supports an environmental factor model.

Leakey's *A. boisei* was found at Olduvai Gorge, a shallow canyon cut into uplifted lake-bottom sediments, stream deposits, volcanic ash and lava flows as well as older beds that had been the land surface during Pliocene time. Reck had mapped the geology and stratigraphy in 1951 and recognized four principal layers or formations (Beds I through IV). The fossil skull of *A. boisei* was found in Bed I at an ancient campsite on the shore of a lake (see Fig. 7.8). Also found in Bed I were a variety of other fossils, nine Oldowan stone tools, 176 stone flakes, and a

120

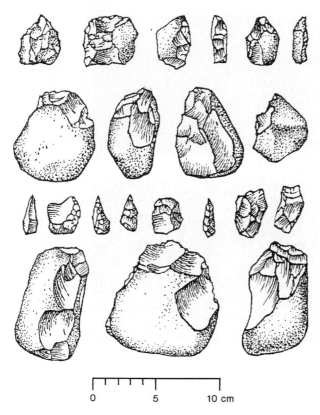

Figure 7.9 Oldowan stone tools. Adapted from *Human Origins,* ed. by Isaac and McCown. Copyright © 1987 W. A. Benjamin, Inc. Used with permission of The Benjamin/Cummings Publishing Company.

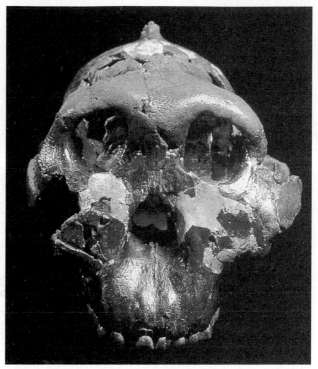

Figure 7.10 *Australopithecus boisei* skull.

hammer stone. The rock materials of these tools were from outcrops some four to nineteen miles distant and clearly represent selected materials that had been carried to the area. The fossil bones were mostly broken and cracked and represented small or young animals. Leakey believed that these hominids had lived along the shore of a lake, made and used tools, stalked these small animals, killed them with the tools and ate their meat. This site and the fossils have been reliably dated at about 1.75 million years.

Eventually more fossils were found at a lower and therefore older horizon. Included were foot bones, finger bones, fragments of two skulls, a collar bone, and bone and stone tools. A left foot consisting of twelve bones was an obvious walking foot, not the grasping one of an ape with an opposable large toe. One skull was that of a child of perhaps eleven or twelve years. Both skulls, however, resembled the gracile australopithecines more than the robust *A. boisei*. More importantly, the dome of the child's skull had a gash and radiating fracture lines caused by some type of blow. This may be another opportunity for recognition of murder in ancient humans.

In 1964 after several years of careful study, Leakey announced the discovery of another new species at Olduvai Gorge from about the same level as *A. boisei*. This form was based on a mandible and some teeth and bone fragments (Fig. 7.11) but was not as robust as *A. boisei*. It was also more human than the gracile australopithecines. Leakey realized that he had finally found the craftsman who made the Oldowan stone tools, the fossil human for which he had spent so many years searching. He christened these remains as *Homo habilis* (handy man). Its brain size and those of three others found later at Olduvai average about 650 cc, a little smaller than the minimum of 700 cc some experts previously considered necessary to be a human. Accordingly, some workers today believe this specimen is too small for human status and identify this species as the gracile *Australopithecus* rather than

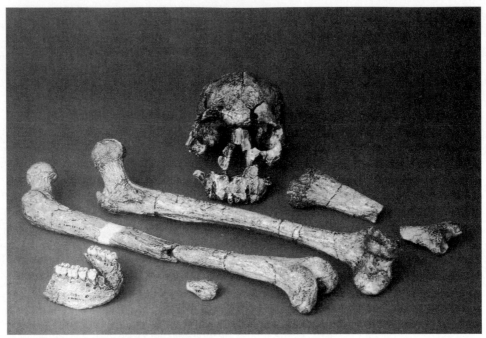

Figure 7.11 Specimens of *Homo habilis*.

Homo. Even so, they do not question its importance as a stage toward modern humans. Most experts, however, support it as the oldest true human, *Homo.*

THE IMPORTANCE OF OLDUVAI FOSSILS

The importance of the Olduvai Gorge hominid fossils was not so much in themselves, because some of them were similar to other specimens already in existence. Some of the stratigraphic units at Olduvai are volcanic tuffs and lavas, materials that contain radioactive minerals and these could be accurately and precisely dated. This meant that accurate ages for any fossils in these particular strata could also be obtained. **Hay** (1967) conducted stratigraphic mapping over a span of fifteen years beginning in 1962 and generally agreed with the Reck's stratigraphy (Fig. 7.12). Samples of rocks from the key volcanic beds were collected and analyzed for age dates by means of newly developed techniques that measured minute quantities of radioactive substances. Bed I, which contained *A. boisei*, was dated by three different mineral methods that agreed with remarkable consistency. The lowermost hominid fossils and tools are now dated at nearly 2.0 million years, more than triple what Leakey originally though it might be

and much older than what other experts had expected for the oldest hominids. Marker Bed A, which overlies the original *A. boisei* fossils, is dated at 1.7 to 1.75 million years and a layer above it dates from 1.6 to 1.65 million years. All of these dates supported the original contention made previously by Dart and Broom that *A. africanus* was at least two million years old.

OLDUVAI PALEOENVIRONMENTS

The important and richly fossiliferous strata of Olduvai Gorge were deposited during Pliocene and Pleistocene times in the East African Rift Zone, an unstable fault valley system produced by tectonic forces associated with movements of Earth's plates. Such rift zones are characterized by volcanism and earthquakes that accompany the spreading apart or rifting of opposite sides of the area. In consequence, the central portion drops to produce a valley. This rift system stretches through Tanzania, Kenya, Uganda and Ethiopia and other countries in Africa. It also connects with the worldwide rift system that encircles the globe, mostly in the oceans, and separates Earth's different crustal plates.

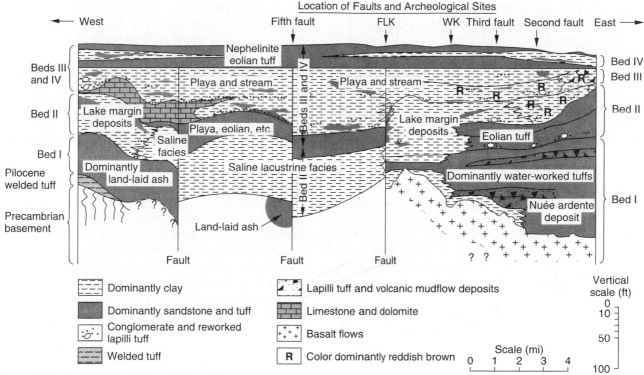

Location of Faults and Archeological Sites

West · Fifth fault · FLK · WK · Third fault · Second fault · East

Nephelinite eolian tuff

Beds III and IV — Playa and stream — Beds III and IV — Playa and stream — Bed IV / Bed III

Bed II — Lake margin deposits — Playa, eolian, etc. — Lake margin deposits — Bed II

Bed I — Saline facies — Eolian tuff

Pilocene welded tuff — Dominantly land-laid ash — Saline lacustrine facies — Dominantly water-worked tuffs — Bed I

Precambrian basement — Land-laid ash — Nuée ardente deposit

Beds II

Fault · Fault · Fault

Legend:
- Dominantly clay
- Dominantly sandstone and tuff
- Conglomerate and reworked lapilli tuff
- Welded tuff
- Lapilli tuff and volcanic mudflow deposits
- Limestone and dolomite
- Basalt flows
- R Color dominantly reddish brown

Vertical scale (ft): 0, 10, 50, 100

Scale (mi): 0 1 2 3 4

Figure 7.12 Stratigraphic section at Olduvai Gorge.

The East African rift valleys have always had a good supply of water that drains into the lowland from the hills and mountains on either side, though most of the region is quite dry today. The central portions of the rift valleys are down-dropped faulted blocks, or blocks flanked by high, uplifted escarpments. In the past, this rift system contained an almost continuous chain of lakes, rivers and swamps. A great many animals still inhabit some of the region today because of the abundance of food. Herbivores eat the lush seasonal grasses and carnivores prey on the herbivores. In the past, animals flourished in vast numbers across the rift valley floor in savannas that had more abundant vegetation and water than today. The australopithecines also lived in these environments, undoubtedly never very far from life-giving water. Because the other animals also depended on these watering places for their needs, the australopithecines undoubtedly congregated there attracted by the ready supply of prey animals and the water. The herd animals and the australopithecines together had spread up and down the rift valley system following the watercourses upon which their very lives depended. Where they lived,

they also died. Only a scant few of the australopithecines who died were not destroyed by predators, scavengers, weathering, or some other posthumous factors, and these few became fossils. Some of these have now been found and provide evidence of this important early hominid type that gave rise to the genus *Homo*. Olduvai Gorge was obviously one place where they had lived in the distant past, some 1.8 to 2.0 million years ago. Where else did they live? Where were the other sites where the australopithecine fossils could be found?

OTHER EAST AFRICAN SITES
THE OMO VALLEY

In 1959 F. Clark Howell was seeking sites where fossils of early humans could be found. He conducted a reconnaissance of the Plio-Pleistocene strata that stretched along the East African Rift Zone northward past Lake Turkana (formerly L. Rudolph) and through the Omo River Valley into Ethiopia. He found complex geology and abundant fossils of the right age in a region that was ripe for discoveries of ancient humans. In 1967 an international expedition of French, American and Kenyan teams traveled north of Olduvai Gorge into this little known portion of the rift zone, to explore for fossils of early humans in the Plio-Pleistocene sedimentary rocks of the Omo Valley. The Omo River is the major watercourse that dominates the drainage of this region and has been doing so for thousands of years. It still flows southward from the Ethiopian highlands through the Omo area to Lake Turkana, a large body of water nearly two hundred miles long.

The Omo region is immense, more than 50,000 square miles. It is also a very arid, inhospitable desert with little vegetation to support animals or people. River courses tend to be dry until seasonal rains fill the channels and produce short-lived streams that flush and shift great quantities of sediments. In the days of the australopithecines, conditions were radically different, with many rivers flowing through forests that covered the region. Abundant wildlife roamed everywhere then and thrived on the abundant vegetation. But the climate was changing and as the precipitation lessened, the region began to dry out slowly and become savanna prairies with grass and brush. Such grasslands and brushlands were able to support vast numbers of herbivorous herd animals, much like some areas in the rift zone farther south today.

The australopithecine-bearing beds at Omo are much thicker and record much more time than the strata at Olduvai. Omo has more than 3,000 feet of sedimentary rocks of Plio-Pleistocene age compared to the mere 350 feet thickness of strata found at Olduvai Gorge. The Omo beds are exposed totally at the surface because of their gentle dip and because of faulting that has raised some of the lower strata to the surface. These strata began accumulating about 4 million years ago and deposition continued for 3 million years. In comparison, the much thinner beds at Olduvai were deposited in only about 800,000 years. The Omo beds are also interlayered with dozens of beds of volcanic ash that spewed from numerous active local volcanoes, like the strata at Olduvai. Because volcanic ash commonly erupts and spreads over a broad area within a few days, each ash layer accumulated essentially instantaneously and provided excellent time-reference markers for stratigraphic correlation and radiometric dating. Fossils found in these strata can be accurately dated to very precise, short time intervals because the volcanic materials contain minerals with radioactive potassium and argon. Thus, the Omo region, like Olduvai Gorge, had several

aspects that made it a very attractive site to search for early hominids. Actually, it was much better than Olduvai because it encompassed a much greater span of geologic time over a broader area.

The three groups of scientists began to work separately in the lower Omo River Valley. Two of the three groups began to find fossils in great quantities, mostly non-primates, but also a variety of hominid fossils including teeth in great numbers, pieces of jaws, two skulls and limb bones. In eight years they collected almost fifty thousand fossil specimens. The mammal fossils provided an important record of more than 140 species that evolved through more than a million years of geologic history. Such fossils are crucial to studies of early hominids because they provide a reference index and timescale that can be used to correlate with any region where these animals are found. The abundance of these fossils indicated that large numbers of these animals had originally lived here.

The American group under F. Clark Howell found four kinds of hominids in these beds. Among these were the two different types of australopithecines that inhabited East Africa, the super-robust *A. boisei* that lived from 2.0 to 1.0 million years ago and a gracile one that lived here from about 3.0 to 2.5 million years ago. In addition, the expeditions found traces of two types of more advanced hominids. There were teeth of *Homo habilis* that are dated at about 1.85 million years and fossils of a younger type, *Homo erectus*, from about 1.1 million years ago. This latter type is well known as **Java man**.

THE LAKE TURKANA FINDS

Richard Leakey, leader of the Kenya group at age 23, was not particularly pleased with his assigned site in the lower Omo River valley because of the age of the beds and the sparse fossils. He shifted his efforts farther south to the eastern shore of Lake Turkana. There, in northern Kenya with the Leakey luck intact, he landed by helicopter at **Koobi Fora**, a locality where the ground was littered with excellent fossils. This turned out to be a remarkable discovery, and within a few years Leakey's team recovered more than one hundred hominid specimens from this site and area. These fossils included three excellent skulls, numerous lower jaws, isolated teeth and limb bone fragments. The fossils represented *A. boisei*, a gracile australopithecine, and the most famous fossil of the region, **KNM-ER 1470** (Fig. 7.13). This specimen is a magnificent, nearly

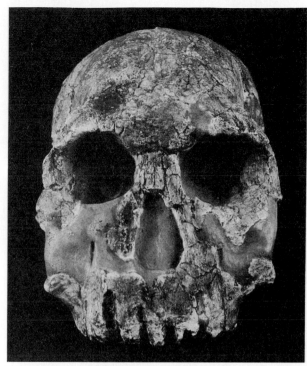

Figure 7.13 *Homo habilis* skull KNM-ER 1470.

complete, toothless skull of an individual with an exceptionally large and modern-looking cranium. It was initially dated at 2.9 million years, which created some consternation. Most experts classify this specimen as the oldest type of human, *Homo habilis*, though a few believe it to be another gracile australopithecine. The date of 2.9 million years initially created difficulty with this specimen of *Homo* because such an age meant it would have predated its presumed ancestors. More carefully collected mineral samples have since been analyzed and the radiometric age has been revised downward to less than 2 million years ago. This age fits ideally into the lineage of humans from the australopithecines.

THE AFAR TRIANGLE OF ETHIOPIA

The richly fossiliferous beds of East Africa, Olduvai Gorge and the Omo Valley have been powerful magnets that have attracted many individuals to field work there, particularly graduate students for whom such field training is an essential precursor to professional careers. One such student was **Donald Johanson**, a doctoral candidate in

paleoanthropology who spent three field seasons in East Africa learning the routines of field organization, financing, and collecting as well as the techniques of identifications of the many different types of fossils present. Few academic institutions provide adequate field training in such bone identifications given the many different kinds of now extinct animals that were associated with fossil humans. Such experience is, therefore, of great academic and practical value to any paleontologist or paleoanthropologist.

Johanson had learned from a French geologist of presumed Plio-Pleistocene deposits at **Hadar** in the **Afar Triangle** region of Ethiopia near the Red Sea, a region 1,600 kilometers north of Lake Turkana covering many thousands of square miles. He believed these deposits were equivalent in age to the older deposits in the Omo Valley that contained only poor fossils. These older beds at Hadar could contain fossil evidence, perhaps even important intermediate specimens that might link older Pliocene deposits with the younger Pliocene ones in which abundant fossils had been found. Johanson argued that the deposits in the Afar region would produce hominid fossils better than those found at Omo and was able to raise enough money for a brief reconnaissance to this extremely hot and inhospitable area during 1972. Fortunately, this expedition found abundant mammal fossils in well-stratified layers, an ideal combination for dating any prospective fossil hominids. Using these finds, Johanson secured funds to support a small expedition to Hadar in 1973–74.

Toward the end of the first field season in 1973, Johanson found two incomplete leg bones that fitted perfectly together at a joint. These were a proximal tibia or upper shinbone and a distal femur or lower thigh that together formed a knee joint. They were of great importance because the knee joint clearly belonged to a hominid that walked upright (Fig. 7.14). Nearby the party also found two proximal femurs, the upper part of thighbones that articulate with the pelvis. To verify the interpretation that these bones were from walking legs, Johanson called on **C. Owen Lovejoy**, an authority on locomotion, to examine the knee joint. Lovejoy concluded that the bones were from an individual who could walk upright, a bipedal hominid. These bones established the nature of this specimen as a gracile australopithecine that used a walking or bipedal lifestyle. The only remaining questions about the bones were their age, their size and their status as hominids.

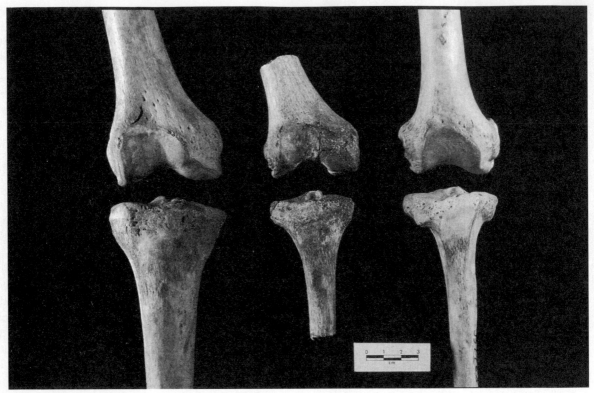

Figure 7.14 Knees of modern human, *Australopithecus afarensis* and ape.

A key basalt layer was found in the Hadar section about 75 meters above the bed that contained the knee-joint. It was analyzed using radioactive potassium-argon dating methods and found to be about 3 million years old with an error of ±200,000 years. Because the basalt was much higher in the section than the leg bones, the fossil knee-joint was believed to be at least 3 million years old, probably older. Later, better samples of the basalt provided an age of 3.75 million years with an error of only ±100,000 years. Further analysis of these strata using paleomagnetic dating techniques substantiated this date. The second aspect of the knee-joint, its size, was equally informative and important; it came from an individual that stood only about three feet tall. No other hominid remains were then known of adults so small in size.

During 1974, the expedition found two hominid jaws that had teeth of different sizes. These jaws were apparently alike and from the same species. The teeth were large incisors and small molars, the opposite condition for similar teeth from known australopithecines. Even more im-

portant was the discovery by Johanson one day of the fossilized remains of what appears to be a single gracile australopithecine. These bones included the back of a skull, several limb bones, a partial pelvis and some vertebrae. Within three weeks the party had collected several hundred pieces of bone that represented about 40 percent of this individual, including parts of arms, legs, pelvis, ribs, vertebrae, shoulder and collar bones, skull and the lower jaw. This individual was ultimately identified as a female (AL 288-1) and given the name **Lucy** (Fig. 7.15). Eventually, Johanson decided this gracile individual was a new species and named it *Australopithecus afarensis* (southern ape from the Afar region).

In 1975 this expedition found another remarkable collection of hominid bones grouped close together at Hadar. These consisted of more than 350 separate pieces of fossil bones, the remains of at least thirteen men, women, and juveniles who had died together in a sudden event, perhaps a flood. These individuals were geologically younger than Lucy but of the same species, and they have become

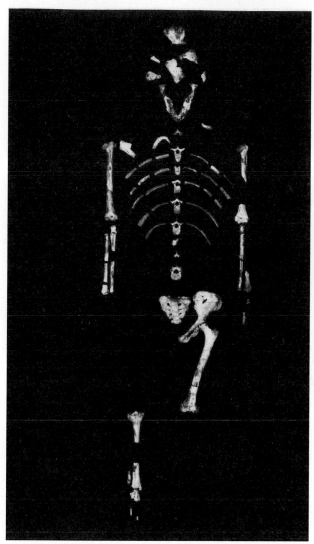

Figure 7.15 The skeleton known as Lucy.

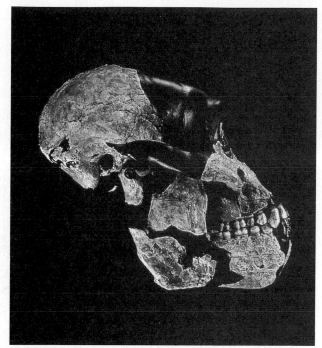

Figure 7.16 Reconstructed skull of *Australopithecus afarensis*.

popularly known as the **First Family**. In 1976–77 the expedition also found stone tools which were tentatively dated to about 2.5 million years. Eventually, enough fossil fragments were found to allow for the reconstruction of *A. afarensis* (Fig. 7.16). The implications and potential for these discoveries at Hadar were far more than Johanson had ever hoped possible and added vital information on additional hominids to the rapidly growing mass of data of human origins.

LAETOLI

Johanson and his associate, **T. D. White** believe that the hominid fossils of their new species from Hadar, *A. afarensis*, are similar to fossils that Mary Leakey found at Laetoli, a site just thirty miles south of Olduvai Gorge. Mary Leakey, however, disagreed with their assessment. Nonetheless, specimens from Hadar and Laetoli are not significantly different and may represent a single gracile species. They also exhibit some differences that can distinguish them from the gracile species from South Africa. The Leakeys originally had gone to Laetoli seeking fossils in Pliocene deposits older than 3 million years but had little success. Returning in 1975, Mary Leakey found a variety of hominid jaws, teeth, and other parts of australopithecines. The greatest prize, however, was the discovery in 1976 of hominid footprints (see Fig. 5.10) in a layer of fine volcanic ash dated between 3.6 and 3.8 million years ago. She reported that the tracks were made by two individuals, one large and one small, who walked across the ash bed while it was still a soft, powdery material. These individuals walked "upright with a bipedal, freestriding gait" and had a foot shaped "exactly the same as ours." The evidence further suggests that the larger individual, presumably male, was about four feet eight inches tall and the smaller individual, presumably female, was just about four feet tall. The smaller individual stopped while walking to the north, turned to the west and apparently gazed in that direction, then resumed her northward pace. Dare we speculate about what this individual saw or thought?

The discovery of an australopithecine walking knee-joint at Hadar and the footprints at Laetoli forcefully demonstrate that upright bipedal walking behavior was in full use by hominids about 3.7 million years ago, nearly a million years prior to the appearance of any specimens assignable to *Homo*. This behavior clearly demonstrated that the hands of these hominids were freed for many tasks, including tool use, and they had a very long time in which to find ways to use these versatile hands. However, no identifiable tools of any type have been found in the Laetoli beds and the consensus is that these australopithecines, about 3.5 million years ago, had not yet reached the stage of development here or elsewhere in which they could use tools or make them to any significant degree. The presence of stone tools in younger Hadar beds that are dated at about 2.5 million years ago demonstrate that the hominids present there, whether australopithecines or *Homo*, had finally learned how to make and use stone tools during the one million year interval after the Laetoli footprints.

In 1985 **Alan Walker** found another robust specimen in the fossil-rich strata on the west side of Lake Turkana. This form (KNM–WT 15000) is dated at about 2.5 million years ago. Though it has primitive characters of the cranium, it also has a "more advanced" face. The cranial characters relate it to the robust form *boisei* but the facial features are too advanced for these robust types. The placement of this specimen thus confuses the hominid evolutionary line from the australopithecines to *Homo*.

The presumed original pathway of human evolution after Dart's discovery of the australopithecines proposes that *A. africanus* gave rise directly to *Homo*. The subsequent discovery of *A. robustus* shows that stock as a second branch from *A. africanus* and required modification of this lineage chart. *A. robustus* was a very conservative stock that had developed very specialized feeding structures and other characters. As such it does not fit in the evolutionary pathway to humans and is considered a dead end. The eventual discovery of these robust australopithecines in East Africa significantly extended their geologic and geo-graphic range, though it did not basically affect the hominid lineage as then understood. *A. boisei* from East Africa, slightly younger than the South African forms, occupied a different environment than their southern cousins; and these environmental differences likely influenced its development. It represents a further specialization of the robust form that developed into a super-robust herbivore. Its close similarities to the South African *robustus* questions the separation of the robust forms into two species.

The discovery of the older gracile australopithecines at Hadar and Laetoli have provided much fossil material over which workers have argued as to which branch provided the direct ancestral stock for *Homo*. The presence of the australopithecines farther to the north along Lake Turkana and in the Omo and Hadar regions of Ethiopia indicate that these hominids had spread widely throughout suitable African environments, especially along the Rift Valley. Johanson's discoveries of the older gracile species at Hadar permitted a revision of the lineage chart to identify this type, *A. afarensis*, as the founder or ancestor of the australopithecine stock.

The presence of large incisors and small molars in *A. afarensis* suggests that it may be more closely related to *Homo* than the other, younger australopithecines. These other australopithecines typically have smaller incisors and larger cheek teeth, a reversal of dental pattern that suggests that the younger australopithecine species, *A. africanus* and the robust types, developed larger cheek teeth in response to harsher or tougher foods than those utilized by *A. afarensis*. Larger back teeth suggest more power was needed for chewing, a character that does not reflect a diet of meat or relatively tender plant materials. These smaller incisors in *A. africanus* reflect the greater power inherent in the larger cheek teeth and imply a reduced dependence on the incisors. *A. africanus* and the robust types thus may not be in the direct line to humans but may be specialized offshoots that became dead ends.

The lineage of hominids, especially that leading to *Homo*, is still being refined. Several interpretations of evolutionary paths from *A. afarensis* to *Homo sapiens* have been proposed (see Chapter 12). They will probably be revised many times as the crucial fossils are found to fill in the gaps. All these lineage charts should be considered scientific progress reports, eligible for revision as new information becomes available. Though the hominoid and hominid fossil finds in recent years have filled some of the gaps in the record and the evolutionary picture has become much clearer than ever before, the nature and characteristics of these fossil species and specimens have generated numerous additional questions about the details of their population variations and how and why they evolved so rapidly. These questions are not concerned so much about the lineage, however, as they are about the details of descent. More fossils must be found to answer these questions, but the places where new fossil hominids might be discovered are not easily found or accessible. The Omo and Afar areas are critical to such discoveries, but because of the political unrest in Ethiopia, the region is unsafe and researchers have stayed away from field work that could perhaps answer some of these questions about man's origin.

THE AUSTRALOPITHECINE TYPES

The general consensus of scientific opinion is that the australopithecines are divisible into two different types, the **gracile** and the **robust** forms. The gracile forms are *A. afarensis* and *A. africanus*, though a few experts prefer to combine these into a single gracile species, *africanus*. The robust forms are *A. robustus* and the super-robust *A. boisei*. Some experts include *A. boise* in *A. robustus* because they consider the size differences between them to be essentially minor and probably related to environmental factors (Fig. 7.17). Others feel that *A. boisei* is sufficiently distinct and has enough different features to be identified as a separate species. The important discovery by Walker (Brown, et al., 1985) of a super-robust type about 2.5 million years old in Kenya has demonstrated that this East African type predates the typical robust forms from South Africa. This fact evokes additional questions about their evolutionary history and the lineages of all the australopithecines.

Over the years, different generic and specific names were proposed for these hominid types as specimens were discovered. After reevaluations these types have been included in the four principle australopithecine species as **synonyms**, which are different names for the same species. Although the relationships of these hominids to each other and to *Homo* has been clarified in some ways, in other ways they are more complex and difficult to assess. Many explanations are simply not yet available or, perhaps, even possible.

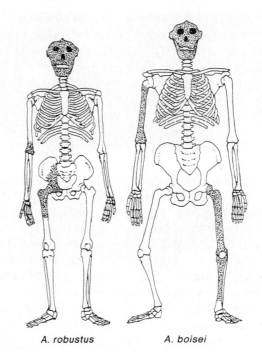

A. robustus   A. boisei

Figure 7.17 Skeletons of the gracile and robust types of australopithecines.

THE GRACILE AUSTRALOPITHECINES

The gracile australopithecines are represented by numerous fossils from many localities in South and East Africa. These specimens include complete and nearly complete skulls, jaws, teeth, long bones, hands, feet, ribs, vertebrae, pelvic and shoulder structures and more. The most important features for identification and separation of these types are the skulls (Fig. 7.18), the jaws and the teeth. These underwent rapid and significant changes as the gracile types evolved on the savannas. The other bones also provide evidence of major importance for differentiation, but overall they did not change as rapidly nor in as great detail as the skulls, jaws and teeth.

The skulls of the gracile australopithecines differ from those of both apes and humans in many respects. In comparison to the gorilla and other apes, all the australopithecines have larger braincases in calvaria with higher vaults. The gracile australopithecines had brains that range in size from 380 to 490 cc. Some of these fall within the normal brain variation of modern apes, but ape brains are much smaller relative to body weight (see Fig. 4.10). The

or the robust australopithecines. Their jaws became smaller and retreated to produce a shorter and more flattened face with reduced prognathism. Their diet relied less on rough and tough plants as they evolved into hominids, and the jaw became even smaller. At the same time the cheek at the base of the zygomatic arch began to shift from above or behind the rear molars, as found in apes, to a more forward position. In humans the base of the zygomatic arch is located approximately above the first molar.

The reduction in size of the jaw bone also had a direct effect on the chewing muscles (temporalis and masseter) that attached it to the top of the skull and the cheek. These muscles, too, became increasingly reduced in size from apes to australopithecines to humans with the changes in diet that required lessened chewing force. In apes the temporalis muscle is anchored on the sagittal crest atop the skull and the jaw bone. This muscle exerts great force in closing the jaws during chewing and its development in apes reflects their diet of tough foods. When the diet changed, the temporalis muscle became smaller and there was no need for an attachment to a sagittal crest; this crest disappeared. The masseter muscle, too, became smaller in the same reduction. To accompany these muscular changes, the zygomatic arch and cheek bone became smaller.

A further reduction in size took place in the supraorbital crests of the gracile australopithecines. This was probably due to two factors; the braincase enlarged to accommodate the larger brain and the muscles that attached to this crest or ridge became reduced in size because of dietary changes. Though the supraorbital torus was not eliminated in the gracile australopithecines, it became much smaller and far less prominent than the same structure in apes.

The jaws of gracile australopithecine underwent other changes as they became smaller. Their dental arcades became more arched into a parabolic or elliptical V-shape that is similar to that of humans. This is in contrast to the dental arcade of apes that has parallel sides and is U-shaped (Fig. 7.19). With changes in diet, the australopithecines no longer needed the large teeth required by apes. Gracile australopithecines have smaller teeth than apes and are essentially human in character though slightly larger. The cheek teeth of the gracile australopithecines tend to be smaller than those of the robust australopithecines, a character in keeping with the differences in diet postulated between the two types. The canines of the gracile forms are almost always the same size as the other teeth, except for

Figure 7.18 Skulls of the gracile and robust types of australopithecines from South Africa.

gracile australopithecines had brains of about the same size as gorillas, but male gorillas can weigh more than six hundred pounds; these diminutive hominids weighed only about ninety pounds and had much different brain size: body-weight ratio.

The gracile australopithecines have less pronounced prognathism than apes, a function related directly to the jaw-face muscle structure evolution and foods consumed. The gracile australopithecine have jaws that fall in between humans and apes in size; they are smaller and thinner than in apes and shorter and wider, with thicker bones, than in human counterparts. This evolutionary trend is directly related to differences in diet. The ape diet is very rough and requires both sturdy jaw and skull bones operated by powerful muscles. As the gracile australopithecines shifted to a diet of less rough and tough plants, they had little need for the powerful chewing apparatus like those of the apes

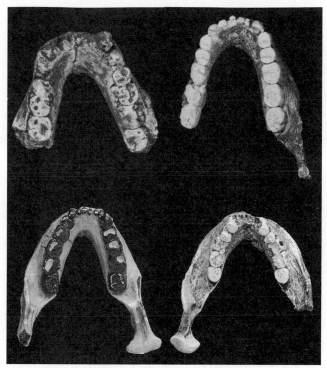

Figure 7.19 Australopithecine jaws. (top left, *A. afarensis;* top right, *A. africanus;* bottom left, *A. boisei;* bottom right, *A. robustus*).

A. afarensis whose canine teeth are slightly larger than the others and appears to exhibit slight sexual dimorphism. Weiss and Mann (1985, p. 321) state,

> The canines of modern humans and other australopithecines show no sexual dimorphism. The apparently male *Australopithecus afarensis* canines are similar in shape to those of the hominids but larger, sticking out slightly above the level of the other teeth, while presumed female canines are somewhat smaller.

This smaller size of australopithecine teeth, compared to that of apes, reflects the australopithecine diet which was significantly less rough than that of the apes; it also indicated a harsher diet than modern humans, hence the comparatively larger size. As these teeth became smaller with time they experienced modifications that are very important in the identification and separation of these ape and hominid species. The slightly larger canines of *A. afarensis* may simply reflect the earliest stages of general australopithecine evolution in which teeth became smaller in size from the typical

pongid condition to accompany the diet changes. Perhaps the canines of *A. afarensis*, though smaller than ape canines, had not yet made the complete reduction in size to the human condition. The still greater reduction in size of canines in the descendant species, *A. africanus*, suggests that this trend continued. The abrupt absence in the australopithecines of the prominent diastema found in apes is additional evidence that distinguishes and separates the apes from the australopithecines. In apes this space accepted the fit of the canine teeth in closed jaws and when the canine teeth were reduced in size in australopithecines, the diastema became redundant.

Further evidence that these gracile australopithecines moved closer toward the human rather than the pongid conditions is, of course, seen in the more delicate character of post-cranial bones. Particularly important are those structures that provide for bipedalism, the walking foot and leg, the pelvic bones and the associated musculature. Johanson's discovery of the walking knee joint (see Fig. 7.14) and Mary Leakey's discovery of footprints at Laetoli have adequately demonstrated that these types utilized bipedalism. These indicate the significance of the knee and foot structures found in certain fossils.

Australopithecus afarensis. The oldest of the gracile types is *Australopithecus afarensis*, clearly a primitive hominid. At Hadar this earliest australopithecine lived from 3.2 to about 2.8 million years ago. If the Laetoli gracile specimens are indeed *A. afarensis*, as argued by Johanson and White, then this species would have inhabited that area at least 3.9 million years ago. *A. afarensis* males weighed from 60 to 80 pounds and stood from four to four and one half feet tall; females were between 25 to 50 percent smaller. A large jaw (MAK-VP 1/12) found recently at Maka, Ethiopia (Gee, 1993) is larger than the jaw of Lucy and demonstrates the variability of the species. Lucy was quite tiny, perhaps only about three and one half feet tall and weighing about sixty pounds, about the same size as an *A. africanus* female. She was between twenty-five and thirty-five years old and suffered from early arthritis. Her brain size was about 400 cc, about that of a chimpanzee and just slightly smaller than that of *A. africanus*. Facially, *A. afarensis* tended to resemble apes with broad faces that included wide, flat noses, prognathus jaws that lacked chins, prominent brow ridges, and low foreheads that retreated into very small, elongate calvaria. Their bodies were likely covered with hair, though probably not as thick as gorilla's pelts. Their skin color is unknown but it was probably dark like that of the African apes. Their bones are thick and sturdy with large muscle scars and they were probably quite strong for their size.

The jaws of *A. afarensis* clearly demonstrate their hominid nature, although those of *A. africanus* and *A. robustus* have more arch to the dental arcade, as shown in Figure 7.19. The teeth of *A. afarensis* (Johanson, p. 359) show three kinds of wear, micro-flaking, pitting and scratching. This wear is probably due to a variety of eating patterns. The micro-flaking and pitting are caused by the biting of objects, possibly bones, that were hard enough to chip off tiny flakes of enamel and expose the softer underlying dentine. The scratches are caused by the abrasive actions of sand grains on plant materials that were chewed or by the silica that is inherent in plant cell structures. These types of wear strongly suggest that these hominids were omnivores that ate virtually anything they could find, fruits, roots, tubers, berries, seeds and meat.

Australopithecus africanus. *A. africanus* was the first australopithecine identified but it is the younger of the two gracile types. It is another relatively small hominid similar to *A. afarensis* in size and physical appearance. Males stood from four and one-half to five feet tall and weighed sixty to one hundred pounds; females were smaller in height and weight. *A. africanus* also had slightly larger brains than their older relatives; their brain case volumes ranged from 428 to 480 cc and averaging about 450 cc. Their diets likely were similar to those of *afarensis* and had the same general effects on teeth, though the slightly larger incisors of *A. africanus* suggest increased dependence on tough foods. In some specimens, the cheek teeth are fully as large as those of robust australopithecines, probably also because of a harsher local component in their diet. These larger incisors and cheek teeth may indicate that *A. africanus* was a more conservative species than its predecessor, *A. afarensis*, and could indicate that *A. africanus* may have been a dead-end side branch of the hominids, perhaps not directly on the main line to humans.

THE ROBUST AUSTRALOPITHECINES

The South and East African robust and super-robust hominid types *A. robustus* and *A. boisei* were clearly larger than the more delicate gracile types. *A. boisei* was somewhat larger than *robustus* with a broader, more massive face and larger jaw and teeth (see Figs. 7.18 and 7.19). *A. boisei* males probably stood about five and a half feet tall

and weighed about 150 or more pounds. The females were somewhat smaller, demonstrating typical sexual dimorphism. The robust type facial features were almost caricatures of the gracile australopithecines with very wide cheeks that emphasized their flat faces, flat (?) noses, more prognathus jaws that were large and powerful, and very large cheek teeth that were intended to grind tough foods. Their chewing apparatus was operated by very large temporalis muscles attached to a prominent sagittal crest that gave their skulls a distinctive, apelike appearance. All these skull and jaw features suggest that these hominids were well adapted to feeding on the harsh and tough vegetation available on the savannas.

Both the South African and the East African robust australopithecine types had clearly developed into highly specialized vegetarians. Their features, structures and characteristics strongly suggest that both types had become very conservative hominids that did not change or evolve much during the more than two million years before the australopithecine lineage became extinct. Compared to the gracile types, however, none of the robust types had features to suggest that they were likely ancestors of humans. Instead, their larger and coarser structures and features support an alternate interpretation, that these hominids had moved into specialized ecological niches that became evolutionary dead ends for them. Nevertheless, these robust types survived for more than 700,000 years and the longer-lived, super-robust *boisei* lived for perhaps 300,000 years concurrently with the earliest human, *Homo habilis*. Whether the presence of this early human had any influence on the demise of *A. boisei* is unknown.

THE AUSTRALOPITHECINE WAY OF LIFE

The numerous finds of australopithecine specimens in East Africa and Ethiopia have added immeasurably to the already impressive collection of data available from the South African specimens. From all these hominid remains, associated fossils, and the sedimentary rocks in which they are found, scientists have drawn a rather impressive picture of the australopithecines and their way of life. Water was absolutely vital to their lives and all of them undoubtedly lived within walking distance of a river, lake or spring, probably within one mile. All frequented the grassy and bushy plains that stretched beyond the thick forests that grew along the river courses. Their foods were the edible products of these environments, materials that they could find and easily collect with their hands. These would have been primarily the plant materials that grew across the savannas, although the gracile type probably occasionally ate meat that was scavenged or caught.

The environments to the north in East Africa at both Olduvai and Omo were similar to those in South Africa. Permanent streams flowed through a dry landscape. Lush forests lined the riverbanks and the lands beyond graded into woodlands and tall grass savannas. The super-robust type, *A. boisei*, was originally found at Olduvai Gorge with scattered stone tools at an apparent campsite on the shore of a lake. Presumably, these robust hominids lived in the same fashion as their cousins in South Africa, opportunistically gathering foods from the plant materials of the area and possibly scavenging occasional meat for variety. Living in this area at the same time was *Homo habilis*, the earliest of the large-brain hominids and the individuals believed to be responsible for the Oldowan tools found at this campsite, throughout the Olduvai Gorge area, and elsewhere. It is difficult to imagine that these three types of hominids, gracile and robust australopithecines and early humans, *Homo habilis*, lived through the same times and were not in contact.

To the north around Hadar in the Afar triangle of Ethiopia, the environments probably consisted of woodlands and savannas flanking permanent streams that were lined with thick woods. This mosaic of varied environments provided a wide choice of foods to the hominids, including meat. It is unlikely that the robust types here or elsewhere were anything more than occasional meateaters. Their powerful jaws and large teeth reflect and were well suited for a specialized herbivore diet and feeding behavior. The gracile types, however, probably ate whatever meat they could find or catch. Their teeth are much smaller than the robust types' teeth and this reflects their dependence on less harsh or tough foods. Presumably this reduction in tooth size would have been selectively favored as they included more and more meat in their diet. It is possible, perhaps even probable, that these gracile australopithecines behaved like modern chimpanzees, who are known to hunt and eat meat on occasion. These australopithecines, however, would probably have hunted or scavenged meat more frequently than the chimpanzees, possibly even more effectively.

Campbell (1976, p. 108) pointed out that South Africa, has been a dry, undisturbed veldt region for perhaps 100 million years, just as it is today. This semidesert region was largely covered with relatively dry, thick brush and grass prairies that contained many types of herbivorous herd animals, the predators and scavengers that fed on them, and many others. The vertebrates are represented by all sizes from elephants, hippos, rhinos and giraffes to birds, reptiles and small mammals. Included in this fauna are the two australopithecine types, the small, gracile hominid, *A. africanus* and the larger *A. robustus*. In these grass and brush prairies, both types of australopithecines found abundant quantities of foods in the form of seeds, roots, tubers, nuts, leaves, berries, fruit and possibly meat. These were all small foods readily available the year round for gathering or picking, something quite feasible for these australopithecines who had the manual dexterity of apes and humans. Some of these foods, like tubers and roots, grow underground and had to be dug out by hand or some other instrument. Though there is no real evidence of tool use by the australopithecines, they clearly had the abilities to dig for foods.

The ancestral hominid stock made the move to the savannas from the forests for a good reason. We do not know whether they were safer from predators in the savannas than in the forests, but the visibility was better on the savannas. If they had been safer in the forests than the savannas, then something must have drawn them to the savannas despite the dangers. The great selection of foods was one goal that could have attracted them, the ease of food gathering may have been another, and the variety of high-energy foods still a third. Obviously, the shift to the savannas from the forests did take place and they became bipedal to facilitate their lifestyles in these new environments. The larger robust type, *A. robustus*, apparently preferred the relatively drier bush country studded with thorn trees over the grassy savannas which were favored by *A. africanus*, the gracile type. This is supported by stratigraphic evidence which indicates *A. robustus* may have moved in to replace the gracile form when the grassy savanna environment dried out and became harsher.

Survival on the savannas with the many predators and scavengers would have been a frightening task for the australopithecines, especially if they did not have any real abilities with tools. If they could have matched the abilities of modern chimpanzees to flail branches and toss an occasional stick or rock at threatening intruders, they might have gained a slight edge in survival. If they had recognized the effect such an activity would have had on predators, they probably would have used it for defense and they would have discovered a valuable survival mechanism. Some experts have suggested that the australopithecines also banded together for mutual support and protection against the meat-eating animals of the savannas, a very likely behavior for these early people who had probably already practiced cooperative behavior. If they lived in large groups, like chimpanzees and baboons, their numbers alone might have been enough to drive off lone predators. Successes in these group actions would have had positive feedback and in time, with more practice, these abilities would have become established as behaviors that increased in frequency and skill. They may even have learned to use this ability to obtain meat by challenging lone predators at kills and occasionally driving them away. If the predator's appetites had been largely sated with meat when these hominids arrived, such attacks might have had reasonable chances for success and meat-eating could have become a more regular practice.

Australopithecines, like all other wild animals, obtained their foods opportunistically as they roamed the countryside gathering whatever plant materials they could find and probably meat whenever they could get it. They collected the plant items by simply picking them from bushes, trees, grasses and shrubs of the savannas and woodlands. They probably also observed other animals digging up underground foods like roots and tubers and began to copy this activity. There is no evidence to support a model for wooden digging implements, only a reasoned belief in their abilities to find and use implements whenever and wherever possible and the knowledge that primitive peoples regularly use them. One has only to watch Goodall's films of chimpanzees using sticks to attempt to pry open a locked box containing bananas to accept such a talent in an early hominid of greater intelligence.

Wooden tools are relatively fragile instruments not easily preserved as fossils like stone tools and would not likely be found even if used by the australopithecines. Further, there is very little one can do to improve a simple stick used for digging and such tools are difficult to recognize even if found. These hominids, therefore, probably did not attempt to make wooden tools but would have picked them up opportunistically whenever and wherever they needed them. Why carry a stick when there are plenty to be found everywhere? Why attempt to fashion it for use when any stick will do? The early hominids may also have learned to use sticks to crack nuts and other tough materials; and it is only a short step from using sticks as hammers to using rocks for the same purposes, something seen in chimpanzees. These are speculations without evidence, though such activities would have been possible in these able hominids that we believe were advanced beyond the capabilities of modern chimpanzees.

The teeth of the gracile and robust australopithecines were somewhat different, particularly in size, suggesting that they probably did not feed on the exactly the same foods. Plant materials probably formed the great bulk of their food-stuffs and were essential for their survival. Their food-gathering techniques were probably similar if not identical to those employed by the terrestrial primates, baboons, chimpanzees and gorillas. The gracile types had relatively small incisors but large molars and this combination suggests a diet of relatively harsh and tough foods that were cracked open by the larger cheek teeth. The robust australopithecines had even larger cheek teeth adapted for harsher and tougher foods. The smaller teeth of the gracile australopithecines imply that they could not use their teeth for cracking open hard nuts and other power tasks as easily as *A. robustus* and *A. boisei*. This suggests that the gracile types could not capitalize on the tougher foods and, in compensation, may have included meats in their diet. Either the gracile australopithecines ate foods that were not as tough, perhaps meats, or they found some way to prepare tough foods. Unworked stone tools used in the raw state would have been of great value to them in opening tough foods, far more efficient than their smaller teeth or bare hands, but there is no evidence to indicate this level of human achievement.

Sillen proposed at a recent meeting (1992) that *A. robustus* ate more meat than previously believed. Newly developed chemical techniques utilize the ratios of strontium to calcium in bones. Plants are richer in strontium and plant eaters have high strontium:calcium ratios; meat eaters, however, have low ratios because meats have much lower amounts of strontium. Tests on fossil bones of *A. robustus*, prepared in a new way to compensate for losses of these two chemicals due to water and acid actions over the years, indicate relatively low ratios of strontium:calcium and support a meat-eating diet for these forms. This view gains support from earlier discoveries that *A. robustus* teeth show small chips, a result of gnawing on bones.

The earliest gracile australopithecines may have climbed into trees for some foods. They probably ate meats, perhaps more frequently than modern chimpanzees and baboons, but not in the quantities of later humans. The acquisition of meat is more complicated than simply picking plant foods. Much meat could have come from dead animals they found, that they could appropriate from predators, or that they may have actively hunted. Schaller and Lowther (1969, p. 326–9) deliberately spent several days walking and scavenging in the open savanna of the Serengeti Plain of Africa to determine if meat could have been obtained without hunting. They found several new-born gazelles in one area but concluded that finding these young animals was seasonal and inadequate for long-term needs. Such a technique was probably effective, however, over a short season when the females were giving birth. These experts also reported that they stumbled upon about seventy-five pounds of usable meat from scraps left behind at a variety of kills. They also saw a hare that could have been caught and a kill being made by a lone cheetah that could have been driven from the meat. Along a wooded stretch of riverbank they could have collected a much greater quantity of meat from several animals; about 500 pounds from a dead buffalo, whole bones at several kills that contained edible marrow, an abandoned and sick live eighty-pound zebra colt, and a 300-pound live giraffe that was blind and easily caught. Such random discoveries would have presented no danger to the australopithecines in contrast to direct confrontations with predators at their own kills. Such expeditions in search of meat would have meant significant gains for these hominids in learning how and when to find such meats and how to cooperate. May we not speculate a bit more and ask if once they had eaten meat, would they have desired it again? Would they have desired it on a more regular basis and have become hunters as a more effective way to obtain it?

It has been argued that the australopithecines on occasion may have been able to intimidate and drive off predators and scavengers from kills. This is based on observations of solitary large carnivores, lions, leopards, and cheetahs, that are occasionally driven from kills by groups of smaller animals. Solitary lions and leopards presented grave risks because they would not easily give up their kills to any scavengers, including australopithecines, but might turn and attack the scavengers. Cheetahs are not particularly aggressive animals and perhaps could have been intimidated and driven off by bipedal australopithecines using group tactics, especially intimidating actions as they hooted and screamed while jumping and waving their arms. Goodall witnessed modern chimpanzees dealing with an intruding leopard at Gombe National Park by thrashing branches about and throwing sticks and stones. Perhaps the australopithecines did the same.

Predators that hunt in groups, including hyenas and wild dogs, are not easily driven from their kills. On occasion, however, noisy and frightening motions and threats have been seen to intimidate and drive away predators, especially if they were already satiated with the kill and not in the mood for a fight. It is not very likely, however, that lions were driven off easily. When hungry, even lions do not tolerate interference while they feast on their kills, except when they are alone and simply unable to defend it against a number of attackers.

Some workers believe that the gracile australopithecines hunted small animals for food, not regularly as a principal feeding activity, but opportunistically. Chimpanzees seem to enjoy eating meat and take to hunting with determination, even though they do not spend much time at it. Would australopithecines as terrestrial primates continue such practices if they once saw or learned it? If meat-eating had become part of feeding behaviors of these gracile australopithecines, then hunting might also have been relatively attractive to them. These bipedal hominids were more intelligent than modern chimpanzees, walked upright, and had much more effective use of their hands for weapons or tools, or for the carrying of meat. We know that later hominids were hunters. Could general hunting have started with these gracile australopithecines?

The australopithecines obviously must have adopted many techniques for gathering foods because they were successful enough to survive for many hundreds of thousands of years. Certainly the new species on the block, *H. habilis*, was an advanced hominid and the earliest human, must have had at least the same capabilities. We know they made stone tools and ate meats; Richard Leakey and Roger Lewin argue (1977, p. 117) that these early humans probably shared food. They say that

Our ancestors switched from being opportunistic food eaters to being systematic food gatherers. The more frequent addition of meat to the menu was a valuable source of high-quality protein. But the exchange of specific foods between individuals, something that no other primate does to any important degree, has deep behavioral and social implications for the emergence of humanness. . .

Food-sharing and hunting behavior are well documented by the numerous stone tools associating the bones of large mammals with *H. habilis*. Hunting such animals requires cooperative behavior and this is of no value and will not be practiced by an individual who does not get a share of the meat. Even if these are not kills but represent animals that were found already dead, there is simply too much meat for one individual to eat and to defend against predators and scavengers. Concerted group activity in hunting and defense would have been the logical means to provide for the most benefit to the group. Evolutionary mechanisms would have favored those populations that adopted group activities and sharing behavior. Selfish behavior would probably have led to lessened success in hunting and resulting in decline of the group, and this could have meant the difference between survival and death for these hominids during times of crisis.

SUMMARY

Dr. Raymond Dart pioneered the discovery of fossil hominids in Africa with the discovery of *Australopithecus africanus* at Taungs in the Transvaal. Though rejected as an early hominid at first by most experts it was eventually accepted as exactly what Dart said it was, an ancestor of humans that was too advanced to be an ape and not quite human. Dr. Robert Broom later found another type of ancestral human, a form that was more robust than Dart's gracile species. Broom named it *Paranthropus robustus* but eventually it was renamed *Australopithecus robustus*. These discoveries were of major importance, but Broom's estimate of two million year age for *A. africanus* raised many objections; it was later verified as correct.

Other fossils of these were found in South Africa until the 1950s. By then both types were reasonably well known. Tools found associated with some of these fossils led to Dart's assertion of *A. africanus* as a toolmaker and hunter, a claim later disproved.

In 1959, Louis and Mary Leakey made the discovery of another type of robust australopithecine at Olduvai Gorge in East Africa. This type is somewhat larger than *A. robustus* but not significantly different. It was found in beds that contained great quantities of shaped stone tools known as Oldowan. Though originally believed by Leakey to be the toolmaker, *Zinjanthropus boisei*, as he named it, has been reclassified as another specimen of *Australopithecus*.

Of particular importance to the finding of australopithecines in East Africa were two very crucial discoveries. These were the reliable absolute age dates from radioactive minerals in beds above and below the fossil-bearing strata. These have enabled workers to place various fossils into a time frame and see how they relate to each other. The second was the discovery of worked stone tools found in the deposits with the fossils. The sites, the age dates, the stone tools and the associated fossils have all combined to produce a picture of simple hominid life in South and East Africa.

By the mid 1960s additional fossil evidence at Olduvai Gorge determined that the australopithecines were not the toolmakers. That honor is assigned to *Homo habilis*, the earliest member of our species, found by Leakey's team at Olduvai Gorge. Additional finds of this species were made to the north of Olduvai Gorge in the Omo River Valley of Ethiopia. Explorations continued northward into likely beds. In 1974 Donald Johanson found another gracile species, *A. afarensis*, this time even farther north in the Afar Triangle of Ethiopia. This discovery of one, nearly complete (40 percent) skeleton of a female, known as Lucy, was very important for three reasons: it was the oldest of the australopithecines, it demonstrated that there were four types of australopithecines, two gracile and two robust, and the evidence from the bones indicated that she walked erect.

Eventually, the discoveries of leg bones with knees in Ethiopia and footprints at Laetoli demonstrated conclusively that the gracile australopithecines walked with an upright bipedalism. Though of small stature, their post-cranial skeletons were significantly advanced over the ancestral chimpanzee stock and not radically different from modern humans. At the same time, their skulls had undergone some changes, particularly in the jaws and teeth which were becoming smaller; such changes are ascribed to a less harsh diet. Their brains were still relatively small, but their intellectual powers must have been improved over their ape cousins because they survived in a hostile world in the face of major predators.

The very fact that these hominids survived and thrived demonstrates their superior abilities over their ape cousins.

It is not unreasonable to assume that the ancestral stock from which these australopithecines evolved was the pre-chimpanzee line. And that stock probably had behaviors not very different from living chimpanzees. With this as a model, it seems likely that the australopithecines had some similar behaviors and were probably more advanced in some others, particularly intellectually.

The success of the australopithecines was obviously due to their various structures, features and characteristics and the behaviors they employed. The change in lifestyles that took place in their new terrestrial environments resulted in different adaptive pressures. These produced different or modified structural features and behaviors, especially their adoption and development of a bipedal behavior, and the manner in which they survived in the face of new or increased dangers. All these resulted in success in the new environments, and all presaged their evolution as improved creatures, as hominids who retained most of the features, structures and behaviors of their ancestors and added to them. That the australopithecines were successful at their way of life, even with all their handicaps, cannot be doubted. That they gave rise to humans appears without question; there are no other candidates.

STUDY QUESTIONS

1. How do the teeth of the australopithecines differ from those of apes?

2. How do the robust australopithecines differ from the gracile ones?

3. Compare the dental characteristics of the gracile and robust australopithecines.

4. What led Dart to conclude that the Taungs australopithecine was an erect biped?

5. What evidence indicates that the robust australopithecines had a harsher diet than the gracile ones?

6. Why did Louis Leakey believe that *Homo habilis* was the maker of the tools found at Olduvai Gorge?

7. Why were the fossils found at Olduvai Gorge so important?

8. Why did Dart believe that *Australopithecus africanus* was a meat-eating hunter?

9. How can Oldowan tools be differentiated from ordinary stream cobbles?

10. Why was Dart's claim that *Australopithecus africanus* was a hominid too difficult for many scientists to accept?

11. Why were early discoveries of australopithecines identified as so many distinct genera and species?

12. What was the significance of the discovery of *Australopithecus robustus*?

13. Describe the environments inhabited by the various australopithecine species.

14. What evidence led Johanson to conclude that *Australopithecus afarensis* was an erect biped?

15. What skeletal features of the gracile australopithecines differentiated them from the apes?

8 The Humans Appear: *Homo habilis*

INTRODUCTION

The discovery of numerous stone tools at Olduvai Gorge was of great importance in the study of early humans because it demonstrated a significant cultural advance for our ancestors. Though we can concede that the earlier australopithecines may also have used tools on occasion, especially sticks or unworked stones, there is no evidence for any systematic tool use or toolmaking in these ancestral hominids. The Oldowan worked tools were a major technological advance, however simple, that improved life significantly for the earliest humans of our genus who otherwise had to rely, like the apes, on their teeth, hands or whatever natural materials they could find. This step forward took place when they began to use objects of their own choosing, modified them in particular ways to extend their usefulness beyond their own abilities, and were able to transmit this skill and ability to others.

Stone tools created a better way of life for these simple humans directly and indirectly. Stone tools helped them augment their diets with harsh or tough foods that were otherwise unavailable. The time required to obtain some of these foods was reduced, preparation was easier, and some foods could be exploited in new ways. Nuts, for example, could be quickly and easily cracked open using stone hammers as chimpanzees do today, and nutmeats, seeds, roots and tubers could be ground up. The butchering of meats with sharp edged stone tools was also possible, and this technique opened up a vast food reservoir of animals for greater exploitation. Stone tools became available for use as weapons, too, probably for both hunting and defense. That stone tools were clearly of value to these hominids is obvious; otherwise they would not have continued and increased their use. Neither would we have expanded their types and varieties and have become so dependent on them. Of equal importance to the first humans were benefits that developed indirectly. The manufacture and use of such tools created a feedback mechanism that encouraged a mentality for making and using such instruments. Tool use and toolmaking provided an impetus that catalyzed these early humans and forced them to develop the first of complex cultures, a trend that we enjoy today. The evolution of stone-tool cultures in Europe is shown in Figure 8.1.

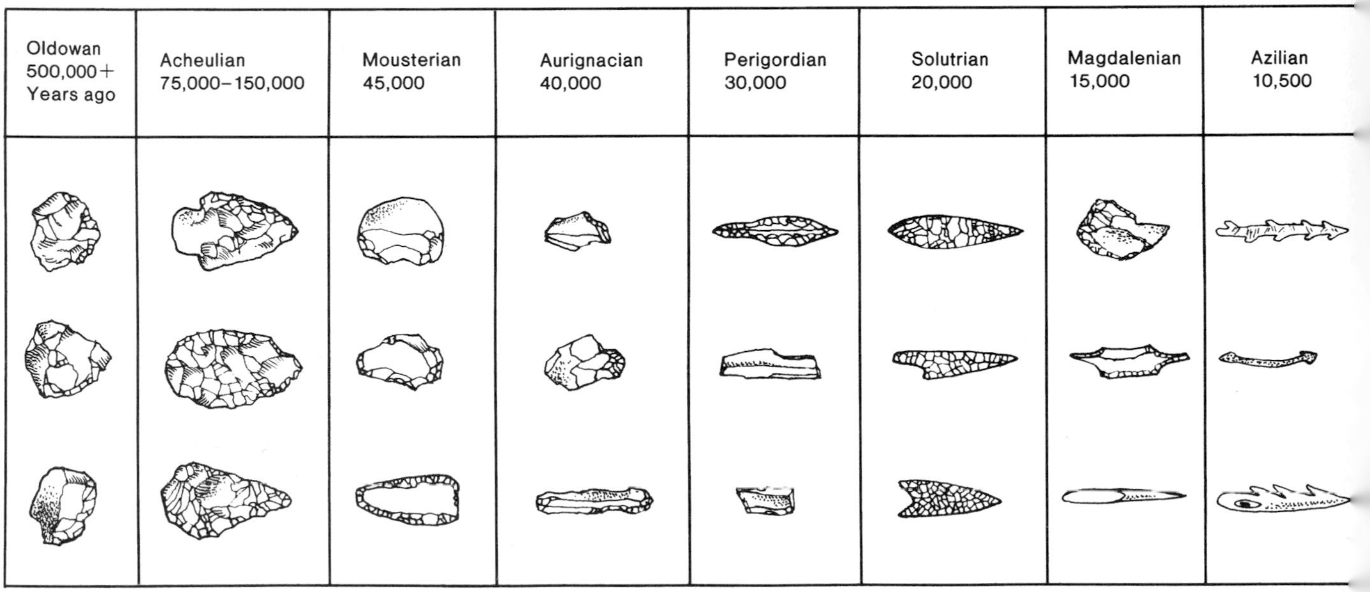

Oldowan 500,000+ Years ago	Acheulian 75,000–150,000	Mousterian 45,000	Aurignacian 40,000	Perigordian 30,000	Solutrian 20,000	Magdalenian 15,000	Azilian 10,500

Figure 8.1 The evolution of stone-tool cultures in Europe.

Figure 8.2 Oldowan pebble tools from East Africa.

The German geologist, Hans Reck, was the first to explore Olduvai Gorge scientifically. He found the first stone tool in 1913; but, although he recovered more than 1,700 fossils from these beds, he failed to recognize these worked stones or rocks as tools. In 1931, Louis Leakey mounted an expedition to Olduvai Gorge and included Reck on the staff. Leakey's aim was to find evidence of Stone Age humans, and he was confident he would find stone tools as proof. Within twenty-four hours he had his first Oldowan tool and additional tens of thousands have since been found throughout the strata of the gorge. Leakey was able to recognize and identify slight differences in manufacturing sophistication and he outlined an evolutionary sequence from simple pebble tools present in Bed I to the more sophisticated tools found in Bed IV. Eventually (1959) Leakey found a fossil hominid in association with stone tools, but it was not the fossilized member of *Homo* that Leakey had long sought. Instead, Leakey found the super robust australopithecine that he initially identified as *Zinjanthropus boisei*. Any disappointment he might have felt at this discovery, however, was not to affect him for very long. By 1964 Leakey had found his fossil human in these beds and named it *Homo habilis*.

The importance of these stone tools in the Olduvai strata cannot be overemphasized. Although geologists routinely provide relative dates for strata using fossils that provided good stratigraphic correlations, such correlations were extremely limited at Olduvai. Radioactive minerals in the strata, however, provided absolute age dates in years, with a precision impossible using fossils. Once the strata had been so age dated, the tools and fossils found in them took the same ages; and once these tools and fossils had been located in time, correlations could be made with fossils and tools elsewhere, even though stratigraphic correlations could not be easily made. Tools similar to the Oldowan type had previously been found associated with fossils in South Africa and precise correlations between these distant sites had been impossible using relative methods. Direct correlations were now possible between the South African and East Africa australopithecines. Tools had also been found at Choukoutien in China in association with a later human, **Peking man**, **Homo (Sinanthropus) erectus**, and these, too, had to be correlated if their position somewhere in the evolutionary tree leading to humans was to have any meaning. The Olduvai stone-tool discoveries and radiometric age dating significantly increased the opportunities for arranging these oldest hominid fossils into some semblance of an order.

OLDOWAN TOOLS

Typical Oldowan stone tools consist of several basic types, though the simplest are choppers (Fig. 8.2). They are

139

easily produced from fist-sized rounded cobbles that have been worked or struck by another stone on only one side. This procedure knocked off chips and commonly produced a rough to sharp, jagged working edge, depending on the intent of the knapper and how the stone had been worked. Such stone tools were typically used to cut, pound or scrape materials. The very common mineral quartz and its mineralogical varieties, flint and chert, are exceptionally well suited to the making of tools in this fashion, as is quartzite, a metamorphosed form of sandstone usually composed of quartz sand grains. Quartz and its varieties are ideally suited for these tools. Their atoms are arranged in a three-dimensional, interlocking network that when struck and broken has no preferred direction of weakness. Instead, quartz breaks with a curved surface, known as **conchoidal fracture**, that has very sharp edges. Small stone implements of this type have been found at Olduvai in great numbers.

In the manufacturing process, stone cobbles or **cores** would be handheld and struck with another stone to remove **flakes** and produce the desired shape and tool. Flakes are commonly found in great quantities at sites where stone tools were made. Some of these flakes have sharp edges and probably also have served as small cutting knives. Their razor sharp edges, which rival those of modern steel scalpels, were undoubtedly used by early humans to cut various materials, especially meat. Such stone knives are easily dulled by use, and these are commonly found with dulled edges. In addition, some impact-marked stones were also found. These are believed to be **anvils** upon which the cores were supported and struck with other stones, **hammers**, to produce particular shaped flakes and cores. With time and increased skill and knowledge, more sophisticated tools production was possible.

The materials from which the stone tools were made and how they were shaped indicate that their makers understood the properties of particular rocks and minerals, how they could be shaped, how they could be used, and the value of certain shapes for particular purposes. Hay (1971) surveyed the geology around Olduvai Gorge and located the sources of many of the materials of the stone tools found there. The early human craftsmen at Olduvai clearly selected certain rock materials over others and were willing to travel great distances to obtain these. In Beds I and II, the lava used for most of the tools was found within two kilometers of the campsite. In Bed III, materials came from outcrops eight to ten kilometers away. In Bed IV, the tools were fashioned from a much greater variety of rock materials that were collected from as far as twenty kilometers away. They had obviously developed preferences for particular materials.

DEVELOPED OLDOWAN TOOLS

As might be expected for such simple tools, the Oldowan tools underwent some slight improvements and evolved into a type that is known as **Developed Oldowan**. They do not differ greatly from the original versions but are just slightly more advanced. This advance was accomplished with additional retouching on the flake tools in an apparent attempt to develop better and more specialized tools. Such changes in technology were hardly indicative of high intelligence because the initial appearances of the two types are about 150,000 years apart; only minimal additional intelligence can be credited to individuals that made the improved types. It would also seem likely that, if the humans were sophisticated enough to develop stone tools and improve them, they also used wooden sticks as implements. There is, however, no direct evidence of such wooden tools, undoubtedly because of their fragility in most environments and rare preservation as fossil tools.

THE FIRST HUMANS
OLDUVAI GORGE

In 1964 Louis Leakey's team of scientists (Leakey, Philip Tobias and Alan Napier) announced the discovery of a new type of hominid at Olduvai Gorge from about the same level as the super-robust *Australopithecus boisei* that had been found in 1959. This new hominid was based on several different specimens (see Fig. 7.12) found from 1960 to 1964, and was about the same size and shape as the gracile *Australopithecus africanus*. It was decidedly more like a human than either the gracile or robust australopithecines. The original fossils were poorly preserved material: a crushed cranium, some skull fragments, two mandibles with some teeth, part of an upper jaw, and some loose teeth. These represented at least two adults and one juvenile. They were found with more than two thousand nonhuman fossils and stone tools scattered about a surface that had apparently been occupied as a living floor some 1.75 million years ago. Tobias reconstructed the skull and obtained a brain size of about 680 cc, clearly larger than any known australopithecine and nearly that of *Homo*.

Figure 8.3 Fossil site at Lake Turkana.

Leakey's team weighed the possibilities of species to which they would assign this type of hominid and considered the fossils that had been found prior to 1964. The brain was slightly smaller in size than one assumed necessary for *Homo*, but after all, there are no absolute criteria for human brain size; the figure of 700 cc as a minimum for humans was simply an educated guess. The skull bones were thin, like humans, and the hand bones included an opposable thumb. An almost complete left foot demonstrated erect posture and bipedal gait; this foot correlated well with leg bones found nearby. The teeth were decidedly human with relatively large front teeth and small cheek teeth. Ultimately the team decided that the fossils did not fit well with any of the known australopithecines. They decided it represented a new type of gracile species which they christened *Homo habilis*. Louis Leakey believed that he had at last found the maker of the abundant Oldowan stone tools that he had found so many years before in these strata. The age of this species, at about 1.75 million years, makes it the oldest true human.

Homo habilis was found to range from Bed I to mid-Bed II at Olduvai and exhibited little change during the almost 750,000 years of its tenure there. Eventually, parts of four different skulls were found and, because it has been a common practice to use informal names or terms for important individual fossil specimens, these were eventu-

ally named Cindy, Johnny's Child, George and Twiggy. Tobias reconstructed these skulls and was able to obtain measurements from three that indicated they had brains that averaged 642 cc, somewhat smaller than the previously considered minimum of 700 cc. Though many experts questioned Tobias' estimate of brain sizes from such fragmentary remains, Holloway (1974, p. 110), an expert on hominid braincases, later independently agreed with the initial measurements and provided his own estimates for the *H. habilis* skulls from Olduvai. However, because of the question of brain size and a difference of opinion about the teeth, some workers still maintain this species is not *Homo*, merely another *Australopithecus*. Most experts have accepted this type as a new species, *H. habilis*.

KOOBI FORA

This momentous discovery at Olduvai Gorge was followed soon afterward by another important find of *H. habilis*. **Richard Leakey**, the third of the famous family to become involved with fossil humans, had been exploring for fossils in the Omo Valley of Ethiopia just north of Kenya. With little success and less interest in his assigned beds, he boldly moved south into Kenya to the eastern shore of Lake Turkana (formerly L. Rudolph) and promptly discovered perhaps the richest site for fossil hominids in the world (Fig. 8.3). There, in 1972, at **Koobi Fora** he found an almost complete skull that has become known around the world by the simple title **KNM-ER 1470**, shorthand for its acquisition label, Kenya National Museum, East Rudolph #1470 (see Fig. 7.13). This remarkable skull, which is about 1.8 million years old, differs from australopithecine skulls in several ways. It is smaller and much more delicate than the skull of the robust *A. boisei* and yet larger and considerably more modern looking than that of *A. africanus*. It differs from both in having a high rounded cranium with a pronounced forehead. It lacks the very prominent brow ridges of both types and has a much larger brain capacity than *A. africanus*. Holloway (1974, p. 110) measured this skull and obtained a brain size of 775 cc, noting,

> Not only had the skull contained a brain substantially larger than the brain of either the gracile or the robust species of *Australopithecus* (and that of *Homo habilis* too) but also this very ancient and relatively large brain was essentially human in neurological organization.

This clearly was no australopithecine and, though its brain was about 140 cc larger than the Olduvai examples of *H. habilis*, it was too small to be a representative of a more advanced species of *Homo*.

KNM-ER 1470 is a highly evolved gracile hominid and more closely related than the older australopithecines to the fragmentary remains of *H. habilis* found by the elder Leakey and his team at Olduvai. A few experts have resisted this assignment, however, and prefer to identify it as *Australopithecus habilis*. Richard Leakey himself preferred not to make identifications to species; he simply identified the hominid fossils from East Lake Turkana as two distinct forms, the robust *Australopithecus* and the gracile *Homo*.

Most workers today believe that the larger brain of the Lake Turkana specimen simply represents the normal variability found in any species and consider it to be *H. habilis*. Though initial tests at Lake Turkana provided dates for KNM-ER 1470 at about 2.9 million years, better samples have provided a revised date of about 1.8 million years, the same general age as the original *H. habilis* find at Olduvai Gorge to the south.

H. habilis is still not a well characterized species; it is relatively new to science and has not been found in any numbers that could allow adequate understanding of its

variability. As a result we know little about the physical nature of this species save the general interpretations that have been made from the fragments of skulls, teeth, jaws and other bones that have been found since 1960. Some of these, however, have provided valuable information from a few crucial fragments. They demonstrate, in keeping with the general rule of evolution, that different anatomical structures evolve at different rates, some more rapidly than others, and that some primitive features are able to relate new species to older ones. At Olduvai Gorge, a partial foot of *H. habilis* reveals that this species walked in the same general manner as modern humans; it was fully bipedal. A partial hand shows that the species had about the same manual dexterity as living humans. This hand ability means that it was capable of manipulating stone to create tools.

Additional evidence from the fossil sites allows some interpretations of their cultural lifestyles and some behaviors. The beds containing *H. habilis* are characterized by Oldowan industry stone tools and this species is generally accepted as the craftsmen who made these implements; no other species from these beds are seen as the likely creators. These earliest humans had a slightly different way of life than the australopithecines, perhaps inhabiting slightly different environments, yet in many respects they were similar. Both of the australopithecines and *H. habilis* undoubtedly lived in small bands or groups of mostly related individuals, the usual pattern practiced by virtually all primates. These bands may have been simple family groups consisting of a male and female pair with offspring, or they may have been extended families with parents, offspring, grandparents, aunts, uncles and cousins. As a maker and user of stone tools, *H. habilis* was able to live a somewhat more comfortable existence than the australopithecines in the hostile environments of the East and South African savannas. Toolmaking and tool using enabled this intelligent hominid to capitalize on its structural and mental abilities to move beyond the level of the gracile australop-ithecines, and this move made it human.

ARCHEOLOGICAL EVIDENCE FOR HOMO HABILIS

The archeological and paleontological evidence collected with great care from hominid field sites has provided information that we use in making interpretations about the lifestyles and culture of these various peoples from the distant past. European hominid sites were well studied during the nineteenth and twentieth centuries. Hominid sites elsewhere were discovered and thoroughly studied later, beginning with the Java sites in the late 1890s. In China, many of the sites are still being worked today. Initially these excavations were crude efforts with little regard for detailed measurements and accurate records, but later workers realized that much important material was lost through haste and carelessness and they began to use improved methods in their field work. Today such "digs" are large operations and models of scientific exploration. Typically, experts from different disciplines examine the sites and contribute their skills and interpretations to the overall analysis. African prehistoric sites, in particular, have been examined in recent years by a variety of cooperating experts and their combined specialized knowledge has provided information previously unavailable. We now know a great deal about these hominids.

OLDUVAI GORGE

A number of the archeological sites at Olduvai Gorge provide evidence about these early humans. Mary Leakey recognized (1971, p. 432) four types of sites found in Beds I and II. She divided these into:

(a) Living floors—the occupation debris is found on a paleosoil or old land surface with a vertical distribution of only a few inches (0.3 ft.).
(b) Butchering or kill sites—artifacts are associated with the skeleton of a large mammal or with a group of smaller mammals.
(c) Sites with diffused material—artifacts and faunal remains are found throughout a considerable thickness of clay or fine-grained tuff.
(d) River or stream channel sites—occupation debris has become incorporated in the filling of a former river or stream channel.

Thousands of fossils and artifacts have been recovered from twenty-three sites dated at from 1.8 to 1.7 million years ago. The geological evidence indicates that the occupation sites were always situated close to streams or on the shore of the lake. They were probably inhabited for only a few days to a few weeks before being abandoned. It is possible that some sites were reoccupied each year. Only the living floors at these sites, however, are capable of providing much information about the behaviors of *H.*

habilis. Site FLK, where *Zinjanthropus* was originally found, is interpreted as a well preserved living floor area of about 3,400 square feet. Louis Leakey initially and erroneously attributed this site to *Australopithecus boisei* because of the magnificent skull found there. He said (1959, p. 491)

This skull was found to be associated with a well-defined living floor of the Oldowan . . . culture. Upon the living floor, in addition to Oldowan tools and waste flakes, there were the fossilized broken and splintered bones of the animals that formed part of the diet of the makers of this most primitive stone-age culture.

This floor had probably been occupied for many seasons because it contained nearly 2,500 artifacts and more than 3,500 large fossils. It consists of a central area that was characterized by a circular zone densely concentrated with very fragmented fossil remains surrounded by a relatively barren zone, especially on two sides, and encircled by another zone of abundant and relatively complete fossil fragments. Interpretations suggest this may have been a structure or dwelling, a windbreak, or even a hunting blind. At site FLK, the central habitation space would logically have contained the debris from meals and other activities, typically small materials not worth picking up and discarding. The encircling barren zone would have been a protective thorn fence where no activities could take place but which would contain a few recoverable materials. The outermost zone with more complete fossils would have been the garbage dump where refuse was simply tossed from the center of the enclosure. This site resembles enclosures constructed by living Bushmen in the Kalahari Desert of southwest Africa. They protect themselves from large predators by constructing a thorn bush fence around a small living space. They create a distribution of waste, bones and other materials almost identical to that found at site FLK.

The stones and fossils at two other sites also suggest circular patterns similar to the habitations used by modern Bushmen. At site DK, the oldest hominid living floor in Olduvai (dated to about 1.85 million years) Mary Leakey (1976) found a rough circle of stones about 4 meters across with some piled into groups around the margin. It is clearly some type of structure and she speculated that it served as a windbreak or perhaps an artificial shelter, similar to those

Figure 8.4 Possible windbreak of stones at Olduvai Gorge.

built by bushmen. This site has hard ground with little or no soil into which branches could be stuck to form a framework to support a canopy or roof of grasses or leaves. However, the piles of stones collected around the margin could have supported branches that did frame such a structure. Another site at Olduvai, from beds about 1.75 million years old, contains a semicircle of heaped stones that could also have been a type of hunting blind or windbreak (Fig. 8.4). Some of the Olduvai sites have provided exceptional information regarding the hunting abilities and techniques of these early humans; this interpretation also implies a division of labor between males and females. Mary Leakey stated (p. 434) that

> Hunting and fishing were unquestionably practiced in view of the remains found on the living floors, but it is probable that scavenging from predator kills was also a method of obtaining meat.

One site, FLK North in Bed II, contains the remains of a *Deinotherium*, an early relative of elephants, and abundant stone tools that were used to butcher it (Fig. 8.5). Fossils of this large mammal and two herds of antelopes

embedded in clays with associated tools at three other sites suggest these animals may have been driven into a swamp where they could be easily killed. An alternate model assumes that these animals became entrapped naturally, died or were killed by the hominids and then exploited. In either version, these early humans were obviously organized and cooperated in these endeavors. The tools are too numerous to have been brought and used by a single hominid. The bones and stone tools, thus, offer ample evidence of success by multiple individuals acting in cooperative fashion.

The evidence of tool use seen in some fossils at these sites is also unmistakable. The bones of three antelopes at one site show that they were killed by blows. Mary Leakey stated significantly (1976, p. 438),

> The depressed fractures on the three frontlets of *Parmularius altidens* from FLK North can be taken to indicate that the animals were killed by means of a blow, delivered at close quarters, since the fractures are accurately placed above the orbits, on the most vulnerable part of the skull.

Figure 8.5 Mary Leakey with J. Desmond Clark at butchering site in Olduvai Gorge.

the amount of work required to produce them, they must have been of great value to the user.

The oldest of the Oldowan tools are probably the basalt stone implements found at Hadar in 1976 by Roche and some flakes found later by Harris; both are dated at about 2.6 million years. The most common tools are **choppers** made from blocks of stone that were worn by streams into rounded, oval or pear shapes (see Fig. 8.2). These raw blocks were simply struck on one end to break off a chip and produce a ragged surface. If struck again adjacent to the first break, another chip would be removed and, with some luck or skill, a reasonably useful jagged edge could be produced. The untouched rounded portion of the stone would be conveniently held and the edges would be sharp enough to cut or tear through flesh much easier and more efficiently than an individual's teeth.

With time, these craftsmen began to improve on their choppers for different purposes. Bed I, which contains the oldest tools, has only six different types. Bed II, however, has ten different types of slightly different style. The relative value of these types may be determined from the numbers of each found at the different sites. Some were made with sharp edges on the sides to form double-edges. Others were made with sharpened ends to form points and chisels that are especially abundant at some sites. **Scrapers**, including both a push type and side-scrapers, are more numerous at some living floors than choppers. Narrow pointed tools, **burins** and **awls**, are useful for making holes but are not very abundant except at one site. More abundant and of great utility were the sharp-edged **flakes** produced as chips from the edges of stones being worked. These probably served as small instruments for several purposes, particularly as knives used to cut the meat.

Mary Leakey makes an important distinction between the two types of tools found at Olduvai Gorge. Beds I and II contain Oldowan and Developed Oldowan tools that were typically worked on only one side. Some other tools found in Bed II are a different style known as **bifacials**. These bifacial tools were often **knapped** or worked all around the edges to produce two working surfaces; they have been variously interpreted as **cleavers** and **hand axes**. Such bifacial tools are given the name **Acheulian** after a particular stone age culture industry identified at a small French village, St. Acheul, where their teardrop shapes were first noted. This industry equates with **Chellean**

The depressed fractures on these antelope skulls may well be related to the **spheroids**, enigmatic spherical stones that were found at these and other sites. They were clearly and accurately worked into spherical shapes for some purpose. Louis Leakey believed that they had been made into **bolas** for hunting. These are simple weapons formed by tying two or three spheroids together with a cord and casting it with a whirling motion. The whirling motion entangles the legs of prey and prevents its escape. Bolas are simple to construct, easy to use and very effective. Spheroids, however, do require much working of stones to achieve the shape. Still other explanation can be offered for these spheroids. Whatever the use for them, considering

or **Abbevillian** cultures, whose names it supplants. Acheulian industry tools are characterized by much finer flaking than the Oldowan tools and required greater skill to produce. They clearly represent an evolution in craftsmanship.

The sudden appearance of Acheulian tools at Olduvai Gorge in an area characterized by Oldowan tools indicates the introduction of a new culture. It is unknown whether this reflects an intrusion by different hominids with new tool types from elsewhere or their innovative development by a small portion of the local hominids. It is possible that the two different industries, Oldowan and Acheulian, were simply utilized by different groups or bands according to their preferences and skills. This view suggests that some of the original toolmakers learned how to make better types of tools and passed this information to their descendants. Some other groups simply did not learn to make these newer types, or perhaps they had a different kind of tool kit (raw materials) which restricted the types of tools they could make. Unfortunately, the lack of evidence precludes a clear explanation for these patterns of tool distributions but it is not unexpected. In our modern world, all people do not use the same style of tools for their needs. Many peoples retain primitive instruments simple because they have not yet come into the modern industrial age or because they do not know how to make more modern instruments.

Leakey originally considered *Australopithecus boisei* to be the maker of these tools because of their presence at the original *Zinjanthropus* discovery in site FLK at Olduvai. Now, however, *Homo habilis* is considered to be the craftsman. Given the choice between these two types of hominids, the larger-brained *H. habilis* is the only acceptable maker. As further evidence, tools have been found associated with *Homo habilis* in at least six sites at Olduvai and elsewhere, including several living floors and kill or butchering sites. *A. boisei*, the vegetarian with powerful teeth, is not known nor thought to have used any stone tools. The first humans, *H. habilis*, therefore, are the people who camped along the lake shore that existed in the Olduvai area millions of years ago. They made and used stone tools, gathered plant foodstuffs, probably hunted and scavenged animals for their meats, certainly butchered them, and produced the debris discovered by paleontologists. The excellent skull of *A. boisei* found in these deposits has not been explained, however, and we may freely speculate on its presence.

Mary Leakey is not alone in her observations on the hunting and meat-cutting tools made and used by *H. habilis*. Merrick (1976) reported on the variety and quantities of artifacts recovered from the Omo Valley excavations that supported these same arguments. Isaac (1976, p. 483) reviewed the archaeological evidence from across Africa and identified toolmaking as a cultural industry that began perhaps 2.5 million years ago. These craftsmen slowly acquired the necessary simple toolmaking skills and eventually developed the ability to improve this tool industry as needs, knowledge, and materials became available to them. Isaac also developed a model of comparisons of behaviors of chimpanzees, protohumans and humans and speculated that the very early humans had "rudimentary sound communication signal systems beyond those of other primates." This would seem likely given the greater abilities and behavioral demands upon these hominids.

To account for the significant step to humans, Isaac specified that these developments took place in two phases that also required assigning these hominids, *H. habilis*, to the human family. The first phase of their humanization involved improvements in their locomotion and subsistence. These were apparent in the development of bipedal gait, something humans share with the earlier australopithecines, and the acquisition of an omnivorous diet with a gradual increase in the consumption of meat. Though we cannot determine how much meat was obtained and eaten by the australopithecines, it was probably enough to encourage a demand for these high energy foods that were available year-round. These dual aspects of phase one would probably never have taken hominids beyond the australopithecine stage of development without the second phase. The second phase was highly significant because it involved toolmaking and food sharing. These improvements were of such a magnitude that they were sufficient to enable these hominids to make the jump from improved australopithecines to humans. Taken together these advances are virtually indicative of humans.

Toolmaking and food sharing are not positively identified with the australopithecines despite arguments by Dart and Ardrey, although they are not negated. Though chimpanzees practice occasional tool use, food sharing is not common in the apes and must be a unique development, perhaps begun by the australopithecines and continued as a practice by the early humans. Isaac and others believe these latter two behaviors, tool use and sharing, are the likely products of larger brains. These brains had the capacity to identify and accept both tools and food sharing as better ways of surviving in their harsh environments. Tools are directly associated with *H. habilis* and so is food sharing; they are not positively identified with any other older species.

OMO VALLEY

Archeological sites with tools and other evidence are present in the Omo River Valley in the middle part of the **Shungura Formation** which dates from about 2.0 to 1.8 million years ago, somewhat older than the original Oldowan sites in Olduvai Gorge. Louis Leakey had predicted in 1936 that Oldowan stone tools would someday be found in these beds at Omo; the large-scale expeditions of the 1960s and thereafter recovered abundant stone tools. These are typically worked quartz pebbles that were recovered from stream deposits, perhaps dried streambeds and muddy streamside environments. These artifacts make a sudden appearance in these strata, rather than a gradual increase in numbers through the time of the deposition. No definitive evidence demonstrates the maker of these tools, because an australopithecine and *H. habilis* are both present in these beds. *H. habilis*, however, is the only acceptable toolmaker.

EAST TURKANA

Three sites with tools from the east side of Lake Turkana are younger than the Omo sites and about the same age as those at Olduvai Gorge. They contain far fewer tools than the Olduvai sites but also demonstrate that large animals were butchered. At one site in East Turkana, a hippopotamus skeleton was found with stone tools, three cores, about 150 flakes and flake fragments, and a hammer stone. Apparently, *H. habilis* took some rock materials with him and made tools on the spot for butchering this hippo. The butchering process would involve much cutting and hacking to free the meat from bones, and sharp edged tools would rapidly dull or break in such use. Cut marks have indeed been found on some of these bones and others from elsewhere and suggest that either reshaping of tools was needed as they dulled or that new ones be made. Thus, a prudent practice would have been to carry raw materials to make new tools.

CHESOWANJA

Midway between Olduvai Gorge and Lake Turkana is the site at **Chesowanja**, an area of about one square kilometer. Two aspects make several places at this site a very important archeological locality. Thousands of Oldowan and Acheulian stone tools have been found in the layers of sedimentary rocks that had been deposited at this site along an ancient lake. The tools were made from the basalt of local lava flows that had poured over the area and from basalt river cobbles. The great number of tools and their variety throughout the strata record the presence of numerous camps occupied by groups of hunters and butchers from 1 million years ago to the present. From Oldowan times to the later Stone Age, hunters and butchers apparently worked here and left the debris of their toolmaking. Flakes accumulated in great quantities in the camps, which have been described as factories for the manufacture of tools.

Of special importance at this site are the lumps of burnt clay that have been found with stones arranged in hearth fashion. These clays are dated at about 1.4 million years ago and technical measurements indicate that the clay had reached temperatures of 400 to 600° C, about that expected for a campfire. If these clay lumps are indeed the remains of campfires, and this interpretation is disputed, they are the oldest known evidence of humans with fire.

STERKFONTEIN

Mary Leakey believes the tools found at Sterkfontein and originally associated with the gracile australopithecines are typical Developed Oldowan hand axes that should be related to *Homo habilis*. They correlate with the Developed Oldowan industry in the upper part of Bed II at Olduvai Gorge. Lewin (1984, p. 48) reports,

> Hominid remains that might well be representative of the Homo habilis grade have been found in South Africa, specifically in the Sterkfontein cave, but they are rather fragmentary and difficult to date with any accuracy.

AUSTRALOPITHECINES AND HOMO HABILIS COMPARED
ANATOMY

H. habilis may be distinguished from the gracile australopithecines in several aspects. Though each aspect may not be conclusive by itself, collectively these produce a forceful argument for separation of these fossils into two different genera. The first and most obvious method involves direct comparisons of the respective skeletal materials, principally the skulls and teeth. *H. habilis* has a high

Figure 8.6 Jaw and teeth of *Homo habilis*.

vaulted cranium with a considerably larger brain that average about 650 cc and ranges to nearly 800 cc (see Fig. 7.13). Their faces are narrow and clearly less ape-like with reduced prognathism and less broadly flared zygomatic arches and cheeks. The gracile australopithecines in contrast have rather low vaulted skulls with smaller brains that range from 400 to 500 cc and average about 450 cc. Their faces are also broader with more pronounced zygomatic arches and cheeks. The teeth of the two types are different, as expected of animals with different diets. *H. habilis* has relatively small cheek teeth but large front teeth (Fig. 8.6) compared to those of *Australopithecus africanus*; these reflect a somewhat less harsh diet of tough plant materials. The australopithecines have truly massive cheek teeth, up to double the size of modern human molars, but small front teeth. The upper jaw of an australopithecine also has a slight diastema; *Homo* has no gap.

BEHAVIORS

Another way to distinguish between australopithecines and the first humans involves analyses of their behaviors, especially toolmaking and their methods for obtaining foods. The systematic making and use of tools strongly indicates an organization recognized only in humans, especially a social structure that probably included their use in hunting and preparation of foods. Stone tools have not been found in close and unequivocal association with *A. africanus*, but they are commonly found with fossils of *H. habilis* at several sites. The conclusion is that the *H. habilis* made and used stone tool; *A. africanus* did not.

Foods are obtained in three ways: gathering, hunting and agriculture. Agriculture is not known to have been practiced or developed until about 10 to 12,000 years ago and it is eliminated as a possible source of food for both the australopithecines and *H. habilis*. Modern primitive people who do not practice agriculture engage in hunting-gathering behavior to obtain their foods. They gather plant foods most of the time and hunt whenever possible. It is likely that early humans lived in family or extended family groups and used a similar pattern for obtaining foods, though we cannot determine the relative percentages of time spent in hunting vs. gathering or the relative success of each.

HUNTING AND GATHERING

Berries, fruits, seeds, nuts and other plant materials can be picked at will by individuals without social organization or cooperative behavior. All primates engage in this type of behavior and in modern hunting-gathering peoples it is usually carried out cooperatively by females and their young. Complex social organizations are not required because the individuals and family groups generally gather enough for their own needs. Hunting, too, can be an individual activity but it is more successful if conducted by several individuals cooperating together. Hunting, therefore, tends to be a much more sophisticated endeavor. Hunting also carries with it the risk of danger from animals who can protect and defend themselves. Animals suitable for food are wary, both individually and collectively, because of the normal fear of predators. These animals, typically herbivores, must somehow be killed or seriously wounded before they are alerted and flee. Successful hunting of this prey requires some degree of ability and skill in planning. The early humans were not as swift as the

148

larger predators and probably could not individually subdue prey animals. But cooperative hunting could be productive and, if successful, the game meat could be shared. Such cooperative behavior, with each individual playing a role in the hunt, was much more productive than individual hunting. It was also essential if large animals were the prey. Even so, the risk of injuries and fatalities from many different animals was always possible. The use of stone tools for hunting, with all its implications, therefore, indicates a high level of sophistication in hominids, much higher than that of other contemporary hominoids and australopithecines. Hunting, whether solitary or cooperative, was not an early behavior in these hominids, but was developed subsequently by *H. habilis*.

It seems likely that such cooperative hunting was performed largely or exclusively by men, although females may have cooperated in hunting roles that exposed them to little danger. This is the typical pattern used by modern hunting-gathering peoples for very good reasons. The danger to females and offspring engaged in hunting was liable to be too great to risk. These females and offspring represented and were the future of the group. To risk their lives would have meant a risk to the group's future. Children, thus, were probably also excluded from hunting because of their inexperience. Thus, gathering was a more likely activity for both females and children. Few were at risk in gathering fruits, berries and other materials, so these females and the offspring concentrated on these methods.

Scavenging was an alternative to hunting as a means of obtaining meat. It may have even been the principal method utilized until they developed adequate hunting skills. Even scavenging, however, was more effective if the individuals cooperated. Cooperative behavior is, therefore, probably a key mechanism in the survival and success of early humans, particularly if they practiced any form of hunting.

The association of fossil bones and tools at particular sites clearly indicates active hunting. The manufacture and use of stone tools by these hominids indicates a relatively high level of intellectual development. Stone tools are utilitarian and made with objectives in mind. The manufacture and use of such tools implies the mental ability to recognize the function and purpose of the tools beforehand. These hominids were practical people, however primitive; their survival depended on their ability to obtain food with as little risk to themselves as possible. A stone tool would be a useful and desired implement if the connection was made between its shape and its use. If one individual could

kill an animal with a stone tool, others would soon adopt that technique and learn to make and use stone tools and weapons. It is clear that they learned how to make and use stone tools, activities we accept as characteristic of humans. We, as their inheritors, and all intervening hominids have continued and improved upon the process by simply expanding on their methods.

CAMPSITES

H. habilis and the gracile australopithecines differed in their camping behavior. There is no evidence yet to support a belief that australopithecines lived in permanent or even semipermanent encampments. Presumably they roamed in family groups over the countryside at will, typically living near the watercourses, seeking food when hungry and sleeping or relaxing at other times. Several sites associated with *H. habilis* have been found, including both living floors and kill sites that have abundant stone tools. The association of tools with *H. habilis* at obvious kill or butcher sites strongly argues for them as the maker of the tools. Thus, *H. habilis*, not only was the craftsman who made the first stone tools, but lived in the first camps or home bases from which they went out to find foods.

STONE TOOLS

From the abundant evidence at Olduvai Gorge, we know that *H. habilis* made and used stone choppers for cutting up meats but we do not yet know conclusively that they actively hunted animals for meat. It seems illogical to assume they could make stone tools to cut up animals for food but were incapable of using them as weapons. If modern chimpanzees can use sticks as weapons as observed during Goodall's research at Gombe National Park, then *H. habilis*, with its larger brain, better hands, and upright bipedal posture, must have had an even greater ability. We must assume that many of the stone tools found with *H. habilis* could have been used as weapons. The distinction between sticks, that are used as tools or weapons, and stones cannot have been so great that it eluded these early humans. Because wooden sticks are not likely to be preserved as fossils, we cannot presume that these humans did not make and use them. Stone tools and fossils are, therefore, the only direct evidence that we can use for understanding of these activities.

Several sites at Olduvai, Omo and East Turkana have skeletons of large mammals in direct association with numerous stone tools. These are significant associations because the stone tools were obviously used to butcher these animals, perhaps even to kill them. Near the base of Olduvai Gorge, for example, one layer contains the bones of an elephant and more than two hundred stone tools restricted to one small area. There is no evidence, however, to indicate that this is a campsite or dwelling place. Instead it probably served as a butchering site. The elephant may have been killed there, or it may have been found dead; then it was butchered. Sites like this demonstrate an important way in which *H. habilis* differed from the gracile australopithecines. *H. habilis* made and used stone tools in systematic fashion and then used them to kill large animals and/or butcher them. None of the strata restricted to the australopithecines has provided evidence of stone tool-use behaviors. We are forced to conclude that australopithecines never learned about stone tools.

COOPERATIVE BEHAVIOR

Richard Leakey and Roger Lewin argue (1977, p. 117) that the early humans owed much of their development to cooperative behavior, probably initiated by the sharing of food. They said that

> Our ancestors switched from being opportunistic food eaters to being systematic food gatherers. The more frequent addition of meat to the menu was a valuable source of high-quality protein. But the exchange of specific foods between individuals, something that no other primate does to any important degree, has deep behavioral and social implications for the emergence of humanness. . .

Howell and Isaac (Coppens, et al., 1976, p. 474) were more specific and noted that

> Human origins are distinguished more by changes in behavior than by changes in gross anatomy, so the archeological traces of characteristic human activities are a crucial complement to the fossils in reconstructing the trajectory of evolution.

Cooperative behavior as demonstrated by the large numbers of stone tools associated with the bones of large mammals would have provided substantial advantages to any hominid group's survival. Typically they would have used stone tools as they united in defense against predators or in hunting. But cooperative hunting and defense are much more efficient when practiced by a group rather than by an individual. Cooperation in hunting also meant increased benefits to individuals who shared in the rewards with less risk than if they practiced solitary hunting. This reward had to be in the greater quantities of meat, skins and other materials each would receive if successful. Such cooperation is of no value if an individual does not share the spoils of the hunt. Therefore, they cooperated and they shared. We could make the same arguments for cooperation for defense.

Even if these early hominids did not hunt and kill the large fossil animals that have been found, they must have somehow cooperated and shared food with each other. A large animal, even scavenged, simply has too much meat for one individual and too much for a single individual to defend against predators, scavengers and other hungry hominids. Concerted group activity is the only sensible activity that would have provided recognizable benefits to the individual and the entire group. Such cooperation would have also encouraged **bonding** between the hunters. We need only consider the bonding that has been reported among modern hunters or others who engage in group activities; it is strong because these individuals depend upon each other for success in their mutual endeavors. Consider the winners of a team sporting event who gather together, hug and slap each other to exhibit their happiness at winning a game.

Both of these activities, cooperative behaviors and the making and using of stone tools, provided benefits of immediate value to these humans and at the same time conferred indirect benefits for their future. These abilities to make and use stone tools for cooperative hunting or defense enhanced their survivability over other groups that lacked similar behaviors and abilities. Because they were better at surviving, these cooperating toolmakers would naturally have increased their numbers in the population. In reality, their better genomes increased in relative percentage over the less-suitable genomes. Eventually, the cooperating and tool-using humans dominated the populations until all individuals were practicing cooperation and stone toolmaking and use to one degree or another. Evolution provided the mechanism for these shifts in the early human population, and opportunity allowed the better-adapted portions of the populations to be more successful than the others. The early humans inevitably followed these trends as they began to evolve into modern humans.

SUMMARY

Although the Leakeys discovered *Australopithecus boisei* at Olduvai Gorge and initially believed that this species was responsible for the great many shaped stone tools found there, the subsequent finding of *Homo habilis* in the same beds identified that species as the toolmaker. From the original discovery at Olduvai Gorge, *H. habilis* has been found across East Africa and in South Africa. This evidence of the earliest human shows that they had the same general physical stature of the gracile australopithecines but their brains were considerably larger; they were more intelligent. Both types probably lived in similar kinds of social groups and probably followed many of the same general behavior patterns.

H. habilis exhibited an important advance over the australopithecines. They discovered how to shape stone tools to improve their lifestyles and left thousands of these artifacts behind. The tools are of two types, Oldowan and Developed Oldowan, and were made of very resistant and durable materials. They can still be found essentially unchanged. The differences between the two types are slight. Developed Oldowan tools have additional retouching on the flake tools, perhaps for specialized use. Together, these two types have left us with a record of *Homo habilis'* advances in tool cultures and a mechanism to identify and correlate them in time and place.

Archeological evidence at several key sites in Olduvai Gorge, the Omo Valley, East Turkana, Chesowanja and Sterkfontein demonstrate the sophisticated making and use of tools that assisted *H. habilis* lifestyles. It is clear these early humans hunted and butchered animals for their meats and other parts, fished and may even have used fire. To accomplish all these activities, they learned to manufacture different kinds of stone tools, including choppers, scrapers, awls, burrins and flakes, and spheroids, perhaps used as bolas.

Analyses of the archeological evidence have enabled workers to draw many conclusions about the behaviors and lifestyles of *H. habilis*. Once toolmaking and tool use entered their lives with any degree of regularity, it led to some new activities and enhanced some they already practiced. Isaac believes that hunting and gathering, which could have been derived from the australopithecines and, perhaps, ultimately from their ape ancestors, became more common. These in turn probably led to a division of labor. Food sharing, too, had to be practiced. This is not a regular routine with living chimpanzees and it is likely that food sharing developed either in the australopithecines or in *H. habilis*. All of these activities require cooperation, an essential behavior of humans that places great emphasis on leadership and following.

Stone tools provided *H. habilis* with much greater ability to survive and improve their conditions than unworked stones or wooden tools. Their use, especially to obtain foods, both meat and plant materials, offered at the same time benefits to those who could use their bigger brains. Feedback between the making and use of stone tools and their increasing intelligence enabled them to develop the ability, however infrequent or rudimentary, to become cooperating hominids, the behavior that characterizes later species of *Homo*. With stone tools and successful planning these humans could survive on the East and South African savannas. They could cooperate in hunting and gathering, use stone tools for coordinated defense, and pool their abilities and tools to better their conditions. Sharing was an essential ingredient; without it, why cooperate. And without cooperation, survival was at increased risk. They obviously survived, and only a few more improvements were needed to turn these early humans into modern humans.

STUDY QUESTIONS

1. What anatomical differences separate *Homo habilis* from the gracile species *Australopithecus afarensis* and *A. africanus?*

2. What behavioral differences separate *H. habilis* from the gracile australopithecines?

3. What evidence supports *H. habilis* as the maker of the oldest stone tools?

4. Why is knowledge of the australopithecine anatomy important in concluding the *H. habilis* was the first true human?

5. Describe the relationship between toolmaking and cooperative social behaviors.

6. Why should the teeth of *H. habilis* have become reduced in size?

7. What is the difference between an occupation floor and a butchering site?

8. What advantages would cooperative hunting have for *H. habilis*?

9. Why is it likely that *H. habilis* developed a division of labor?

10. Why are sex roles likely to have developed from a division of labor?

11. Distinguish between Oldowan tools and Developed Oldowan tools.

12. How do scientists differentiate between stone tools and nonworked stones?

13. What evidence supports the argument that *H. habilis* used fire?

14. Differentiate between the gracile and robust australopithecines and *H. habilis* by brain size.

15. Differentiate between the teeth of gracile and robust australopithecines and *H. habilis*.

9 The Great Migration: *Homo erectus*

INTRODUCTION

During the several million years of their existence, the australopithecines spread gradually through the Rift Zone of East Africa and into South Africa. There is no evidence, however, that they migrated beyond the African continent. Their descendants, *Homo habilis*, also spread along the Rift Zones of Africa and they, too, apparently were restricted to Africa; none have been found elsewhere in the Old World. Both of these hominids had successfully adapted to the environments of their African birthplace, and both were destined to remain there forever. But the descendants of *H. habilis* did travel out of Africa, through the Middle East into Europe and across Asia, places throughout the Old World where its fossils have been found. These descendants were *H. erectus*, and their widespread distribution is ample testimony of the great success they had in adapting to many different environments. Their successes were not simply in adapting to various habitats, however. They were responsible for developing and inventing many of the devices that we still use today. Our search for these wandering humans began with an idea proposed during the middle of the nineteenth century, about the same time that the original Neandertals were being debated.

HAECKEL AND THE MISSING LINK

One of the earliest scientists to support Darwin and his concept of evolution in Europe was **Ernst Haeckel** (1834–1909), a German evolutionary theorist, biologist, ardent nationalist and the first great popularizer of evolution. Though many Europeans and Americans had accepted Darwin's ideas, Nordenskiöld (1932, p. 505) pointed to Haeckel and said

> . . . there are not many personalities who have so powerfully influenced the development of human culture.

Despite his prominence in biology and evolution, Haeckel made no important scientific discoveries of his own, and his influential ideas were largely borrowed from others. Haeckel had, however, greatly impacted European social thought for many years because he had become deeply committed to liberal social reforms like so many of Germany's leading scientists of the day. This attitude colored his scientific views and carried over into his writings on human evolution. In particular, he had been struck by Darwin's omission of humans from his *Origin of Species* and thereafter began a life-long interest in the evolution of mankind. Unfortunately, he developed many unsubstantiated views that were uncritically accepted by followers who compounded his errors. In this regard, he created some misconceptions that have continued to this day, especially among creationists who cite him and his errors as evidence that evolution is a false concept.

Haeckel's principal scientific idea was his **biogenic law** in which he erroneously argued that an individual's embryonic development (ontogeny) passes through the evolutionary stages of the species (phylogeny). His views of evolution, especially as it related to the social organization of humans, led him to argue for the superiority of an Aryan race of humans. He believed that these humans were better fit biologically for survival than other races. Some of these misguided ideas, including those involving eugenics, were unfortunately later adopted in part by the Nazi government in Germany during the 1930s and 1940s and used as justification for its murderous programs of genocide carried out in the name of racial purification. Nonetheless, Haeckel is today recognized as a principal driving force in modernizing biology. He is particularly known for his views on evolution that began the search for the origin of humans.

From our modern vantage point, it is difficult to imagine the impact that Darwin had on the entire scientific world after *Origin of Species* was published. It is even more difficult to imagine how influential Haeckel was in Europe, particularly because most people today are not really familiar with his scientific views. But Haeckel's prominence and the times combined to provide him with an unequaled opportunity to promulgate his ideas. In 1856 the famous Neandertal man had been found in Germany, the first of many celebrated fossils. It generated much debate in the scientific community and the general population about its origin. But these important fossil bones were shelved without resolution of their nature or status. Scientific Europe had been given its first hints about human ancestry, but failed to capitalize on these ideas. Haeckel, fortunately, realized that Darwin's book neglected to consider the origin of humans and that these newly discovered ancient "Neandertals" must somehow be fitted into an evolutionary pattern.

Although the efforts to find fossil men had not yet begun, Europeans had for many years been making discoveries of what appeared to be stone tools. Unfortunately, these were discounted as the products of human craftsmanship and relegated to the status of simple, natural objects, perhaps shaped by lightning or some other natural force. Boucher de Perthes had published the results of his study of chipped stones and reasoned that such shapes were not formed by natural forces; he concluded that these could only be implements of great antiquity shaped by men. Constable reported (1973, p. 14) that de Perthes was correct,

> ... more right than he knew; the stone tools that
> he had found were at least 300,000 years old.

These stone tools were, in fact, hand axes. Some group of very ancient people had made these tools, but who?

THE MISSING LINK

In this milieu of 1863 Haeckel studied man's place in nature. He proposed that humans descended from the apes and was most closely related to the gibbons. He theorized that modern humans and apes are different but are linked evolutionarily by an intermediate stage that is neither human nor ape, but a man-ape or ape-man. He described this hypothetical and undiscovered ancestor as a **missing link** between hominids and apes and named it *Pithecanthropus alalus* (ape-man without speech). He believed that humans had evolved from the higher primates (gibbons) still living in Java, and proposed that fossils of the ancestral human species would be found on that island in the Pacific Ocean. To demonstrate this he drew the first family tree of mankind (Fig. 9.1). Later (1868) he refined the site of man's origin as **Lemuria**, a hypothetical continent that had disappeared by sinking into the Indian Ocean. In effect, this tropic paradise island was identified as the Garden of Eden.

EUGENE DUBOIS AND JAVA MAN

The year 1863 was important for human paleontology for two reasons. Not only did Ernst Haeckel propose that humans were descended from apes but Thomas Huxley wrote an essay that posed the critical questions:

> Where, then, must we look for primeval Man?
> Was the oldest Homo sapiens Pliocene or

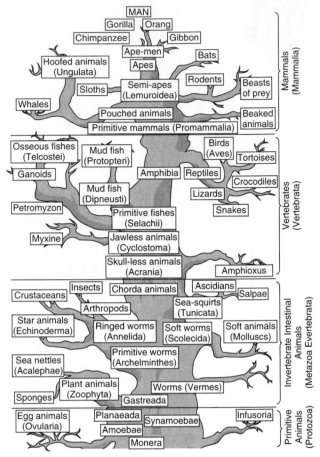

Figure 9.1 Haeckel's family tree of man.

Miocene? In still older strata do the fossilized bones of an ape, more anthropoid, or a man, more pithecoid than any yet known await the researches of some unborn paleontologist?

These ideas of ancient humans, the debates and the fossils profoundly impressed one individual, **Marie Eugène Francois Thomas Dubois** (1858–1940). He was a Dutch physician who took Huxley's challenge seriously. Dubois had taken a degree in medicine at the University of Amsterdam where he studied under some prominent scientists, including the botanist Hugo de Vries, but declined to practice as a physician. Instead, he preferred the academic life and lectured as a professor's assistant for six years before becoming deeply interested in ancient humans. He was an ardent evolutionist and the Neandertal discovery, though made before he was born, had filled him with interest in ancient humans. He knew the works of Darwin

154

and Wallace, and both of these suggested that humans had originated in warm or tropic lands. Then, inspired by Haeckel's writings of early mankind in the tropics, Dubois announced in 1886 his determination to find the missing link, that hypothetical intermediate stage proposed by Haeckel to lie between modern humans and the human-like apes. Dubois accepted many of Haeckel's views, especially the idea that hominids were most closely related to the great apes and had descended from them. He believed that hominids had either originated in Africa because of the presence of gorillas and chimpanzees or in the Malaysian Archipelago because of the presence of gibbons and orangutans.

Alfred Wallace, who had independently proposed the evolution of species by natural selection from his observations of animals in the islands of Southeast Asia, had suggested that fossil humans should be sought in places where apes live today. The Malaysian region was one such tropical region, warm with a favorable climate and abundant food. Life there could be very pleasant without much effort and to some people this typified the Garden of Eden as reported in the book of Genesis. Dubois knew that Haeckel's proposed ancestral stock for hominids, the gibbons, lived in this region, and he concluded that the area would be a likely place to find ancient humans. Accordingly, in 1887 he accepted a position in the colonial service as an army surgeon assigned to the Dutch East Indies, then a part of the Dutch colonial empire. He even announced to friends that he would travel there and find the missing link, that first humanoid described by Haeckel to link apes and humans in an evolutionary pathway. Today the islands are part of the island nation of Indonesia, but in the late 1800s they were a far-off Dutch colony with many poorly explored regions. Eventually Dubois was successful, but not in finding the mythical creature conjured up by Haeckel. There is no single intermediate stage or missing link to connect humans with apes, but a number of stages in which each type improved and contributed aspects that ultimately led to modern humans. We humans are, accordingly, only the latest version of the hominid lineage that began with the australopithecines.

Dubois conducted expeditions from his base in Sumatra to likely sites seeking fossil humans but without immediate success. His discoveries did not happen with the speed of Richard Leakey's hominid finds at East Rudolph, or of Johanson's at Hadar in Ethiopia. Dubois searched for two years without success, then heard that someone had found a fossil skull in 1890 at Wadjak on the island of Java. He immediately obtained government funding to seek such fossils there and, with a convict crew, set out to find his missing link. Dubois found another fossil skull similar to the first Wadjak find, but both were relatively modern in appearance and of no discernible antiquity. He continued these explorations and at Trinl on the Solo River dug into thick riverbank deposits of sandstone and volcanic ash. There, in 1891, he found the fossils that were to become famous as **Java man**.

The first find was a tooth followed by the famous skullcap of Java man (Fig. 9.2). At that point, however, rising waters forced suspension of the digging for several months and Dubois spent the time studying the skullcap and tooth. Though he could not decide their taxonomic position, he reported,

> That both specimens come from a great man-like ape was at once clear.

In 1892, almost a year after the first discovery, he found a left femur (Fig. 9.3) within fifty feet of the first find. It was straight and human in character and modern in every respect except that it was sturdier and heavier. The next month he found an apelike tooth similar to the one found the previous year. All of these fossils were recovered from the same beds, the Trinl layer. All were found (Dubois, 1899, p. 447)

> . . . at exactly the same level in the entirely untouched lapilli stratum.

And now Dubois' expertise as an anatomist became extremely useful in evaluating these remains.

The skullcap of Java man was much lower and more flattened than a modern human skull which typically has a high vault. The Java skullcap was, therefore, clearly more primitive. But the femur was human and modern in all respects, except that it was heavier. Obviously, this femur was from an individual who had walked erect like modern humans, but how did this leg bone relate to the individual who had the low and elongated braincase? Both were found in the same stratum about fifteen meters apart on a sharply defined surface and may even have come from the same individual. If so, then they presented some problems to the idea, then current, of what ancient humans looked like.

Many scientists of that day subscribed to Haeckel's missing link as an actual stage intermediate between apes and humans. Therefore, they expected to find a type with intermediate characters. The skullcap of Java man had very

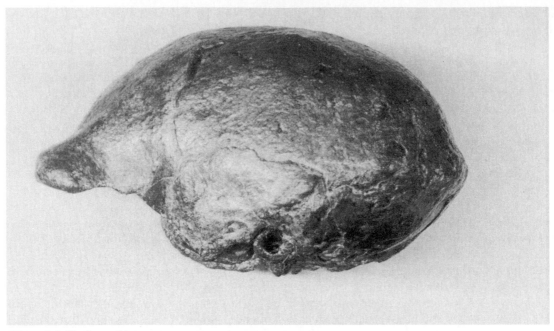

Figure 9.2 Skullcap of Java man.

thick bones, virtually no forehead because of the low slope to the vault, and an exceptionally large and prominent brow ridge that projected outward from a narrowing or constriction of the cranium. However, it possessed some of the appropriate features of the expected missing link. Dubois had enthusiastically accepted Haeckel's missing link stage and, to him, this skull verified Haeckel's predictions. He announced his discovery by publishing a report in 1892 in a mining bulletin (an unfortunate choice) and claimed,

> ...through each of the three recovered skeletal parts, and especially by the thighbone, the *Anthropopithecus erectus* (Eugene Dubois) approaches closer to man than any other anthropoid.

He later changed his mind and renamed this specimen stating (1899, p. 459)

> I believe that . . . this upright-walking apeman, as I have called him, and as he is really shown to be after the most searching examination, represents a so-called transition form between men and apes . . . and I do not hesitate now...to regard this *Pithecanthropus erectus* as the immediate progenitor of the human race.

Dubois had accomplished his life goal; he believed that he had found Haeckel's **Übergangsform**, his missing link and this is indicated by the name *Pithecanthropus*. It was the original name proposed by Haeckel for the missing link and meant ape-man. Dubois' great find showed the world that there were indeed more fossils of ancient men to be found and that the recorded history of humans could be recovered from Earth's strata. Scientifically, and in the public's eye, the search for the path to humans had begun in earnest; there were fossil humans to be found. Today, this skullcap and thighbone are dated at between 800,000 and 900,000 years old making them truly ancient humans. They are also considered fully human and have been reassigned to our own genus *Homo* as the species, *Homo (Pithecanthropus) erectus*.

THE CONTROVERSY

Dubois cabled the results of these Java finds to his friends in Europe and promptly initiated a great controversy over the interpretations. We must remember that the Neandertal fossils were not yet broadly accepted as ancient humans, and scientists continued to argue about their merits and taxonomic positions. Java man did little to settle these arguments and instead generated criticisms from

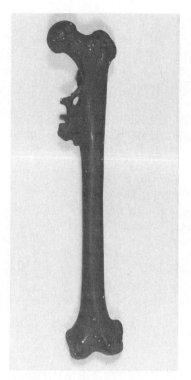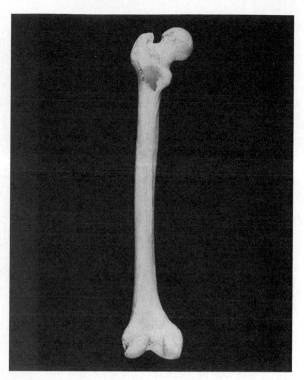

Figure 9.3 Cast of Java man femur (left) with left femur of modern human.

scientists and others who had not even seen or examined the fossils. Without direct study, they argued that the primitive skullcap and modern femur were probably from two individuals with different characters. Dubois was then quickly subjected to ridicule from individuals who used arguments based on preconceived ideas, not on the facts. Clergy denounced him from pulpits as if he had created the fossils and was, thereby, personally responsible for the controversy. Though Dubois allowed any and all to examine the fossils and had published photographs and prepared detailed descriptions, he had little success in convincing the scientific world of his interpretations. Campbell (1976, p. 208) reports that European scientists generally objected to the fossils on nationalistic lines; the

> . . . Germans believed that Pithecanthropus was an ape that had manlike characteristics; most Englishmen thought it was a man that had apelike attributes; and the Americans tended to lean toward a transitional form more along the lines Dubois had suggested.

Of a group of nineteen scientists gathered to examine the fossils, five considered it to be an ape, seven thought it human and the remaining seven thought it was intermediate between these two types. There was clearly mixed support for the human nature of Java man from these scientists who voiced no concensus. Unfortunately, the nonscientists took this opportunity to voice much criticism.

Dubois was deeply hurt by these rejections and withdrew the fossils from public scrunity, even going so far as to bury them in a box beneath his dining-room floor. By 1920 some scientists asked that he again show the specimens, but Dubois refused. Then, apparently in a pique, he announced that he had long ago obtained other important fossils, including the Wadjak skulls that had originally been found in 1890, but had kept them hidden from public or scientific view during all these years. He added that he would never again allow them to be examined. In addition to the original specimens and the Wadjak skull, he also had found (McCown and Kennedy, 1972, p. 190) four other femora in the Trinl Beds in 1900 and an important mandibular fragment elsewhere. Clearly Dubois was seeking revenge of sorts for his humiliations. In 1923, however, he relented and allowed **Ales Hrdlicka** of the Smithsonian Institution to examine the Wadjak skulls. Thereafter he opened all these original finds of Java man to the scientific world. Some materials, however, remained hidden from scientific knowledge and view until 1932.

The *Pithecanthropus* braincase was shown to have a volume of about 900 cc, much larger than living apes and smaller than living humans and the fossil Neandertals. Today, some experts still argue whether the femur and the skullcap are from the same species. Whatever the relationship between these two fossils, Dubois had opened the door for more scientific inquiry about ancient humans. Thereafter it would be much easier for later discoveries of fossil hominoids and hominids to be accepted by the scientific community and the public. It is no wonder that Dart's discovery and explanation of *Australopithecus* created such a stir among scientists and the general public; it is even more surprising that *Australopithecus* generated such negative reactions in face of the potential for other early humans implied by Java man. Apparently, the views of early humans had crystallized sufficiently over the years within the scientific community, and Dart, not a member of the insider clique, was dismissed without much difficulty because his finds did not fit the characteristics "accepted" and expected by the experts.

Dart's discovery did, however, solve some of the problems for Java man and changed the scientific climate in the search for ancient humans. *Australopithecus africanus* was clearly more apelike than Java man and seemed to demonstrate a less human and more apelike form that could be placed evolutionarily between apes and *P. erectus*. This possible relationship made *P. erectus* more human in appearance and more acceptable as a primitive human. Additionally, by this time a variety of hominid fossils of undetermined affinities had been found in different parts of the world and the general public and scientists had become further conditioned to the idea of ancient humans.

One of these important hominid fossils was a jaw found in a sandpit at Mauer, Germany in 1907 and examined by experts at Heidelberg University (Fig. 9.4). There, **Prof. O. A. Schoentensack** decided that it was so wide and thick that he would have considered it from an ape,

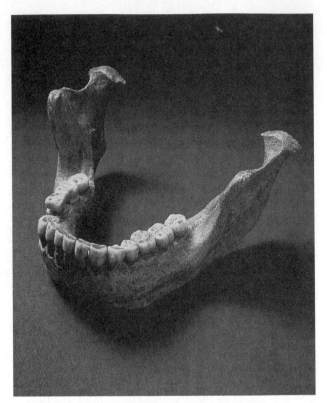

Figure 9.4 The Mauer or Heidelberg jaw.

except for the teeth which, though very large, were remarkably human. He decided it was a new species and named it *Homo heidelbergensis*; today it is considered another specimen of Java man. This fossil is exceedingly important not only because it demonstrates that these humans had spread to Europe, but also because it was found in strata along with a great variety of other well known animals that had lived during the early Pleistocene, about 500,000 years ago. Java man was becoming so well known that scientists could identify the species of animals that these early humans had hunted for their meats, hides and other parts. The critical fossils of Java man had been found, and the first steps were being taken to determine their distribution, history and culture.

PEKING MAN

During the original Dubois excavations in Java, the native digging crews were discovered to be surreptitiously smuggling some of the fossils and selling them to Chinese merchants. These "**dragon bones**" were then ground and made into various medicines, especially aphrodisiacs, for sale within the Chinese community. Such tragic sales still continue today with illegal ivory, horn and other materials from animals poached in Africa and elsewhere. It is unknown how much valuable fossil material has been lost in this fashion over the years, but apparently much is gone. Ultimately, the Dutch government made such sales illegal in Java, though the ban was circumvented and fossils continued to be so absurdly destroyed. This ban on traffic in fossils in the Dutch East Indies, however, had no effect in China.

For many years, occidentals were aware of such dragon bones being used for medicinal purposes in China. Probably the earliest recognition was in 1899 when a European physician found a fossil tooth in a Peking pharmacy, and it was identified as being from an ancient human or some unknown type of ape. This tooth was finally identified in 1935 as coming from a very large animal, an herbivorous hominoid that **G. H. R. von Koenigswald** named *Gigantopithecus*. von Koenigswald then found in a similar shop in Hong Kong an enormous tooth also destined to become medicine to relieve someone's ailment. von Koenigswald was aware of such practices and had even purchased a collection of fossil teeth that had been gathered earlier in the 1900s. Another individual who knew of such practices was **Dr. Davidson Black**, a Canadian physician who had gone in 1919 to the Peking Union Medical College, then being established and funded by the Rockefeller Foundation.

THE DISCOVERY

Black was very much interested in such fossils but was dissuaded by the school authorities from searching for them. In 1921 **John Andersson**, a Swedish geologist, began to dig for fossils at the village of **Zhoukoudian** (Choukoutien), about twenty-five miles from Beijing (Peking). He dug in **Dragon Bone Hill** beside an abandoned limestone quarry from which fossils had been dug for medicinal purposes for hundreds of years. There, in the limestone beds, he found only one fossil tooth of any potential value, but it was a molar from an ape. With the

tooth Andersson also found many pieces of quartz, materials not normally associated with limestone. He speculated that these quartz rocks might have been brought into the area by ancient men and used as tools. Black ultimately examined some of the fossil teeth and was finally able to persuade the Rockefeller Foundation to support excavations at Dragon Bone Hill. These began in 1927 and the diggers quickly encountered solidified cave-fill, materials that had filled ancient caves that honeycombed the hill and then became cemented into rock. One tooth was found in this material and Black decided after extensive examination that it was the molar of a very ancient human. He named it *Sinanthropus pekinensis* (Chinese man of Pekin), though it became more popularly known as **Peking man**.

In 1928 a fossil jaw of this species was found, and in 1929 a skull was found by **W. C. Pei**, a Chinese paleontologist. Black determined that the skull had a brain capacity of about 1,000 cc, slightly larger than Java man and closer in size to modern humans. He also had the good sense to make excellent plaster casts, photographs, measurements, and detailed descriptions of these fossils and others that were found later. Eventually the entire hillside was cut away revealing more than 160 feet of cave deposits arranged in irregular layers, each with important fossil evidence. Cave bears, hyenas and more ancient human skulls were found, plus tools and ashes. All told, these layers recorded more than 200,000 years of almost continuous occupation of these caves by humans beginning about 460,000 years ago and lasting until about 230,000 years ago. After Black's untimely death in 1934, **Dr. Franz Weidenreich**, a noted anatomist, replaced him as director of the excavations and continued the work. Weidenreich was the world expert on the evolutionary changes in the pelvis and foot that had enabled upright posture in hominids and was eminently qualified to conduct this new research. His efforts produced excellent monographs of the fossils that were recovered, and these contributed significant information to our knowledge of these ancient humans.

By 1937, the cave deposits at Dragon Bone Hill had yielded fossils of more than forty individuals, men, women and children, represented by five complete and nine fragmentary skulls (Fig. 9.5), fourteen lower jaws and more

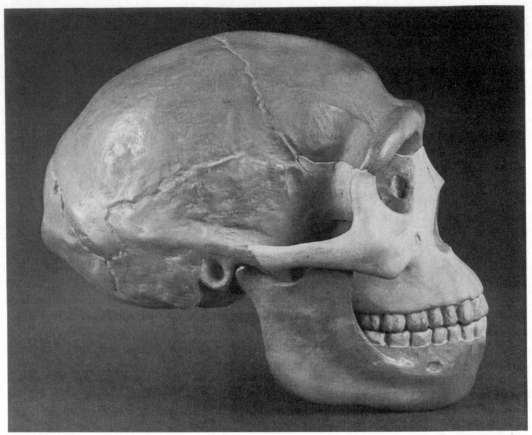

Figure 9.5 Peking man skull.

than 150 teeth. In addition, one cave alone yielded more than 100,000 stone tools and fragments, tools of animal bones and antlers, charred pieces of wood and bone, and baked zones in the clays found beneath the carbonized wood. There is little doubt that these were humans, however primitive. These fossils allowed experts to understand and accept Peking man as a likely ancestor of modern humans.

Weidenreich's research and monographs on Peking man supported Black's contention that this species was clearly a primitive human and not a missing link. Unfortunately, in the same year of 1937, the Japanese Army invaded and took control of northern China. Shortly thereafter, digging for fossils at Dragon Bone Hill ceased, but there was ample fossil material to continue the laboratory analyses.

THE LOSS

In 1941 the political situation in China was deteriorating rapidly and war between Japan and the United States was imminent. Weidenreich took a set of plaster casts of the fossils to New York City to prevent any possible loss of these valuable scientific materials. In August, Dr. W. H. Wong, the director of the Geological Survey of China and the administrator responsible for the fossils of Peking man, attempted to get the American ambassador to accept and remove these fossils to the United States for safekeeping. It was mid-November before the political decisions were finalized and arrangements were made to send the fossils by U.S. Marine Embassy guards to the port city of Chingwantao where they would be taken aboard the liner SS *President Harrison* to the United States. Unfortunately, war broke out just then and the fossils were lost sometime after December 5 when the marine guards left by train for the port and were taken prisoner by the Japanese.

Shapiro provided a detailed account (1974) of the search for these fossils of Peking man after the war's end. Regretfully, they have never been found, though a reward stands for their recovery. The loss of their value to science cannot be estimated. Fortunately, this major scientific catastrophe was lessened somewhat, because Black had the foresight to make excellent plaster casts of the originals. Today, these casts, the photographs, measurements and detailed descriptions are the only evidence of the original specimens of a major segment of human evolution. Though casts are very useful in making comparisons between fossils, the original specimens are much better for determining relationships. All was not a total loss for science, however, because these original specimens of Peking man were not the only ones of this type that have been found.

THE RELATIONSHIPS BETWEEN JAVA Man AND PEKING MAN

Black had concluded in 1929 that Peking man and Java man represented the same type of early human, a view that contradicted Dubois' idea that Java man was an intermediate stage between humans and apes. These interpretations meant that the Far East was inhabited by a wide-ranging group of humans who had spread across eastern Asia as far south as the Malayasian Archipelago. Black noted that Peking man, with its slightly larger brain of 1,000 cc, was slightly more advanced toward modern humans than Java man. That was precisely the situation to be expected in an evolutionary sequence, because Peking man was slightly younger than Java man. Clearly, these different Asian populations of *Homo erectus* would have experienced slight variations due to evolution in different environments and with time. We recognize similar variations today in modern populations; but an understanding of variations in populations due to different environment and time was not often considered by many scientists then studying ancient humans. If these differences in populations of this fossil species could be recognized, what other discoveries about fossil humans were yet to be made?

In 1931, more skull fragments were found near Ngandong along the **Solo River** in Java not far from the

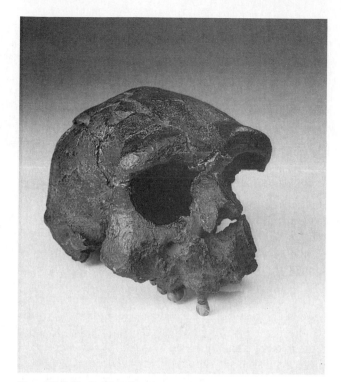

Figure 9.6 Skull from Sangiran.

original Dubois discoveries of Java man. In 1936 and 1937, von Koenigswald found three fragmentary skulls of two adults and one child at **Sangiran** (Fig. 9.6). These ultimately proved to be virtually identical with the original Java man specimens. Ultimately, more than thirty fossils, including skulls and several shinbones, were found at this rich site. von Koenigswald developed a unique method for finding fossils in Java. He offered a trivial monetary reward for each fossil brought to him. Van Bemmelen reported (1960, personal communication) that von Koenigswald bought many fossils in this way. Soon, however, he began to receive numerous broken pieces from children who had realized that they could break the specimens into pieces and get a greater reward. Van Bemmelen said that von Koenigswald did not object because the amount of money was trivial, and he could always glue the fragments back together.

Weidenreich and von Koenigswald brought together their fossils from China and Java for direct comparison in

161

1939 at an historic meeting in China. They decided that the two types were identical and no more different than two races of modern humans. Le Gros Clark, the brilliant anatomist, also agreed with this assessment and proposed that the generic name *Sinanthropus* be dropped in favor of *Pithecanthropus,* which had priority. Accordingly, the two types were then renamed and identified as *Pithecanthropus erectus* and *Pithecanthropus pekinensis*. This plan was acceptable to both scientists but not to Dubois, who objected strenuously. **Ernst Mayr**, the evolutionary taxonomist, then argued that the fossils could not be separated from the genus *Homo* in view of zoological standards. Howells (1966, p. 49) stated of Mayr that

> In his opinion the amount of evolutionary progress that separates *Pithecanthropus* from ourselves is a step that allows the recognition only of a different species.

Accordingly, both types, Java man and Peking man, have been reassigned to the genus *Homo* because of the validity of Mayr's arguments. The differences between these two types and the living species, *H. sapiens*, are not more than what we should expect between two species; thus, Java and Peking men are grouped into a single species, *H. erectus*. The distinctions between the two are, however, sufficiently important to rank them as subspecies. In such a situation, the rules of zoological nomenclature specify that priority is given to the original valid species. Java man is this original species, and it has become *H. erectus erectus*; Peking man is now identified as *H. erectus pekinensis*.

ADDITIONAL FOSSIL EVIDENCE

The widespread distribution of *Homo erectus* throughout the Old World was not fully demonstrated even though many similar fossils were in South Africa, East Africa and North Africa. These specimens along with others found in Java, China, Europe and the Middle East both before and after World War II were not originally identified as *H. erectus* but were also variously identified by different generic and specific names. Most of these specimens could not be otherwise assigned, because they were incomplete and the concept of the species was not well known. As these specimens began to accumulate and the ranges of specimen variability became known, however, these were shown to fit into the defined species and subspecies of *H. erectus*. Today, these additional specimens are considered to represent widely scattered populations of *H. erectus* that vary only slightly from the original specimens.

INDONESIA

Indonesian geologists and paleontologists have found additional fossils of *H. erectus* and have worked out much of the crucial stratigraphy and associated faunas from the original discovery site. These studies show that fossils at three principal sites, Trinl, Sangiran and Solo, are found in three distinct layers, the **Djetis**, **Trinl**, and **Ngandong Beds** in ascending order. The Djetis Beds are the oldest and are dated at about 1.3 million years. They contain a fauna that is related to that of Bed II at Olduvai Gorge in Africa in which *H. habilis* has been found. One skull found in Djetis strata has a relatively small brain for *H. erectus*. The overlying Trinl Beds contained the original Java man specimens and are dated at about 600,000 years, much younger than the Djetis Beds. The overlying Ngandong Beds are unreliably dated and may be of middle Pleistocene age, perhaps somewhat older than 125,000 years. Specimens from the Ngandong strata at the Solo locality have a larger brain size than the Djetis and Trinl specimens. This trend in increasing brain sizes suggests, to some extent, that *H. erectus* evolved very slowly from *H. habilis*.

ZHOUKOUDIAN

The Zhoukoudian cave site (Zhoukoudian Locality I) was eventually reopened (Rukang and Shenglong, 1983) after World War II by the Chinese government, and additional fossils of *H. erectus* have been found there. More than 120 different scientists from a broad range of disciplines were employed to recover all possible information from the geology and the fossils at the cave. They reconstructed the cave's geologic history, from the first solution of the limestone bedrock by percolating groundwaters to its final end when the roof collapsed and buried the debris of human occupation and other materials that had washed into the cave. Excavations of these cave fillings have provided a detailed picture of the occupation and culture of the *H. erectus* inhabitants. We now know when these people inhabited the cave, how they lived, what they ate, the nature of their tools and much more.

The fossils were recovered from more than fifteen places in the cave and comprise an impressive population. They total (Rukang and Shenglong, 1983, p. 89)

> ... six complete or almost complete skulls and 12 other skull fragments, 15 pieces of man-dibles, 157 teeth, three fragments of humerus (the bone of the upper arm), one clavicle (collarbone), seven fragments of femur (the thighbone), one fragment of tibia (shinbone) and one lunate bone (a wristbone shaped like a half moon).

From this evidence a good physical reconstruction of *H. erectus pekinensis* was made. Peking man was much like modern humans skeletally but with limb bones that had thicker walls and smaller marrow cavities. Their skulls are more elongated and have a pronounced angular rear portion and a much lower vault (see Fig. 9.5). The skull also has thicker bones, protruding brows that are relatively flat and an obvious narrowing behind the brow ridge. The teeth are larger than those of modern humans and have a **cingulum**. This feature is a ridge or collar of enamel around the crown that is an ancient trait common in apes but relatively rare in humans. The mandible of *Homo erectus* lacks both a chin and a simian shelf, but is very thick to provide the necessary strength to cope with their rough diet. Coon (1962, p. 349) noted that this simian shelf also took up space that was needed for the tongue, and that as the mandible in hominids became smaller, the simian shelf shifted to an external position where it evolved into the chin. He added,

> The absence of a chin in *Homo erectus* does not tell us whether or not he could speak, or, if he could, when he began to do so.

Peking man's elongated skull accommodated a brain slightly larger than that of Java man, averaging about 1,050 cc vs. the 900 cc found in the older variety *H. erectus erectus*. This slight difference of only about 150 cc is not particularly significant in total volume but it may well explain many of Peking man's accomplishments that advanced them beyond their earlier cousins and marked them clearly as humans.

The hominid remains found in the Zhoukoudian caves delineate a primitive people that lived in groups and, presumably, had a relatively simple social system. They were likely the first true cave dwellers and went out from these shelters to hunt meats and gather plants, often bringing them back to the cave to eat. The climate of the region in their day was temperate, probably similar to the climate in this area of modern China. The vegetation of the past was also similar to the vegetation of today. Caves were obviously desirable as shelters during the cold winters, and four very thick layers of ashes indicate that Peking man used fire, a welcome addition to life in a cold, damp cave. The evidence does not, however, indicate whether *H. erectus* made the fire or merely brought it into the cave from some natural fire, perhaps one caused by lightning. One ash layer, more than six meters deep in places, demonstrates their lengthy dependence on this important discovery or invention.

The thick piles of ash strongly suggest that fires were concentrated and controlled in specific places, probably where they provided the most warmth or were best for cooking. Charred bones and charred seeds found in the ash suggest a crude form of cooking, roasting or barbecuing, the same methods we still use and enjoy today. This is not a sophisticated cooking technique and could easily have been discovered unintentionally. One of these people merely had drop a piece of meat by accident into the fire, then retrieve it and resume eating. Eating such burned or roasted meat would have allowed them to discover that this meat was tastier and easier to chew than raw meat, a pleasant and desirable activity even today. In the absence of definitive information from elsewhere, therefore, Pekin man may be regarded as having invented cooking.

AFRICA

Africa is widely regarded as the birthplace of humanity and where the lineage can be traced back to the hominoids. Africa also has provided all the evidence of the ancestral australopithecines and the oldest specimens of *Homo erectus*. In South Africa at Swartkrans, jaw and facial fragments of *H. erectus* have been found in beds that also contain *Australopithecus robustus*. These finds are dated to at least 1.8 million years ago, though some workers believe them to be several hundred thousand years younger, perhaps only 1.2 million years old. A variety of specimens of *Homo erectus* have also been found at Olduvai Gorge. These include: a skullcap found in Upper Bed II; some fragmentary materials from Beds III and IV, including part of a lower jaw; and a partial pelvis from Bed IV. Some of these exhibit remarkable similarities to the Zhoukoudian specimens. Stone tools ascribed to them and dated at about 1.5 million years offer further evidence of *H. erectus* at Olduvai. At Lake Turkana in East Africa an important *H. erectus* skeleton (KNM-WT 15000) was found. It was of a youth about twelve years old and 5'4" tall (Fig. 9.7). It is described as the most complete early hominid skeleton ever found and is not equaled in the fossil record by Lucy or others until Neandertal burials one million years younger. Brown, et al., (1985, p. 792) emphasize

> . . . it is also the first fossil hominoid, let alone hominid, in which brain and body size can be measured accurately on the same individual. These two crucial variables, on which so much speculation about human origins and behaviour has been based, can now be determined for at least one individual early hominid.

Day (1986, p. 239) supports this view and notes

> It can be expected that further study will contribute significantly to the debate that currently surrounds the status of the taxon *Homo erectus*; the relationships between its African and Asian examples, its ancestral relationship (or not) to modern *Homo sapiens*. . .

At East Turkana, two skulls of *H. erectus*, KNM-ER 3733 (Fig. 9.8) and 3883 are similar to the skulls found at Zhoukoudian, China. They are, however, dated to more than 1.5 million years, making them about one million years older than the Zhoukoudian remains. As such, they are strong evidence that *H. erectus* originated in Africa and spread to the rest of the world. Less complete remains of *H. erectus* are also known from North Africa in the Atlas Mountains. One site at Ternefine, Algeria contained three jaws, a parietal skullbone and numerous stone tools. Fragments of lower jaws have been found also in Morocco at Rabat and Sidi Abderrahman. All of these resemble the specimens from China.

EUROPE AND ARCHAIC SAPIENS

The European specimens of *H. erectus* are sparse and less clearly defined than many other specimens. The first specimen was a jaw found at Mauer, Germany in association with a rich fossil fauna. The associated fossils date the *H. erectus* specimen at about 500,000 years ago. The jaw was turned over to geologists at Heidelberg University and is popularly known today as **Heidelberg man**. It is

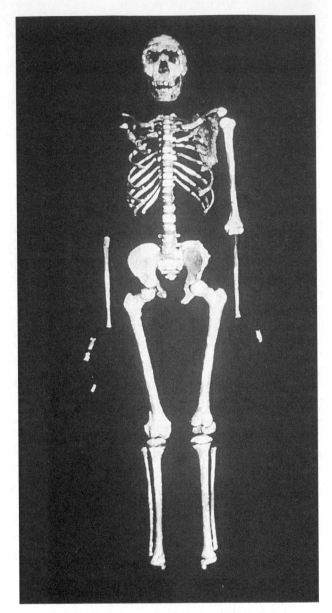

Figure 9.7 *Homo erectus* skeleton, KNM-WT 15000.

considered by some to be *H. erectus* but the identification is clouded by some disagreement; no firm conclusions can be drawn because the remainder of the skull is missing. Scattered stone tool chips at two other localities, Le Vallonnet in France and Stranská Skála in Czechoslovakia, suggest the presence of *H. erectus*, but these are totally inconclusive except for their ages which are 1 to 2 million

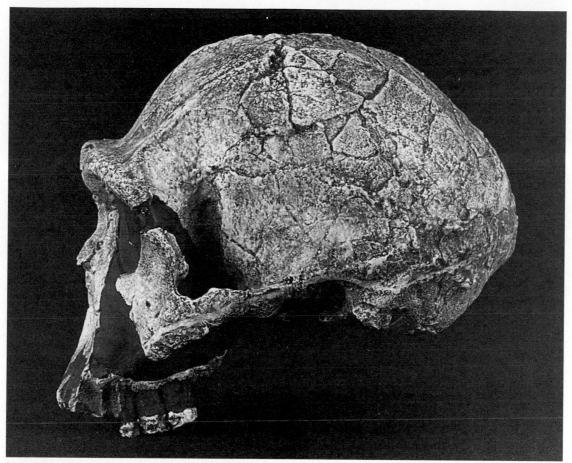

Figure 9.8 *Homo erectus* skull, KNM-ER 3733.

years ago in France and 700,000 years ago in Czechoslovakia. Similar unclear situations exist for incomplete fragments of several skulls found at Bilzingsleben, Germany, and a nearly complete skull from Petralona, Greece. Both sets of these specimens have affinities with both *H. erectus* and *H. sapiens*, especially the Petralona skull. These characters and their ages could indicate very late *H. erectus* or even very early *H. sapiens*. If these are very early *H. sapiens*, they would demonstrate that the modernization of humans was already beginning to take place many hundreds of thousands of years ago. Further, these specimens would also necessitate redefinition of the nature and taxonomic position of later *H. sapiens* and how modern humans evolved. Because of such difficulties, all of

these specimens are commonly identified as **archaic sapiens**, a category that refers to the very old and more primitive examples of our own species, *H. sapiens*.

The recent discovery (Dorozynski, 1993) of a complete skeleton from a cave near Altamura, Italy offers some clues to archaic sapiens development. The skeleton, still in place, is partly covered with cave dripstone deposits and is not totally visible. Features of its face are visible, and these suggest a stage of development between *H. erectus* and Neandertals. The skull has a prominent supraorbital torus, a Neandertal feature, but other aspects of the face suggest it has not yet evolved to the Neandertal stage of development. These aspects of the Neandertal and pre-Neandertal stages suggest that the specimen may provide

Figure 9.9 Acheulian tools.

important information about the transition from *H. erectus* to archaic sapiens, but it has not yet been dated.

There is one well-recognized *H. erectus* site in Europe at Isernia in central Italy. It is dated at about 730,000 years ago. Thousands of stone tools have been found at this site which bordered a lake. Fossils recovered from here are of butchered animals and include both elephant and bison. These animals indicate that the climate was wet and cold, more a steppe than forest. The stone tools are simple choppers and flint flakes that included scrapers and retouched points; no hand axes have been found. Remarkably, the general appearance of these tools is Oldowan industry rather than the more advanced Acheulian industry typically associated with *H. erectus*. This pattern suggests that, although the people at this site in Europe were the newer, smarter type, they had not yet learned to make

Acheulian style tools (Fig. 9.9). One explanation for this anomaly is that this site in Italy was far removed from contemporary African centers of innovation where Acheulian tools were being developed. Though these people in Italy had evolved into a new species from *H. habilis*, they continued the tradition of their ancestors and made the only tools with which they were familiar, the simpler Oldowan tools. Communication of the new advances in tool manufacture had evidently not yet reached them from centers of development. This and similar evidence in Spain leads to the tentative conclusion that the first humans in Europe were *H. erectus* who began their migration from Africa before Acheulian tools had come into general use. Their basic tools were Oldowan tools and they carried this manufacturing style with them. But, their migration took so long that advances in tools in Africa had already progressed to the Acheulian stage by the time the site at Isernia was in use. Subsequently, Acheulian tools made their way to Europe in a later wave. A simpler hypothesis proposes that this difference in tool manufacturing was due to the lack of availability of materials from which more advanced tools could be made, though this seems difficult to justify given the rock materials available throughout Italy.

All of these younger specimens of *H. erectus* present a dilemma for taxonomists in making species assignments. The transition to *H. sapiens* had to come at some point in time, but when and how? Did *H. sapiens* indeed evolve from *H. erectus*? Was the transition abrupt or was it gradual? Over how many years did it take place? Did some features of the older humans change more quickly than others and, if so, which features?

Debate swirls around *H. erectus* today: questions raised about its morphology, cranial and dental aspects, and more have led many workers to accept it as a "true" taxon even as others are objecting to that position. Day (1986, p. 409) asks "Does *Homo erectus* exist as a true taxon or should it be sunk into *Homo sapiens*?" He adds (p. 411)

> It is clear, therefore, that the taxon *Homo erectus* is under intense debate in terms of its geographic range, its temporal range, its origin and its evolutionary fate.

BEHAVIOR
DIET

The fossilized bones and the stone tools found at *H. erectus* sites offer important information about their lifestyles. In addition, the evidence of different foods also found at sites provides us with eloquent testimony of their varied diet. Plant pollen from the layers in cave floors reveals that walnuts, hazelnuts, pine, elm and rambler rose were present. These are all plants that have edible berries and seeds and were likely to have been included in their food supply. Charred hackberry seeds suggest that they may have even roasted at least some of these nutritious materials. Such fossilized plant materials also demonstrate the considerable abilities and success of these people in gathering a wide range of simple and nutritious foods which were brought back to the cave where they were eaten and, in all probability, were shared.

Vegetable or plant matter could not have been totally satisfying to these humans because only certain types of plants could be eaten. Also, many of these materials could only have been obtained seasonally. The most abundant materials eaten by many animals, grasses and leaves, were available as food over much of the year in many regions, but these may have been indigestible by these early humans. Fortunately, they were able to obtain quantities of energy and nutrients from the meat of animals. Typically, meat will yield as much as 550 calories per 100 grams, about five times as much as the equivalent weight of most plant foods available to *H. erectus*. Thus, the practice of eating meat, probably begun as a rare event in the hominoids, is likely to have continued and become increasingly more common in the hominids. In *H. erectus* meat eating could easily have been a common practice. There were additional advantages to eating meat. Animal meat was available year-round to hunters, an important benefit in environments that had seasonal changes in which vegetation was available only in certain seasons. Meat is also much tastier than many plant materials. Meat eating, thus, could have been a logical choice for our *H. erectus* ancestors because it was available year-round, provided greater caloric benefit, tended to be filling, and was tastier.

There is considerable evidence that *H. erectus* used fires. Charred bones have been found recently in Swartkrans Cave, South Africa These are mostly from antelopes, but baboons, warthogs and zebras are also represented, as well as an australopithecine finger bone. There is, however, no way to determine whether these fragments were scraps left from meals or simply trash that was disposed in the fire. Older deposits in the same cave did not contain evidence of burned bones, so these may record the first use of fire at this site. Brain has suggested that these fossil bone fragments could be 1.5 million years old, but others believe them to be slightly younger, perhaps only 1.2 million years old. Another site in Kenya suggests the fire was being used there less than 1.4 million years ago. Some experts, however, are hesitant at accepting this site date. Other, younger sites have provided additional evidence of the use of fires. Thick layers of ash along with charred seeds and bones have been found at Zhoukoudian in China. These deposits are considerably younger than the Swartkrans fire ash and may only be about 500,000 years old or younger. Charcoal was also found with the elephant remains recovered at Torralba in Spain, though this material could have been the result of grass fires unrelated to humans; these fossils are dated at about 400,000 years ago.

Figure 9.10 The kill site of Olorgesailie.

SHARING BEHAVIOR AND HUNTING

Sharing was a behavior commonly used by *H. erectus* to improve their lives. The great number of animal bones found at Zhoukoudian argue emphatically for this view. Some ninety-six different types of mammals, both small and large, have been found in the Zhoukoudian cave layers and demonstrate that these people probably obtained meat from a great many different types of animals. These remains also indicate that Peking man was a hunter competent enough to kill many different animals, including some much larger than himself. Deer appear to have been their favorite prey. The remains of about three thousand deer of two different types have been found at the caves. They were much sought after game animals because they were available, more easily hunted than other species, or were particularly tasty. Hunting of such swift animals like deer or large animals is not usually practical for an individual hunter. These animals are too wary or too strong to kill easily. But the same animals can be taken relatively easily by several individuals hunting together. The American Indians, for example, were able to regularly kill large numbers of buffaloes for their needs long before they had horses. The Indians simply worked together and used fire or noisy commotion to stampede these herd animals over cliffs where they fell to their deaths. Then it was a simple matter to gather the meat in large quantities from the dead and dying animals.

OLORGESAILIE. Ample evidence of the hunting prowess of *H. erectus* is seen at sites in Africa and Europe where fossils of these people are directly associated with numerous examples of Acheulian tools and bones of various animals. One of these is at **Olorgesailie** (Fig. 9.10) in Kenya, about forty miles from Nairobi. There, in 1940 at a site that may be as much as 650,000 years old that flanked an ancient lakeshore, Louis and Mary Leakey found hand axes weathering from the ground. Isaac later excavated site DE/89 from that area and found more than four hundred hand axes in an area of only about three hundred square meters. These stone hand axes were associated with the bones and teeth of many animals, particularly a large extinct species of baboon, *Theropithecus oswaldi*, represented by at least fifty adult and thirteen juveniles; this baboon was about the size of a female gorilla and weighed about 250 pounds.

The animal remains and the numerous tools were not found together in casual association. The stone weapons could only have been obtained from rock outcrops at least twenty miles away. They had to have been brought to this site with deliberation. Further, the great number of these stone tools together weighed more than two thousand pounds and would have required many trips by a number of individuals who were capable of carrying only a few with each trip.

The Olorgesailie evidence indicates that a troop of baboons was hunted and killed by *H. erectus* using stone cobbles and the hand axes. Because baboons are alert and fierce fighters, they are too dangerous to attack directly. Presumably the attack took place at night when these large animals were asleep in trees, their normal sleeping refuge, or the hunters must have quietly surrounded these trees and then began the attack by hurling cobbles at the baboons, probably at dawn. When the baboons descended to the ground to defend themselves, they were beaten with cobbles and hand axes until at least sixty-three of them were dead and dying; others may have escaped. This was followed presumably by a feast on the baboons.

From this scenario we can deduce several interpretations about the abilities of these hunters that must command our respect. The transport of these great numbers of stone weapons from distant sources is very strenuous. To carry this sheer quantity and weight of cobbles this great distance by hand would have required a significant effort by these people. Probably numerous females and juveniles as well as adult males were involved in transporting these weapons. It is unlikely that these weapons were carried in their hands because of obvious limitations. Unless we postulate that each individual made a great many individual trips some twenty miles each way, and this is unlikely, we must consider some other model. It is entirely possible, indeed probable, that they carried these stone weapons in baskets, bags or slings made of animal hides or woven grasses. In this fashion they each would have been able carry a reasonable number of stone weapons per trip. Still, they each would have had to make many trips to accumulate the two thousand pounds of stone cobbles that have been unearthed.

The attack was undoubtedly launched during the night when the baboons were asleep and surprise was possible or when day was breaking. A daylight attack by a single hunter or by numerous individuals would have easily alerted the baboons and they could have counter-attacked or quickly fled the area. Thus, a daylight attack would have defeated the purpose of the assault and negated the massive effort involved in supplying the stone weapons. The hunters must have surrounded the trees during the night and waited until the first light of dawn when there was enough light for the attackers to see their targets. This would also have caught the baboons before they were awake. Obviously, considering the strength and fierceness of the baboons, as Campbell (1976, p. 252) points out,

> . . . there was nothing casual about the hunt. It must also have taken a great deal of courage to bring off, since if it did occur at night, nocturnal predators would have been on the prowl, as eager perhaps to eat the somewhat vulnerable hunters as these were to make a meal of the baboons. By the time of the Olorgesailie baboon ambush, human beings already had become very skilled hunters.

Equally important was the cooperation that this organized attack demonstrates. It is one thing to form a group of hunters to attack a solitary baboon but entirely different for them to attack an entire troop of these very large and dangerous animals. These early hunters must have planned their strategy carefully and have implemented it by arming themselves with adequate supplies of stone weapons. Numerous individuals were necessary to bring the stone cobbles from the distant outcrops. Then the hunters had to select the target troop and learn where they found refuge at night. The weapons then had to be brought into place, perhaps during the daytime when the baboons would be alerted but not aware of their impending doom. This must have taken several trips or required numerous carriers. Finally, the hunters had to plan and organize the actual attack. All of these activities required advance planning and coordination. Not only was planning and coordination essential to success, the entire plan had to be communicated to all the participants so that they would know and understand their roles and carry them out. All this demonstrated a particularly impressive attack, because it succeeded against a formidable foe. We do not know if these *H. erectus* people could talk using some type of language. Regardless, they coordinated the attack by some manner of communication.

TORRALBA-AMBRONA. A second hunting locality attributed to *H. erectus* consists of two adjacent sites, **Ambrona** and **Torralba** in the Ambrona River valley in Spain. These were discovered in 1888 when workers digging a trench accidentally uncovered huge bones of an extinct elephant with straight tusks. A Spanish archeologist, Cerralbo, excavated at Torralba in 1907–11 and found some twenty-five elephants, stone tools, and worked wood. In 1960, **F. Clark Howell** reopened the site, dated at about 400,000 years ago and found more than fifty additional elephants there and at Ambrona. The evidence eloquently identifies the events and the hunting procedures used by *H. erectus*. Hunting elephants is not an easy endeavor for any people, ancient or modern, because of the elephant's size and strength. The scale of the Torralba-Ambrona hunts must have been great as indicated by the size and numbers of elephants recovered from the excavations. In fact, these hunts were probably too large for only one band of these people. It may be that several bands united to successfully carry out this hunt. Certainly, an elephant would have produced enough meat for more than one band, and several elephants would have supplied them with food for a considerable period of time.

This valley was probably a favored travel route used by elephants and other animals as they regularly traveled the region. The hunters knew this and could easily have developed a plan to attack them. The stratigraphic evidence indicates that some of the ground was swampy, especially where many of the bones were found. Apparently the plan was to attack the elephants after they had become mired by their great weight. Pieces of charcoal have been interpreted by some experts as the traces of campfires or grass fires, perhaps both. The hunters could have stampeded the elephants into swampy ground by setting fire to the grass. There, the elephants would have become entrapped by their great weight and size in the soft ground, eventually becoming exhausted, and could become easy prey to the hunters using stone and wooden weapons. These hunters may even have waited until the trapped animals weakened and died. Horses, deer, aurochs and smaller elephants were also caught in this ambush or subsequent ones and were also killed with stone weapons, spears and sharpened tusks. Within a short time and with relatively little physical effort and danger, the hunters had more meat than they could possibly eat. This hunting procedure was used at Ambrona on at least ten different occasions.

The bones, tools and other debris dug from these sites also record some of the procedures in butchering and gathering the meat. The larger bones show where the animals lay after they were killed. The hunters apparently hacked quantities of meat from the great bodies and may have eaten some of it on the spot. Numerous piles of charcoal are associated with the bones and presumably represent campfires for cooking, warmth or light. These suggest that at least some of the meat could have been roasted on the spot; it is likely that, given the long time required to butcher several elephants, that they dined on some of the choicer parts immediately. Quantities of smaller bones and bone fragments are concentrated in small areas and testify that these hunters brought large pieces of freshly butchered meat from the bodies to carving places where they were probably cut up into smaller pieces suitable for carrying. These smaller fragments, thus, may indicate that the major portion of the meat was not eaten on the spot but was taken away.

Surprisingly, only one elephant skull has been found at Ambrona; its braincase is broken open, probably to get at the brain which also may have been eaten. A rather enigmatic collection of five long bones and a tusk are laid out in a straight line for some unknown purpose. Originally these were interpreted as providing a walkway across boggy ground, but this idea has been questioned on geological grounds.

TERRA AMATA. Another ancient campsite of *H. erectus* was discovered in 1965 during excavations for an apartment at Nice, France. This site, known as **Terra Amata**, is located at the mouth of the Paillon Valley. It is dated at about 300,000–400,000 years old and contains strata more than seventy feet thick. These layers record the presence of *H. erectus* in at least twenty-one separate levels that were deposited during thousands of years of time. More than 35,000 objects have been recovered, including fossil bones, tools and imprints, even an adult footprint. The site has been interpreted as a seaside beach or dune area that was subjected to drifting sands. These sands tended to quickly bury and preserve any traces of the human activities.

Apparently *H. erectus* people occupied this site only part of the year, then moved elsewhere. Of particular importance is the clear evidence of twenty-one shelters apparently constructed in three places by the hunting bands during their brief stays here, probably in a span of no more than one hundred years. These were built on a sandbar along the seashore, on a beach, and on the dune belt. Each location appears to represent a series of camps that were used consecutively over the years, perhaps by the same band of hunters.

The shelters (Fig. 9.11) in the dune belt are represented by oval rings of stones found in place. These are interpreted as the foundation outlines of crude huts that were from 26–49 feet long and 13–20 feet wide. Presumably the stones served to anchor or brace walls of sticks or brush. One hut measured 20 by 40 feet and had large circular holes in the floor, possibly to erect posts that supported the roof. The center of the floor of this hut has a baked area of discolored earth, indicating that it held a hearth, a feature also found in other huts. This hearth is partially surrounded by a ring of stacked cobbles that may have served as a windbreak, suggesting that these huts were drafty. Stone core, tools and chips were found to one side of the hearth where they

Figure 9.11 Reconstructed hut from Terra Amata.

were concentrated around a large transported stone that probably served as an anvil; this area is believed to have been a toolmakers area. Another slab of stone in another hut has prominent scratches that have been interpreted as cutting marks produced when meats were sliced with sharp stones; likely it was a butcher's block. There is even evidence of a primitive latrine area.

This series of camps indicates that these were well-experienced and capable hunter-gathering peoples. They wandered through the countryside looking for game, edible plants and drinking water. As the local game became depleted, these people moved elsewhere to find more productive hunting areas. The evidence demonstrates that at Terra Amata they had an ideal site for a hunting camp. There was fresh water in the springs that emerged from the limestone hillsides, fish and shellfish were available in the river and sea nearby, and a great variety of plants grew in the valley and on the hillsides. They dined on turtles, birds, rabbits, rodents, red deer, elephant, mountain goat, boar, rhinoceros, oxen, oysters, mussels, limpets and fish. This variety of foods easily rivals that dined on by the modern pleasure seekers that congregate in Nice today. Pollen in their fossilized feces indicates they ate spring flowers; they undoubtedly also ate other plant materials that did not carry pollen, particularly tubers, roots and leaves. Of great significance is the size of some of the animals they hunted

for meat. Rhinos, elephants and boars are dangerous animals and hunting them could have resulted in injuries and death. Remains of these animals could have been scavenged, but these were capable hunters.

The earlier series of camp huts on the sandbar and beach were inhabited by people who made larger fires and produced less sophisticated tools than those made by the inhabitants of the later sites in the dune area. The tools of the earlier people were simple pebble tools, rough bifacials, scrapers, cleavers and projectile points. The tools of the dune campers were more advanced types that had been made by flaking cores to produce chips that also could be flaked. Some of their materials came from a locality thirty miles away. The dune people also made tools and weapons from bones; a number have been found with sharpened points that are fire hardened.

Three other pieces of evidence allow us to understand some of their culture. At one site, the soft sands have preserved the impressions of animal skins or hides. These may represent skins that were worn as clothing, they may have simply been used to sit on, or they may represent a cover of some type, perhaps a blanket. Worn pieces of red ocher (weathered iron ore) suggest that these people may have made red markings on themselves or other objects. These marks could have been made for ceremonial purposes, or perhaps they marked themselves as protection against sunburn. A final bit of evidence is provided by a rounded impression in the sand that has been interpreted as a bowl. It could have been used to collect plant foods or to serve as a cooking vessel or storage container for water or various other items.

All of these evidences support the view that these individuals had achieved a simple but effective culture that provided them with some comforts. Their level of culture could be compared favorably with that of primitive hunter-gatherers still living among modern humans.

COOPERATIVE BEHAVIOR

Cooperative hunting behavior as practiced by *H. erectus* implies a relatively high level of social organization. These people probably lived in groups or small bands of perhaps thirty-five or fewer individuals; this is a convenient maximum that can be fed by their style of hunting-gathering. Their hunting had probably evolved into several different methods which are still used effectively by modern hunting peoples. A principal method is **ambush**, in which the hunter hides from unsuspecting animals and, when they come near, attacks suddenly before they can flee. A second method is termed **persistence**. This is the continued chase of a quarry until it is too weak to continue flight and becomes an easy victim. A third method is **stalking**, the stealthy tracking of a cautious animal and sudden attack from close distance. All of these methods can be accomplished by the individual hunter or by those hunting in groups. Cooperative hunting, however, differs significantly from individual hunting in that it requires the development of at least three additional behavior patterns in the individuals involved: an organization consisting of a leader and a willing group of followers, some type of communication between the hunters, and the intelligence to organize the individual into a concerted plan of action.

Any organization must have a **leader** who plans a cooperative hunt and coordinates or directs it. The leader must determine the local situation, identify the animals to be hunted, decide on a plan likely to succeed with these animals, assign specific tasks to each individual and supervise to ensure that each task is carried out. Leadership and planning also imply that the members of the group are willing to become **followers**, at least to some degree. They must be willing to accept instructions and to carry them out. Good followers would be willing to participate in such group hunts, because they expect to share in the rewards, the meats, furs and other useful parts of the hunted animals. This share must also be larger than can ordinarily be obtained by solitary hunting, or it must be easier to obtain, or it must involve less effort and risk. Without benefits, cooperative hunting offers little for an individual whose efforts are otherwise exploited.

The natural development of a leader-follower organization required some form of sophisticated communication within the group. There is no way of knowing whether this communication was vocal, by visual signals or by a combination of these. **Vocal signals** are likely to have been in common use, if only as grunts to indicate pleasure or displeasure. Evidence for a **language** in *H. erectus* does not exist though they must have communicated at some level approaching primitive modern peoples. The anatomical structures indicating an ability to talk simply do not lend themselves to preservation, and none have been found. **Sign language** may have been their only effective method of sophisticated communication. Such signing has commonly been used between groups of modern humans who do not share a common language. This method would have been effective enough for their needs.

These ideas suggest that cooperative hunting behaviors would be advantageous to these humans and likely to be reinforced through feedback mechanisms. Once such hunting began and was successful, it would have become more common as individuals recognized and desired the advantages. Small game was always relatively plentiful and available to individual hunters, but solitary hunters were not likely to kill large game. Therefore, there was a limit to the hunting productivity of an individual. Two or more individuals hunting together, however, increased the chances of success for each with less expenditure of time and energy. Larger game animals could be hunted and killed to provide greater quantities of meat to share. With successful hunts, the individuals would have been more willing to cooperate, because they knew that they would share in a greater quantity of meat. The killing of large animals meant that there would be abundant meat for the entire band for several days. Success encouraged increased cooperation. Such organized leader-follower relationships and interactions would have meant a significant advance for survival in those groups that could develop and accept such roles. But more than simply surviving, such groups would have thrived and advanced, because the cooperation learned in hunting could have been transferred to other endeavors with little effort, or vice versa.

INTELLIGENCE

Intelligence and the ability to recognize situations is not unique to humans. All animals experience hunger, fatigue, sexual drive, parental instincts and much more to some degree; they also learn many behaviors if they are to survive, especially how to recognize danger. There is no reason to believe that these are understood in context, except in the higher primates. The type of reactions by animals to particular situations can mean survival or death; animals either survive or they don't. The animals better fit or suited to survive are those that had better behavior "reactions" or patterns and adaptations that they received genetically and by training from their parents; the animals with poorer behaviors and adaptations were less likely to survive. The more intelligent animals tended to develop better behaviors and better adaptations and survived more than less intelligent ones.

Better hunting patterns in the hominids were undoubtedly developed by the more intelligent individuals who were able to successfully meet particular challenges. The smarter individuals who planned and led better hunts were more likely to become the group leaders, and any group with smarter leaders would tend to develop advantages over other groups with less smart leaders. In this manner, the cooperative hunting that required intelligence and leadership was a direct result of particular leadership qualities that appeared through ordinary variation in the different populations of *H. erectus*. Leadership became an advantage to be exploited, and intelligence became a desirable asset related to leadership in these hunting peoples. Working through feedback mechanisms, the various aspects of cooperation, intelligence and leadership qualities tended to be self-perpetuating and self-increasing. With a survival premium attached to these aspects, populations of *H. erectus* undoubtedly recognized the advantages of greater intelligence, and this enabled them to advance further toward modern humans.

TOOLS

Homo erectus has been called a migratory species because they were active hunters who constantly wandered in search of foods, especially following the migrating herd

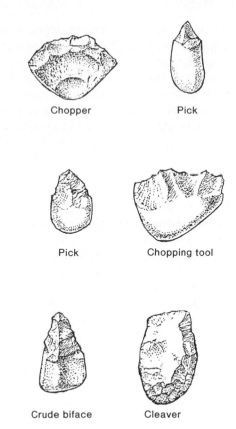

Chopper   Pick

Pick   Chopping tool

Crude biface   Cleaver

Figure 9.12   Acheulian tools from Terra Amata. From "A Paleolithic Camp at Nice" by Henry de Lumley. Copyright © 1969 by Scientific American, Inc. All rights reserved.

animals upon which they fed. As these people increased in numbers, so did their demand for food; plant materials simply could not be collected in sufficient quantity over the entire year to provide enough foodstuffs. On the other hand, one deer could provide much more energy than equivalent amounts of plant foods, and meat satisfies hunger for longer periods of time. To accommodate their appetites, needs and desires, these people probably hunted more and improved upon techniques invented by their ancestors, *H. habilis*. They also improved upon the tools of the Oldowan and Developed Oldowan industries that had been in use for about 1.5 million years. The improved tool that they produced, the Acheulian tool, came into use about 700,000 years ago, even as Developed Oldowan tools continued to be used.

Acheulian tools (Fig. 9.12) are typically bifacial hand axes with teardrop shapes and are formed by flaking along the entire margin. They are commonly composed of fine-grained materials, flint, chert, obsidian, quartzite, etc.; these have no preferred directions of cleavage but fracture along curved surfaces and develop razor sharp edges. The usual Oldowan tool was a chopper made by simply striking one cobble with another to create a rough but serviceable working edge. The Developed Oldowan tool was essentially the same but required more blows that removed additional large chips to produce the semblance of a specialized edge. The fact that it took hundreds of thousands of years to learn to modify these tools is an indication that their intelligence by modern standards was not significant. Yet despite this minimal level of achievement, they were far advanced beyond hominoids. Acheulian tools were the product of an even more advanced culture. They were produced from generally rounded or oval cores by **knapping** or striking purposely with another stone using a delicate touch along the edges that removes small flakes. This technique produced a flat-shaped stone that could produce a handy cutting edge that contrasted with the simple rough chopping-edge tools made by earlier cultures. There was an additional advantage in using this manufacturing technique the knapping produced razor sharp flakes that could be employed as handy small knives.

In time, *H. erectus* learned to use different types of hammers for shaping such tools. Initially these instruments were simple hammer stones that could remove large, deep flakes from the cores and create irregular edges. Later they learned to use bone and wood hammers that were softer; these offered better control for knapping the flakes and edges. These new hammers also produced straighter, sharper edges on both the core and the flakes, making both more useful. The greater degree of control they developed in the knapping process allowed them to make a variety of tools in addition to the better hand axes. They made cleavers, scrapers, chisels, awls, anvils, spheroids and hammers. The scrapers were designed to remove thin shavings or scrapings and suggest they had begun to learn how to prepare animal hides, a significant cultural and social advance. Such hides could have been used for a great variety of purposes—clothing, blankets, containers, and shelters—all extremely useful to any people, particularly those trying to survive in the rigors of an extremely primitive culture.

Rukang and Shenglong point out (1983, p. 92) that the tools of *H. erectus pekinensis* were advanced in two ways over the tools made by the earlier *H. erectus erectus*. First,

H. erectus pekinensis made their tools from a wider variety of rock materials, although most were of a quartz composition. Their raw materials included rock fragments that were not stream-rounded cobbles but were broken pieces obtained from outcrops of rock that contained veins of quartz and quartz crystals. Second, the tools were generally flakes made in three ways: **anvil percussion**, **direct percussion**, and **bipolar percussion**. In the anvil method, a piece of rock was struck against a larger piece of rock (anvil) until flakes were chipped off the original piece. These pieces could be used as formed, or they could be improved by reworking using another stone to strike flakes. In the direct percussion method, a **hammer stone** was used to chip flakes directly from a **core** stone held in the hand. In bipolar percussion, a piece or core of rock, often vein quartz or a related material, was held vertically on a large flat anvil on the ground. This core was then struck from above by a hammer stone and flakes were chipped off both ends, thus producing a bipolar tool.

Such improvements in tools made by *H. erectus* were not uniformly advanced throughout their Old World populations. In some regions very simple stone tools that began in use about 1.5 million years ago continued in use by early types of *Homo sapiens* until perhaps 100,000 or less years ago. In other regions the manufacturing techniques for making Acheulian tools quickly spread through the population until they became common. With such a base of manufacturing skills, these techniques were easily passed on to successive generations. We are simply the most recent beneficiaries of these advancements in toolmaking techniques. Today we make tools that in turn are used to make other tools that are used to make machines that can do wondrous things, not the least of which is to travel in space or to write a book using a computer.

The continued use of simple stone tools by some local human populations even as other populations were using more improved tools strongly suggests that contact and exchange of ideas was not common or uniform throughout the entire population of *H. erectus* or its history. Some populations were undoubtedly isolated, either physically, socially or culturally, and simply did not or would not receive the news of the improved and better methods for making advanced tools. Whether this was by choice, design or by accident is unknown, but many simple peoples have lived successfully into this century using such Stone Age technology.

SUMMARY

Though Darwin provided the idea that humans had probably evolved from other species and the evidence to support it, Haeckel was the one who impelled the search for the origin of humans. He conceived of a missing link, *Pithecanthropus alalus*, as the intermediate form between humans and apes, specifically the gibbons, and believed that this form could be found on the island of Java. This was so appealing in the height of the origins controversy that it caught the attention of Eugene Dubois, a Dutch physician who announced in 1886 his determination to find this missing link. In 1887 he traveled to Java to search for the mythical creature. Even more remarkably, he found humans fossils in 1891–92 that he named *Pithecanthropus erectus*, Java man, after Haeckel's hypothetical intermediate form. Eventually a number of fossils were found that demonstrated this species as an ancient bipedal human that walked erect, used tools, and had a brain size of 900–1,000 cc.

Excavations supported by the Rockefeller Foundation to find ancient hominids in China began in 1927 in Dragon Bone Hill at Zhoukoudian near Peking (Bejing) in a filled-in cave after some shaped stone tools were found. In 1923 and in 1927, three ancient human teeth were found and named *Sinanthropus pekinensis*, Pekin man. In 1928–29 a fossil jaw and skull of the same species were found in the cave deposits. Meanwhile, more skulls and other fossils of Java man had been found in Java. The similarities between the two forms led eventually to their being classified as the same species, *Homo erectus*. Today they are considered separate subspecies, *H. erectus erectus* and *H. erectus pekinensis*.

The cave at Zhoukoudian showed ample evidence of habitation by these ancient humans for more than 200,000 years and included the bones of animals they hunted and plant materials they gathered. One ash layer more than eighteen feet demonstrated that they knew how to use fire; charred bones and seeds strongly support the view that they had also learned to cook foods.

The oldest *H. erectus* specimen is from South Africa where it is in the same beds as *Australopithecus robustus*. These are dated at around 1.8 million years ago. Other remains have been found in East Africa (1.5 million years ago) and North Africa. Stone tools found in East Africa are dated at about 1.5 million years ago and show affinities to those from Zhoukoudian. Other specimens with *H. erectus* affinities have been found at Heidelberg and Bilzingsleben in Germany, Petralona in Greece, Terra Amata in France, and Torralba and Ambrona in Spain.

Toolmaking by these ancient people became a significant activity compared to that of their ancestors, *H. habilis*. Tools of *H. erectus* are known as Achculian after a site in France where they were first described. These are typically teardrop shaped hand axes, bifacials, formed by flaking along both sides. This flaking procedure, knapping, required a more delicate touch than did the previous Oldowan and Developed Oldowan tools of *H. habilis*. Acheulian tools became diversified into cleavers, scrapers, chisels, awls, anvils, spheroids, hammers and anvils. Because of the great distances that the *H. erectus* people had spread, the advances in their toolmaking did not keep pace with their spread. They obviously took their particular stage of toolmaking with them when migrating. Advances that occurred in distant places were slow to spread. As more advanced tools were produced, some areas continued to use the older versions.

The presence of *H. erectus* from Africa to Europe and Asia is demonstrated by their fossils at many sites. These sites are particularly important because of the information they provide regarding the lifestyles of the *H. erectus* people. It is obvious that improvements in culture, language, hunting skills, toolmaking and more contributed to their advancement as humans beyond their ancestors, *H. habilis*. Though their lifestyles still centered around an economy based on the simple gathering of plants supplemented by hunting of animals, there is ample evidence for their cultural advances over *H. habilis*. They were clearly a hunting, fishing, meat-eating people who lived in bands. They used and probably made fire, cooked foods, built shelters, made specialized stone tools, used hides as coverings or bags, and probably used vessels or containers. All these were techniques that assisted them forward on the path toward the modern human condition. Obviously they were successful in these advances.

Much of their success can be attributed to their greater intelligence than *H. habilis*. They were clearly a cooperative people who were successful gatherers and who practiced group hunting. Such hunting requires leadership and follower roles, and success in this endeavor requires intelligence to understand the prey, plan the hunt and carry it out. It also implies some sort of communication, whether visual or vocal or both. Success in communicating produced feedback that emphasized brain power. It is clear that greater intelligence was a highly desirable trait.

Many workers believe that *H. erectus* originated in Africa where their oldest remains have been found. If so, they probably survived there in only modest numbers, gradually spreading northward into East Africa from Olduvai and East Rudolph by following watercourses and hunting animals. There were no natural barriers to prevent their spread northward along the Rift Zone and, at the Mediterranean, they were easily able to spread along both African and European coasts. In North Africa their bones are known from Ternefine in Algeria, and they probably reached the Atlantic Ocean. Walking around the eastern Mediterranean they spread into the Middle East where they found abundant grasslands filled with animals available for hunting. From there they spread to the north and west into Europe. They could have easily have spread along the shore of the Mediterranean Sea through Italy and France as far as Spain, where they hunted elephants. They also spread extensively into the interior of Europe through the great river forests of the Balkans, reaching Hungary, Germany and France, where their important fossils have been found.

The presence of *H. erectus* in Asia is documented by fossils found at widely scattered sites in China and a number of places on Java in Indonesia. Some workers believe that *H. erectus* spread from Africa through the Middle East and traveled into China to Lantian and Zhoukoudian. From there they spread to Southeast Asia, where they populated the island of Java. A variant of this model proposes that *H. erectus* spread to China and Java from Europe through central Asia. Much exploration for additional fossils in Asia must be done before a clearer picture of *H. erectus* dispersal can be determined.

STUDY QUESTIONS

1. How do the Acheulian stone tools differ from the Oldowan and Developed Oldowan types?
2. Explain why stone-tool development often lags behind migrations of populations.
3. What contributions did Haeckel make to the search for human origins?
4. What evidence did Dubois use to establish that Java man was an erect biped?
5. Why were Java man and Peking man grouped together into the genus *Homo*?
6. Cite the evidence that *H. erectus* used fire.
7. Cite evidence that *H. erectus* likely invented cooking.
8. Cite the evidence from Olorgesailie that *H. erectus* actively hunted baboons.

9. Why is Africa regarded as the birthplace of *H. erectus*?
10. What is the evidence that *H. erectus* probably hunted elephants?
11. What evidence from Terra Amata indicates that the *H. erectus* hunting bands had a varied diet?
12. How is the manufacture of core stone tools different from that of flake tools?
13. How did the *H. erectus* people differ anatomically from *H. habilis*?
14. What evidence argues vehemently that *H. erectus* people were socially cooperative?
15. Describe how and why human cooperative behavior, especially as used in hunting, could have led to increases in intelligence.

10 The Neandertals Arrive

INTRODUCTION

The descent of *Homo sapiens* from the earlier species *H. erectus* is neither abruptly delineated nor clearly defined. Instead, the early members of our own species probably evolved gradually during the Pleistocene Epoch. Though scientists have learned much about how this crucial evolution took place, where it happened, and when, many details are still unknown. Different ideas have been postulated to explain this basic pathway, all based on the same evidence, the fossils and the sites, which are not in dispute. It is the way in which these discoveries have been interpreted that create the differences of opinion and allow for alternate explanations of this evolution. If the fossils were scattered evenly through time or in space, or if they were clearly distinguishable as one species or another, these interpretations would be simplified and there would be fewer disagreements. Unfortunately, evolution does not operate for the convenience of scientists but creates new species from existing ones in usually gradual and nondirectional fashion. Fossilization itself is a random process that provides no guarantees for long-time preservation. The result is that we have not found all the hominid fossils and so have an incomplete scenario of human evolution.

The series of critical events in human evolution, including the rise and spread of *H. erectus*, took place during the middle part of the Pleistocene Epoch, a time when great masses of glacier ice had spread across much of the continental landmasses of the Northern Hemisphere, covering vast lowland areas and the high mountain ranges. These continental and mountain glaciers were accompanied by a general worldwide lowering of temperatures that resulted in environmental conditions far from ideal for many species of plants and animals. In some of these changing environments, climatic conditions imposed new and different stresses on the existing flora and fauna, including the hominids. Some species had little difficulty in adapting to these new conditions. Others could not adapt and found survival difficult; many quickly became extinct. New species appeared and evolved as the existing species routinely gave rise to variants, some of which had modifications that allowed for adaptations suitable for the new conditions. Of particular interest among these was the new hominid, *H. sapiens*.

The Pleistocene Epoch (Table 10.1), which began about 1.8–2.0 million years ago, has been recognized by geologists to consist of four principal episodes of major glacial ice advances separated by three interglacial intervals. The four separate glacier stages were first studied in Europe and named **Günz**, **Mindel**, **Riss** and **Würm** for deposits recognized there. In North America, different names for these glacial advances and retreats were used, but the separate stages are essentially equivalent. The European terminology is commonly used in most parts of the world by many workers, especially for events during the Pleistocene that relate to the evolution of hominids.

Dating of the Pleistocene glacial stages has long been a difficult problem for geologists. Studies of magnetic particles in iron-bearing rocks demonstrate that reversals of Earth's magnetic field have taken place repeatedly in the past. A valuable paleomagnetic timescale (Table 10.2) has been compiled using **magnetic reversals** in conjunction with dates obtained from radioactive minerals. This scale provides a means of correlating evolutionary events around the world. Dates obtained from rocks at key localities have been compared to this timescale and have provided reasonably accurate dates of many important sites. Unfortunately, there are not enough fossils to provide complete coverage of the time involved in human evolution; scientists would prefer more dated sites much closer together in time.

Recent studies indicate that these four major glacial stages comprise a series of perhaps seventeen separate minor episodes of glacial advance and retreat. Deep-sea cores taken from ocean-bottom sediments indicate that as many as twenty separate glacial advances may have occurred during the past 800,000 years. This evidence also indicates that climatic conditions alternated back and forth during this long period of time but were prevailingly cold during the four major intervals. Precipitation during these intervals produced great quantities of snow that formed the mountain or **alpine glaciers** in high altitudes and the great **ice sheets**, or **continental glaciers**, that spread across the lowland surfaces, principally in the high latitudes. In between the glacial advances were times of warming, identified as **interglacial stages**, during which conditions

Table 10.1 Pleistocene glacial and interglacial stages with dates and tool cultures.

GEOLOGIC STAGE	GLACIAL STAGE		CULTURE
	N. EUROPE	NORTH AMERICA	
RECENT			
10,000 years———————————————————————————————			
UPPER PALEOLITHIC			
UPPER PLEISTOCENE	**WÜRM**	**WISCONSIN**	(PERIGORDIAN, etc.)
125,000 years————————————————————————			
MIDDLE PALEOLITHIC			(MOUSTERIAN)
		SANGAMON INTERGLACIAL	
	RISS	**ILLINOIAN**	(ACHEULIAN)
600,000–400,000 years————————————————————			
LOWER PALEOLITHIC			
		YARMOUTHIAN INTERGLACIAL	
MIDDLE PLEISTOCENE	**MINDEL**	**KANSAN**	(ABBEVILLEAN)
1,400,000–1,200,000 years ————————————————			
		AFTONIAN INTERGLACIAL	
	GÜNZ	**NEBRASKAN**	
1,500,000 years ————————————————————————			
			(DEVELOPED OLDOWAN)
LOWER PLEISTOCENE			
			(OLDOWAN)
1,800,000–2,000,000 years ———————————————			

ameliorated and much of the ice in these glaciers melted. This melting producing great torrents of meltwaters in some places and flushed great quantities of sediments from the glaciers to be deposited downstream.

The principal events of human evolution leading to modern *H. sapiens* occurred during these later Pleistocene cycles of severe cold and mildly warm intervals. These climatic shifts from cold to warm and from dry to moist, and back again, must have had significant influence on the hominids wherever they lived. That the existing stock of humans, *H. erectus*, were able to survive and prosper during what must have been unimaginably difficult and

Table 10.2 Pleistocene paleomagnetic timescale.

Paleomagnetic Epochs	Paleomagnetic Events	Million Years B.P.
BRUNHES NORMAL		
		——0.69——
MATAUYAMA REVERSED	Jaramillo	0.87–.92
	Olduvai	1.6–1.8
	Reunion	1.95–2.1
		——2.40——
GAUSS NORMAL	Kaena	2.8–2.9
	Mammoth	2.9–3.08
		——3.20——
GILBERT REVERSED	Conchiti	3.7–3.8
	Nunivak	4.0–4.1
	C1	4.3–4.4
	C2	4.5–4.7
		——5.20——

long winter conditions in some regions is ample testimony to their adaptability for survival, specifically their ability to find foods and shelters from the cold. Although the summers were much milder than winters during the glacial stages, they were still cooler than at the present. It was the stresses of life under such cold winter and cool summer conditions that may well have eventually shaped the development of physical and cultural adaptations in humans that facilitated their survival and advanced them toward the modern species, *H. sapiens*. Gradually, these *H. erectus* people changed or evolved under these influences until they gave rise to the modern human species.

The harsh glacial circumstances that prevailed over Europe, where a sizable population of *H. erectus* lived, were not experienced worldwide. In much of Africa and other tropic and equatorial regions with *H. erectus* populations, the climatic conditions were neither as harsh as those in Europe nor perhaps even difficult. Some of these other regions may well have developed into virtual paradises, with temperatures that declined until they became pleasantly warm. In addition, increased rainfall in the Southern Hemisphere is likely to have paralleled the increased precipitation in the Northern Hemisphere that fell as snow and produced the glaciers. Certainly the contrasting conditions in different parts of the world had different effects on the native human populations. In Europe, however, the frigid conditions were far from ideal, and into this setting appeared an important population of *H. sapiens*.

THE DISCOVERY AND THE DEBATE

The first people considered to be a type older than modern humans were the **Neandertals**. Today we are both considered to be the same species, *H. sapiens*, though this assignment was neither easily nor quickly determined. As close relatives of modern humans, Neandertals represent a complex group of people who lived in many parts of the Old World, who devised and improved many of the cultural techniques that we accept as fundamental for successful life in the wild, and whom some consider behaviorally similar to many primitive living peoples. These views are based on hundreds of fossil specimens that have been found, particularly in Europe, Africa and the Middle East. Collectively these fossils provide ample data for analyses of the characteristics that demonstrate their wide range of variability. These finds have also provided substantial evidence that demonstrates the human qualities of these people.

The discovery of the Neandertals was made in 1856 when workmen quarrying limestone in the Neander Valley near Dusseldorf in Germany opened a small cave, the Feldhofner Grotto. Digging into the floor of the cave for limestone they found some old bones that probably came from a complete skeleton, though only the skullcap, ribs, limb bones and a partial pelvis were preserved. Apparently an entire skeleton had originally been buried in the cave but, because of ignorance, carelessness, or lack of interest, only the skullcap and some other bones were saved. The owner of the cave and the bones believed them to be of a cave bear. Eventually, however, they were passed to a local science teacher, J. K. Fuhlrott, who recognized them as being from a very unusual human. However, these remains received a less than enthusiastic welcome from the scientific community.

The bones were seen to be very thick, and the skullcap had a heavy brow leading to a slanted cranium. One initial interpretation proposed that the bones were of an individual from very ancient times who died in the Flood of Noah and had been washed into the cave. A professor of anatomy at the University of Bonn, Hermann Schaaffhausen, even identified the bones as belonging to one of the "most ancient races of man," perhaps a barbarian predecessor of the Celts and German tribes no more than a few thousand years old. The prevailing view of mankind's origin at the time was that provided in Genesis and interpreted by religious leaders. They believed that humans were not a very ancient creature, but came into existence on Earth only about three or four thousand years ago when Earth had first been formed. It was proclaimed in Genesis, and the faithful accepted it without question. Since that time, humans had remained essentially unchanged. In other words, modern humans were the same type as the original man formed in the Garden of Eden and no one believed that other types of ancient people could have appeared in such a short span of time.

There was no jubilation over the sudden arrival of these older humans into the family of humans. In fact, their gradual acceptance as fossil humans closely related to living humans was marked by much controversy and dispute. Scientists were hesitant to accept these fossils as nearly modern humans and their reluctance was, after all, understandable. In the midst of this debate, Darwin had published his volume *Origin of Species* explaining how species of plants and animals evolved from earlier types. This radical idea generated much confusion and controversy regarding the acceptability of such a view for animals and plants, regardless of the facts. Darwin's concept also unwittingly introduced the one question that he wished to avoid: "Could man, too, have been evolved from earlier species as a product of evolution?" At that time there were no acknowledged fossils of ancient humans. There were also no known fossils that could be considered even close ancestors to them. The opponents of Darwinism naturally asked, "If man was a product of evolution, where was the evidence?" However, study of the Neandertals and additional discoveries of other fossils eventually shifted the balance of the situation. It became obvious to many that the Neandertals were a viable candidate for ancient humans, but therein arose much of the controversy of these types.

Though it would have been logical to assume an evolutionary origin for humans given the paleontological evidence of the Neandertals available at that time, the nature of these Neandertal fossils was uncertain and their placement in a family tree leading directly to modern humans was premature. There simply was no precedent for this radical idea that went so firmly against the prevailing religious bias of the time, and as a result most experts could offer no satisfactory explanations for the Neandertals or their origin. **Dr. William King**, however, was one expert not reluctant to make a decision about the Neandertals. King was Professor of Anatomy at Queen's College, Galway, Ireland. He believed that the Neandertals were an extinct type of humans closely related to modern ones and belonged in the same genus. He felt their differences from modern humans warranted their inclusion in a new species of *Homo*. In 1864 he became the first to recognize them as such and formally named them *Homo neanderthalensis* and explained that they represented a brutish type of ancient men who lacked the abilities to think as modern humans. With his proposal, the idea of ancient humans, different from living people and preserved as fossils, was formalized and quickly began to spread.

Curiously, a variety of other humanlike fossil bones had been found previous to the Neander Valley finds, although they had not been considered relevant to the origin of humans. Fossils had been found in Belgium in 1829 and at Gibraltar in 1848, shortly before the Neander Valley find, though neither led to any definitive conclusions of their nature at those times; both of these were later assigned to the Neandertals. Stone tools of apparent antiquity had also been found for many years, but these were considered the product of natural forces not the results of human activity. The real identity of these stone tools as the handiwork of humans was settled in 1858 by a team of British scientists, just in time for the original Neandertal discovery.

Within a few years of this important find, the origin of humans was further complicated by the discovery in 1868 of another new type of fossil human in the **Dordogne** district of France. That fossil find, **Cro-Magnon man**, was immediately recognized as fully modern in all respects and accepted as *H. sapiens*. If this very old fossil specimen was of a modern human, what was its relationship to the presumed very old humans, the Neandertals? Could the Neandertals have been the ancestors of Cro-Magnons or were they contemporaries? There was no way to compare these two types of fossils because there were too few specimens for adequate taxonomic and systematic analyses, and it was unknown when they lived. Further, it was not even known if the available Neandertal and Cro-Magnon fossils were normal or abnormal individuals. Experts could not decide whether there were two separate species or whether Neandertals were simply variants of Cro-Magnon man. Without enough evidence to draw proper technical conclusions, most scientists adopted a prudent attitude and simply awaited more evidence. Many others, however, began to assume that the Neandertals were a separate species of humans that probably gave rise to the Cro-Magnon people.

The study of the Neandertals can be divided into three stages. The earliest stage, from 1859 to about 1921, was one of discovery of Neandertals accompanied by many erroneous interpretations of their nature because of the scarceness of fossils. Especially important was the belief that the Neandertals were a unique hominid restricted to Europe and that they were much older than Cro-Magnons.

The second stage of study, from 1921 to about 1955, was a period in which the Neandertals were recognized as a relatively primitive human species that had spread through much of the Old World. The third and present stage, which began about 1955, focuses on the recognition of the Neandertals as a primitive type of modern humans, fully human in most respects and perhaps the originator of many of the cultural practices of modern humans, but not their direct ancestor.

This most recent idea is based on the discovery of numerous sites with abundant specimens, especially some older fossils that exhibit features with both modern and Neandertal characters. These have led to a revised understanding of these people and demonstrate that the Neandertals had a rich history in Europe and elsewhere. This view of the Neandertals was based on detailed restudy of many specimens and their archeological sites with full discussions of their nature. Today, we can accept the Neandertals as successful members of our own species who survived for thousands of years through the rigors of the ice ages in Europe, Africa and Asia. Their survival was aided to a great extent by their development of many behaviors and survival techniques. The success of the Neandertals can be seen in the relationship between their behaviors and survival techniques and our own modern ones. Their abilities provided much that we recognize in modern human culture.

RECOGNITION OF THE NEANDERTALS AS FOSSIL HUMANS

The exact nature and relationship of the Neandertal fossils has been debated and argued, often heatedly, by scientists and non-scientists alike, from the time of their discovery; these debates continue. One reason for this spirited controversy was the introduction of Darwin's concept just a few short years after the Neandertals had been discovered. Darwin had deliberately excluded humans from his concept of evolution, an omission that was pointedly noted and corrected in his *Descent of Man* (1871). Darwin's writings generated such criticism and debate from opponents in the latter half of the nineteenth century that discoveries of any new specimens of "fossil men" did not often settle the question, but instead only fueled the disagreements. Darwin's two prime works on evolution had a dual effect on the modern world. While his concept revealed the natural relationships between organisms by developing an explanation supported by abundant evidence, it also polarized opinions into two great opposing camps. To propose that evolution had taken place was daring enough but to propose that modern humans were part of this animal evolution was particularly shocking to many people, even as others found it logical and long overdue. His opponents trumpeted the unacceptability of placing humans in the same system with other organisms, even as the proponents were finding evidence to support his views. The rejections, unfortunately, were not based on logic, scientific understanding, or any other rational argument. Instead they were based on religious arguments.

While the original arguments about Darwinism raged across Europe and America during the latter half of the nineteenth century, discoveries of Neandertal fossils were being made; however no scientific concensus about their status as early humans was reached at that time. At least no concensus was reached while preconceived ideas of many individuals held firm and scientists rejected the evidence before them. Darwin's work solidified much of the opposition's opinions into a general denial that such evolution could possibly be true, and that it could never apply to mankind, because that it was contrary to the explanations in Genesis. These arguments were not raised because Darwin failed to provide enough evidence to prove his case; he provided ample data and examples, and many people accepted his evidence and logic. Rather, his opposition developed because the prevailing views and existing opinions were still too firmly established to be dislodged by evidence. The vehemence attached to opposing ideas has not disappeared; it is still present in many groups that continue to voice the same objections to Darwinian Evolution with the same faulty scientific arguments, even going so far as to promote public laws to require schools to teach religious beliefs. Fortunately, the courts have recognized the fallacy of such arguments and have consistently denied attempts to foster such religious beliefs purporting to be science.

During the last decades of the nineteenth century the vociferous public view was in opposition to Darwinism. Dubois' discovery of Java man in 1896, however, provided great support for the Neandertals as fossil humans and for Darwin's evolution as the mechanism of change. Although Java man was initially described as evolutionarily midway between humans and apes (Dubois had at first informally named his discovery *Anthropopithecus erectus*), its dis-

covery and implications could not be discounted as an important landmark in the evolution controversy, and it was demonstrably clear that these erect, bipedal hominids had evolved much closer to the modern human condition than Dubois had initially assumed. It was soon obvious that Neandertals were not the only type of fossil humans or near-humans that had lived in the past. And shortly after the turn of the century, Neandertal man was considered by many to be, at the very least, a close relative of modern humans. With two types of ancient humans now identified, scientists and the public began to wonder if perhaps there were other types yet to be found.

Little changed until 1924 when Raymond Dart discovered *Australopithecus africanus* and truly jolted the scientific world with his interpretations of a man-ape or ape-man, intermediate in general structures and features between apes and humans, and who had given rise to the latter. Though not all workers agreed that this new species was an ancestor of humans, some scientists began to foster the view that there may have been several different evolutionary stages between humans and apes, including this new type. *A. africanus*, for example, was initially considered by many experts to be a variant of an ape, though Dart and others thought otherwise from the very beginning and accepted it as the ancestral stock for humans. It was the turning point, well into the twentieth century, of the successful search for the ancestor of humans. Dart's discovery, coupled with previously discovered types of indeterminate status, provided ample evidence that a variety of truly primitive types of fossil humans had lived in the past. By comparison, the Neandertals were shown to be much more modern looking, and most workers soon began to view them as humans closely related to modern ones. For the first time, a relatively adequate lineage to modern humans with actual hominid stages could be demonstrated.

EARLY INTERPRETATIONS

Most people have viewed the European Neandertals as the typical cave people. We have seen them in museum reconstructions, in movies and in cartoons which depict them as caves dwellers who wore fur loincloths and carried heavy clubs (Fig. 10.1). In one of these popularly seen museum reconstructions, they are shown to be very coarsely built humans with short, stocky bodies that rose directly into large heads without apparent necks. Their slightly prognathous faces were dominated by large, broad noses

Figure 10.1 Reconstruction of Neandertals using Boule's concept.

beneath prominent brow ridges; they exhibit none of the delicacy of modern human faces. Their skulls generally rose from massive sloping foreheads into very low cranial vaults that were elongated and pointed at the rear. They had rounded shoulders which gave them a hunched over, apelike posture. Their arms and legs were thick and powerful looking because their bones and muscles were much sturdier than those of modern humans, a reflection of a lifestyle almost totally reliant on their strength. Their bodies are depicted to be moderately hairy, neither as thickly pelted as apes nor as bare as modern humans, but this is conjecture. Animal skins were worn as clothing, perhaps to compensate for the lack of insulating and protective fur. In general, these reconstructions offer little to suggest that the Neandertals were thinking people of any great achievement. All of these primitive models have given rise to the term **Classic Neandertals**. Figure 10.2 shows a reconstruction of these people based on modern interpretations.

As Campbell (1976, p. 293) notes, these views of the Neandertals were clearly biased. Many of them undoubt-

edly wore fur clothing, lived in caves and survived on great strength in environments filled with hazzards we would hesitate to face. But these characterizations are not necessarily true for all Neandertals. Today, Neandertals are exceptionally well known physically, and the numerous Neandertal fossils demonstrate variations in them as great as in modern humans. In general, however, most of these Neandertals exhibit the same general physical characters, especially their skulls which are very important for identifications. Their faces do have projecting jaws with weak chins, large noses and enlarged cheeks. Above these faces their prominent and curved brow ridges rise through sloping foreheads into low crania that once contained large brains.

Of great importance is the abundant evidence found in excavations of Neandertal sites throughout the Old World that has allowed reconstruction, interpretations and understanding of some of their culture and behaviors. Wear of their teeth, for example, is similar to that of modern Eskimos and suggests that, like the Eskimos, they held animal hides with their front teeth during preparation.

Figure 10.2 Modern interpretation of Neandertals.

These hides were probably used for clothing, tents, blankets, and carrying bags. Though some of their behaviors and practices are thus reasonably well known, there is no evidence to support the caricature of Neandertal males obtaining mates by attacking females with clubs and dragging them off by their hair, or of females enjoying such practices.

BOULE'S RECONSTRUCTIONS

Many of the original misconceptions about Neandertals are due to the reconstructions (1911–13) of **Marcellin Boule**, the great French paleontologist who made the first definitive scientific descriptions of Neandertals from a specimen found in France. This was an exceptionally important task that was intended to provide a firm basis for the scientific definition of the Neandertals. Unfortunately, Boule's work was later shown to be flawed and inadequate. Beginning in 1908, numerous excellent specimens of Neandertal adults and children had been found in caves and rock shelters in France at La Chapelle-aux-Saints, Le Moustier, La Ferassie and La Quinta. These individuals were found in shallow graves, a fact not appreciated at that time. Collectively, these specimens represented virtually complete remains with variations that should have enabled a reasonable characterization. Instead, Boule selected for study the skeleton of an old man from **La Chapelle-aux-Saints** (Fig. 10.3). Though this was a nearly complete and excellent skeleton, Boule made major errors in reconstructing the remains. Other scientists, examining these same bones in 1955, found that this individual had suffered from severe arthritis of the joints and the vertebrae, and from gum disease (he had lost almost all of his teeth). Boule had missed these pathologies and misinterpreted the fossil remains. He is also reported to have incorrectly arranged the attitude of the foot bones and the knee joint which, in turn, led to misinterpretations of the posture and gait which he assumed to be similar to those of the apes. As a

Figure 10.3 Neandertal remains from La Chapelle-aux-Saints.

consequence, Boule reconstructed the spine as a straight rather than recurved column and placed the head too far forward. The brain size was assumed to be very small because of the low, elongated shape of the skull. This led Boule to believe that this was evidence of an intellectual ability that fell between humans and apes. Boule described these features and wrote of the

brutish appearance of this muscular and clumsy body, and of the heavy-jawed skull that declares the predominance of a purely vegetative or bestial kind over the functions of the mind... What a contrast with the men of the next period, the men of the Cro-Magnon type, who had a more elegant body, a finer head, an upright and spacious brow, and who have left behind so much evidence of their material skill, their artistic and religious preoccupa-tions and their abstract faculties–and who were the first to merit the glorious title of *Homo sapiens!*

Boule was aware that the biologist **Rudolf Virchow** had explained the brutish appearance of the original Neandertals as deformities caused by pathological defects of otherwise modern men and may have been unduly influenced by Virchow's beliefs and reputation. Boule reviewed all the known Neandertal remains in three great definitive volumes that were filled with details. These established the lessened image of Neandertals that pre-vailed for many years, even though some scientists had disagreed with it and correctly extolled their true nature. Boule's misconceived view was that this creature was simply too brutish and too near an ape to be a successful human. He ignored the volume of the brain and concen-trated on the shape of the cranium. In effect, he said that the Old Man of La Chapelle-aux-Saints was a localized type, essentially adapted by European environmental conditions and of restricted range.

Boule's reputation and ideas immediately prevailed and greatly influenced paleontologists and others. In par-ticular, these ideas swayed archeologists, most of whom relied on his reputation and expertise and may have been relieved to have evidence that purported to separate Neandertals from modern humans. Archeologists, in par-ticular, failed to make the apparent connections between the cultures of the Neandertals and the subsequent Cro-Magnons, even though cultural evidence found at many sites was available and obvious. Instead, the two types were defined as separate, and it was generally accepted that the Neandertals were an extinct, aberrant species of near humans. These misinterpretations were perpetuated in both scientific and popular works, especially in the illustra-tions made from reconstructions. With such impressive authority arguing for this viewpoint, the general public could hardly be expected to believe otherwise. As a result, many scientists assumed that the Neandertals were not the direct ancestors of modern humans and continued to search for another species.

Boule may have been mistaken in many of his views and his interpretation of the Neandertals, but he was not alone in accepting their status as a near-ape. Others were even more emphatic in their interpretations of these ape-like hominids. One worker insisted that the Neandertal

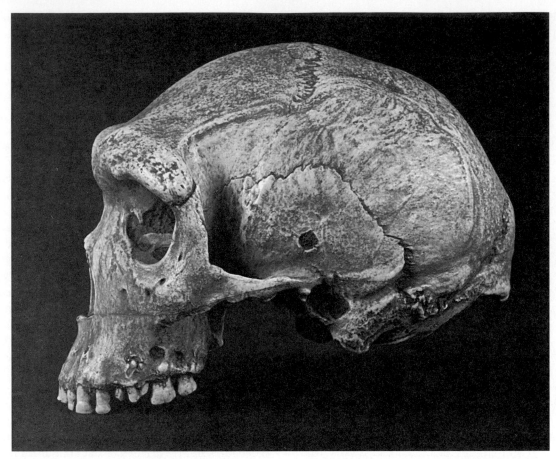

Figure 10.4 Rhodesian man (Broken Hill) skull.

hand was incapable of the dexterity of the modern human hand, even though the two are virtually identical. He also believed that Neandertals probably had a "shaggy covering of hair over most of the body," though there was no evidence to support this idea. It seems that many argued against the true status of the Neandertals with opinions that supported and satisfied preconceived views. As long as the Neandertals were considered strictly an European species, however, these problems were not significant.

OLD WORLD DISTRIBUTIONS
AFRICAN NEANDERTALS

By 1921, a great many Neandertal remains had been found across much of central and southern Europe. They were found in Germany, Spain, France, Rumania and the Crimea. A questionable fossil discovery was made on the island of Jersey in the English Channel. Great Britain and the northern European countries were unrepresented, perhaps because climatic conditions and the English Channel were effective barriers to the Neandertals' spread from the continent. No Neandertals had been found outside of Europe, which supported their status as a localized and aberrant product of evolution in Europe. All this changed, however, in 1921 when a strongly Neandertaloid skull and some other bones were found in Africa by miners in a cave at Broken Hill near Kwabe in Zambia (formerly Rhodesia). This large individual (Fig. 10.4), known as **Rhodesian man**, had a large, typically Neandertal skull with very prominent brow ridges, sloping forehead, and a low-vaulted, elongate cranium. The skeleton did, however, have straight and slender long bones that contrasted with the heavier and bowed bones of the European Neandertals. The combination of straight, slender limb bones mixed with a "brutish" skull created confusion, because the skull features allied it with European Neandertals even as the more delicate limb bones suggested a more modern stage of development. Initially believed to be about 40,000 years old, a new date has been proposed at 125,000 years based on reevaluation of the fauna and tools.

Speculations about the status of the Neandertals based on Rhodesian man appeared with almost as much variety as the number of workers offering opinions. Some experts pronounced Rhodesian man as more closely related to the apes than to Neandertals; others said that he was simply a modern individual deformed by disease. One expert considered Rhodesian man to be ". . . simian, and that he walked with a stoop." Others, however, believed that this was simply an African Neandertal and that the differences were normal for a widely distributed species inhabiting different environments. The only difficulty with this latter view was that if this specimen was a Neandertal, its presence in Africa indicated that the Neandertals were not a purely localized European stock after all. Complicating all these views is the suggested age of 125,000 years for this skull. The Rhodesian man skull is neither *H. erectus* nor a modern human; it therefore must be placed somewhere in-between, and Neandertal is the likely taxon. Some workers believe (Rightmire, 1989, p. 117), however, that this skull and others do not "closely resemble Neanderthals" and instead believe its features give it "an archaic appearance."

African Neandertals have now been found in Morocco, East Africa and South Africa. The specimen from Saldhana Bay in South Africa is similar to Rhodesian man but has some features that are slightly more primitive. The skull from Florisbad also resembles Rhodesian man skull, but it is more modern. Tools found with it are similar to those found at a site that has been dated at about sixty thousand years ago. Two skulls found in Morocco in 1961 present another dilemma. One is clearly a Neandertal, but the other has both modern and primitive features. The Bodo skull from Ethiopia has features that relate it to both Neandertals and *H. erectus* and it was found with Acheulian tools. This skull is remarkable because it bears cut marks similar to those on modern human skulls that had been scalped.

Because Neandertal features were now identified in African specimens and this form was not restricted to Europe, perhaps they were not aberrant humans as origi-

nally suggested, and perhaps they also could be found in Asia. If so, the implications to their status as relatives of modern humans would be enormous.

ASIAN NEANDERTALS

JAVA. Between 1931 and 1933, some 25,000 mammalian fossils had been found by ter Haar in the Ngandong Beds in Java (see Chapter 9). Included in this vast collection were the fragments of eleven human skulls that are known collectively as **Solo man** (Fig. 10.5). These partial skulls had been recovered from obvious burials but were not immediately identified as Neandertals, because their slightly thicker skull bones suggested a somewhat more primitive stage of evolution. Though some experts identified these fossils as Asian specimens of Neandertals, some others considered them to be relatively late specimens of the ancestral species, *H. erectus*. Their brain volume, which is slightly larger than other specimens of *H. erectus* and low for Neandertals, suggests an intermediate stage of development. It may be, therefore, that these Solo fossils represent the Asian variety of the earliest Neandertals with appropriate variations due to the environmental conditions of Java.

It may also be that these Solo fossils are simply more advanced specimens of *H. erectus* that had not yet fully evolved into a Neandertal stage. If these are Neandertals their presence in Java indicates that these humans had either spread across all the lands of the Old World or that they were the result of independent evolution in Southeast Asia from an earlier type. It remains for experts to determine which of these contrasting models is correct after more individuals have been found and examined.

The Solo discoveries in Java further complicate the study of ancient human behavior. These Solo fossils consisted solely of fragmented skulls and two postcranial bones. Such an abnormal distribution was an obvious intentional burial and undoubtedly represented some type of ceremony. Why else bury corpses? Why else bury only some portions of the bodies? The skulls showed evidence of being intentionally damaged, probably by their contemporaries, perhaps by their killers. Four of the skulls had been broken open at their bases and this suggested an intent to remove the brain, perhaps for cannibalism or some ritual. Several of the skulls, all thought to be female, show evidence of blows severe enough to produce scarring of the bones. Whether their sex was a factor in the circumstances

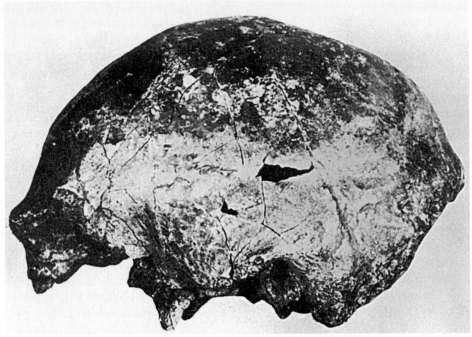

Figure 10.5 Solo man skull, Ngandong calvarium No. 6.

is unknown. The significance of the traumas is also still uncertain, but the implication is that these individuals died from violence, possibly murder. This view also implies that ancient man was fully capable of the same failings as modern humans.

MIDDLE EAST NEANDERTALS

THE MOUNT CARMEL CAVES AND ELSE-WHERE. Explorations for Neandertals in the Middle East were particularly successful and enlightening. Important specimens were found over a number of years in this crossroads region that links Europe, Africa and Asia and support the belief that Neandertals had spread throughout the Old World by wandering from a single birthplace. During 1931 and 1932, an Anglo-American expedition found important hominid remains in adjacent caves on **Mount Carmel** near Haifa, Israel (then Palestine). These caves contained Acheulian (see Fig. 9.9) and Mousterian tools (Fig. 10.6) that recorded cultures extending from the Lower Paleolithic through the Middle Paleolithic (see Table 10.1); other caves nearby showed similar tool histories but extended the tool cultures into the Upper Paleo-lithic. The Acheulian tool culture is considered

Figure 10.6 Typical Mousterian tools from Mount Carmel.

representative of *Homo erectus* but the **Mousterian culture** is associated with the Neandertals. The Mount Carmel remains were associated with Mousterian tools that documented their status as Neandertals. These caves, thus served as habitations for both *Homo erectus* and Neandertals.

The first cave, **Mugharet es-Tabun** or **Cave of the Oven**, yielded two individuals who were obvious Classic Neandertals but who possessed some aspects that were more modern than primitive. One had a somewhat more delicate skeleton than Classic Neandertals, perhaps because it was a female. But this skull also differed from the Classic Neandertals in having arched brows that led into a low cranial vault that was rounded at the rear, not typically pointed. A second specimen, a male jaw, was also Neandertal but with a more modern prominent chin instead of the expected receding chin.

The second cave, **Mugharet es-Skhül**, **Cave of the Kids**, contained the remains of ten buried skeletons perhaps slightly younger in time than those in Tabun Cave. All the individuals had been placed in shallow graves with their knees raised to their chests. One individual, a forty-five-year-old male, held the jawbone of a large boar cradled in his arms. Apparently some ceremony had been involved in these burials, a technique that is now known to be common among the Neandertals. One of these skeletons (Waechter, 1976, p. 106)

> . . . had a deep wound in the pelvis, passing into
> the head of the femur. A cast taken of the cavity
> shows that it was made by a pointed wooden
> stake. . .

At least two other Neandertal fossils, from Europe and Iraq, have such wounds to their left side and suggest that they were assailed by right-handed individuals. We may also freely speculate that this wounded **Skhül** individual, like the other two, may have been the victim of a deliberate act of violence.

Though the Mount Carmel fossils were accepted as Neandertals, there was confusion about their status. Some of the specimens were obvious Neandertals, some looked slightly more advanced or modern, and one was very close in appearance to modern humans. In general, all of the Skhül individuals were tall and more delicate or less robust than the European Neandertals. All the Mount Carmel skulls tend to have highly vaulted crania with flat sides and rounded rather than pointed rears, all traits considered

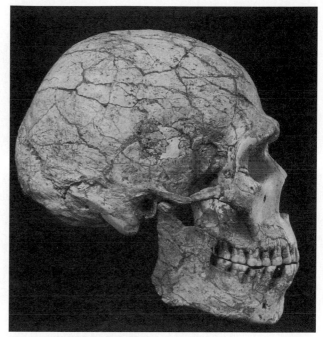

Figure 10.7 Skhül V skull.

relatively modern and non-Neandertal (Fig. 10.7). The faces of these skulls, however, are typically Neandertal but with smaller brow ridges than expected. Their supraorbital tori rise into foreheads that range from low and sloping to high and prominent. The specimens appear to have a mix of characters that relate them to both Neandertals and more modern humans. One skull (Constable, 1973, p. 22)

> . . . displayed a trace of the thick Neanderthal
> brow ridge, but the forehead was steeper, the
> jaw more delicate, the chin more pronounced
> and the shape of the cranium distinctly modern.

Because both the Tabun and Skhül specimens possess some traits typical of the Classic Neandertals and some of more modern people, determination of their taxonomic position was far more difficult than originally supposed. At the same time, this determination was of great importance. Were some of these individuals Neandertals and some Cro-Magnons? If so, which? Perhaps they were a mixed group both of Neandertals and of early Cro-Magnons that lived together. Perhaps they represented an evolutionary stage intermediate between the two types and were undergoing changes that led from one to the other. They were obviously individuals that had been buried, but why? Was this treatment a usual event with Neandertals or special with

189

these individuals? Did they deserve special treatment because of their fate or because of their position among their group or family? There were then, even as now, too many questions about these specimens and not enough answers.

Specimens were found in the cave at **Jebel Qafzeh** (Kafzeh), near Nazareth in Israel during 1934 and 1935. These five individuals were badly damaged but show affinities with the Skhül remains. Tools found with these fossils indicate mixed Acheulian and Mousterian affinities. Several other specimens found in northern Israel in **Amud Cave** by Japanese scientists included a nearly complete adult, a partial adult and two infants. The adults indicate a relationship with the Tabun-Shanidar and Skhül-Qafzeh groups, perhaps intermediate between them. Amud I, for example, has the largest known cranial capacity of any hominid fossil, 1,740 cc. These skulls also support the contention that many Neandertals in the Middle East have both modern and primitive aspects.

The evidence is not only clear that Neandertals practiced burials and had ceremonies or rituals, it is overwhelming. Even as a primitive people leading a simple life and fighting for survival, they were obviously thinking individuals with a sense of the importance of the brain and a respectful view of death. This idea of a people relatively modern in abilities contrasts markedly with the extreme hypothesis of one recent author who proposed that until about 1000 B.C. humans acted unconsciously, driven by inner voices assumed to be divine but in reality simply instructions from the right side of the brain to the left. This radical view is contradicted by too many of the Neandertal's achievements, not the least of which is their stone-tool technology that demonstrates advances over previous cultures and tool industries.

The questions of the fate of the Skhül Neandertal fossils also relates to the fate of all the Neandertals. What happened to them evolutionarily? Were they a distinct species that simply became extinct in the face of a more modern type, the Cro-Magnons? Could they have met their fate through conflict with the Cro-Magnons? Could they have been absorbed by interbreeding with the Cro-Magnons? Perhaps they evolved into the Cro-Magnons? One idea that did arise and gain favor suggested that the Skhül specimens represented an advanced but separate Neandertal population that had been evolving for some time, as demonstrated by their more modern characters. None of these questions, however, could be answered and the speculations contin-

ued. Fortunately, the explorations, excavations and discoveries also continued.

PROBLEMATIC EUROPEAN NEANDERTALS

Additional Neandertals were found in Europe during the 1920s and 1930s and a number of these discoveries were of such characters that they allowed for interesting interpretations. Saccopastore in Italy, Ehringsdorf and Steinheim in Germany and Piltdown and Swanscombe in England all contributed important information. The infamous Piltdown man, however, was a spectacular hoax and the only specimen that was totally anomalous in its characters. The others showed variations in characters that were controversial, some interpreted as very modern, some as very primitive. But they fit into an evolutionary pathway that did not admit Piltdown man. Though these specimens were in sharp contrast to the earliest discoveries of European Neandertals, the primitive-appearing Classic type, they provided little information that could be used to settle the controversy between primitive and modern forms. In part, this inconclusiveness continued between the World Wars because most of these fossils were incomplete fragmentary remains that lacked some definitive features. They provided sufficient evidence, however, to eventually demonstrate that the Neandertals were a much more varied group of humans than the experts originally thought.

Today we can look back at the original specimens and propose evolutionary hypotheses that make much more sense because of the great mass of information that has been accumulated. Like a jigsaw puzzle or mosaic, each specimen was by itself essentially an unknown until it was fitted or compared to another specimen, particularly after they all were fitted together. Today, interpretations of the Neandertals provide a very reasonable picture of their nature and relationships and the general pattern of their evolution.

PILTDOWN MAN. The greatest problem of the European Neandertals was generated by **Piltdown man** found by Dawson in England in 1911. This find consisted of a jaw and several parts of a skull, including a braincase with a vertical forehead and small brow ridges (Fig. 10.8). It was found in a gravel pit along with fossils of other animals and flints that indicated a great age, about 200,000–300,000 years. Its cranium and braincase, however, suggested the enlarged and advanced brain of a modern human, not an ancient one 200,000 years old. The jaw was obviously

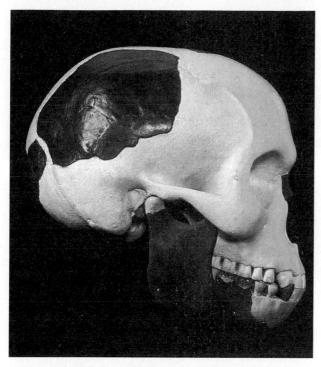

Figure 10.8 Reconstructed skull of Piltdown man.

apelike and offered strong evidence of a very primitive nature. Taken together these contrasting characters depicted a human with a relative modern brain but with the jaw of an ape. They fostered a belief that the human brain had evolved into a modern state much earlier than the postcranial body, even though this was contrary to the evidence already in existence. Java man, with its small brain and essentially modern legs, had already demonstrated that the brain evolved much more slowly than the legs and associated postcranial structures.

Obvious discrepancies with the Piltdown skull were quickly apparent but not fully appreciated. It had elderly teeth set in a youthful jaw that articulated with a clearly middle-aged skull. Many workers, however, began to have grave doubts as to Piltdown's validity because of these problems. By 1953, the evidence was insurmountable that Piltdown man was a hoax, a combination of a modified orangutan jaw and old human skull bones. This remarkable fraud was only discovered with geochemical tests for fluorine not even devised or possible when the fossil was first found. Careful reexamination of the fossil in 1953 even showed file marks on the orangutan teeth which had

been obviously "carved" to the proper shape. Dawson was discredited posthumously, but the perpetrator of the hoax and motives are unknown. The principals are long dead but the speculations continue.

EHRINGSDORF REMAINS. Fortunately, other important European Neandertal fossils were not fakes. The **Ehringsdorf** remains were found in 1925 by workmen at a quarry near Weimar, Germany. These consist of some teeth, an adult jaw, a child's jaws, and several skull fragments that show evidence of at least five wounds. The fossils were found in a layer that also contained (Sauer and Phenice, 1972, p. 60)

> . . . ashes, charcoal, broken and burnt bones, as
> well as implements in the rock, suggesting the
> site of an ancient hearth.

Constable reported (1973, p. 105) that the adult skull was a female who

> . . . had been repeatedly clubbed on the forehead; her head was severed from the body and,
> as at Solo, the foramen magnum had been
> enlarged.

This woman was obviously the victim of foul play, whatever the motive.

SACCOPASTORE SKULL. The **Saccopastore skull** (Fig. 10.9) is another female who died when about thirty years old. It was found in 1929 by Sergi in a gravel pit outside the walls of Rome. A second skull, of a thirty-five-year-old male, was found in the same pit in 1936. Both are badly broken but are typical Classic European Neandertals. Also found in the pit were rhinoceros, elephant and hippopotamus fossils.

STEINHEIM SKULL. The **Steinheim skull** (Fig. 10.10) is a third nearly complete female skull that lacks the lower jaw. It was found alone in 1933 in a gravel pit at Steinheim, Germany, near Stuttgart. This deposit is dated at about 250,000 years, which places it about the time of the Riss-Würm interglacial. Part of the face is missing and the skull base is broken open at the foramen magnum. This skull has an elongate cranium with a low vault and prominent brow ridges, features that are very primitive and suggestive of the more ancestral species *H. erectus*. The rear of the skull, however, is relatively rounded and modern in appearance, as is the articulation with the vertebral column. In addition, this specimen has a cranial capacity between 1,150 and 1,175 cc, a rather inconclusive dimension that does not support assignment to either *H. erectus*

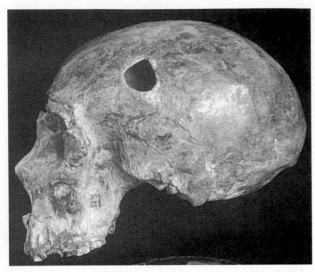

Figure 10.9 Saccopastore I skull.

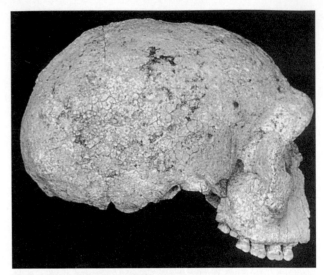

Figure 10.10 Steinheim skull.

or Neandertal man. It is clearly too advanced to be *H. erectus* and too primitive to be modern man. A number of experts believe that it represents a very early type of Neandertal that fits between *H. erectus* and the Classic Neandertals.

SWANSCOMBE SKULL. The **Swanscombe skull**, probably another female, consists of three fragments from the back and top of a skull that were found in a gravel pit along the Thames River in England. Two skull bones were found by Marston in 1935–36 and a third from the same skull was found by Wymer in 1955. This deposit contains abundant fossil remains of animals that provide a date during the Riss-Würm interglacial and it correlates roughly with the Steinheim skull. Numerous shaped tools and flake tools along with small pieces of charcoal and reddened flints, presumably evidence of campfires, were found in the same layer as the skull fragments. The deposit is dated at about 200,000–250,000 years indicating that this skull is actually older than the Classic Neandertals of Europe.

The Swanscombe woman was determined to be about 20–25 years old based on the open sutures between the skull bones. The back of the skull has a relatively modern appearance superficially but the bones are rather thick and suggestive of a more primitive stage of development. Some other details suggest that a prominent brow ridge was probably also present, but the face is missing and cannot be evaluated. The cranial capacity is estimated at about 1,300± 25 cc and compares favorably with modern

Europeans. However, the cranial vault is relatively low and suggests a more primitive stage of *Homo* that fits somewhere between *H. erectus* and Neandertals.

Tools found in the Swanscombe gravel pit have been classed as middle Acheulian industry. They include scrapers, engravers, saw-toothed cutters, flake tools, flakes and the ubiquitous hand axes. Though of the same pattern as the earlier examples of Acheulian tools from Africa, the Swanscombe tools are much better made and indicate a more sophisticated manufacture. The hand axes, for example, include numerous pointed specimens with wide bases. Similar tools have been recovered from another site about sixty-five miles away and support the view that these tools may represent a local improvement on the basic Acheulian style. Such local designs are common in other regions, and they may reflect an innovative way of working the materials developed by the inhabitants of this area or even a skilled artisan who produced a particularly distinctive style.

THE NEANDERTAL PROBLEM

The Swanscombe skull and the Steinheim skull are about the same age and have similarly shaped skull backs that resemble those of modern humans. With Steinheim man, however, there are no other such similarities with modern man because the Steinheim skull has a thoroughly primitive Neandertal face. Though it is not primitive enough to warrant assignment to *Homo erectus*, it is also not modern enough to be considered modern man; it must lie somewhere in between the two and the only reasonable status proposed is as Neandertal man. If the Swanscombe and Steinheim skulls are similar in shape and about the same age, then the Swanscombe specimen also cannot be very modern, nor should it be old enough to be *H. erectus*. It, too, should be considered Neandertal, though as a very early specimen like the Steinheim skull. These are reasonable interpretations given the limitations of this fossil material.

The absence of clear distinctions between the different types of problematic hominids was eventually resolved to some extent by a sophisticated type of analysis that partially answered some of these questions. By the 1960s, computers had come into general use for complex calculations of seemingly insurmountable problems. One was used to determine the status of the Swanscombe and Steinheim specimens and their relationships. Comparisons of seventeen different measurements on the two skulls showed that there was no essential difference between them. Campbell (1972, p. 304) emphasized that

> . . . the Swanscombe fossil was no more modern than the Steinheim skull. Instead of being precociously sapient, both skulls were about as primitive as one would expect from their ancient date.

There is no real evidence for a truly modern skull at such an early date, merely misinterpretation of the assumed modern qualities of the Steinheim fossils and the Swanscombe skull which lacked a face that could have provided important assessment. These skulls are essentially the same type, of similar age, and have some modern appearing characters that contrast sharply with the more primitive characters of the typical Classic Neandertals, which are actually younger in age. These juxtaposed aspects, old and more modern appearing vs. younger and more primitive appearing, created the difficulties that came to be known as the **Neandertal Problem**.

CLASSIC VS. PROGRESSIVE NEANDERTALS

The first recognized Neandertals were those "brutish" specimens from central Europe and Gibraltar. These were the typical cave people that most of us have seen reconstructed, however exaggerated, in museums and photographs and are known as the **typical** or **classic** type. As more specimens of these primitive humans were found in

Europe, the Middle East and Africa, workers began to seriously question these Classic European Neandertals with their primitive nature as representative and typical of all Neandertals. The arguments were based on some of the well-preserved remains found at Mount Carmel and several other sites in the Middle East that have some more modern features. These specimens are considered **Progressive** types (see Fig. 10.7). Other specimens, however, have characters clearly similar to the Classic European specimens. Those specimens found in Europe at Swanscombe in England and Steinheim in Germany also were recognized as having some modern as well as some primitive aspects. The presence of both primitive and modern aspects in different populations of what was considered a single species created all the confusion.

The confusion, however, was not simply the recognition of differences between some specimens that had primitive or modern aspects (**Classic vs. Progressive**), but *when* in time these characters had developed. Scientists had initially reasoned that there would probably be a straight-line evolutionary sequence leading from *Homo erectus* to modern humans, though they did not know how many different species, types or stages might be involved. When the Neandertals were discovered, this sequence included three distinct species in a gradation from the primitive ancestral form, *H. erectus*, through the newcomers, *H. neanderthalensis*, to the final stage, *H. sapiens*. This evolutionary model assumed that the specimens of any one stage would be more primitive looking or conservative than the younger types. Thus, the older Neandertals would be expected to be more similar in appearance to the ancestral *H. erectus*, and the younger Neandertals would be more similar to the descendant *H. sapiens*. The time distributions of these fossils, however, demonstrated just the opposite situation. The chronologically older specimens proved to be more modern looking (progressive) than the classic or more primitive type of European specimens. As the evidence accumulated, the Neandertals were eventually recognized and accepted as having both modern and primitive conditions. Classic Neandertals that were more primitive looking were actually the *younger* types. The more modern looking specimens from the Middle East and Europe were actually older in age; they are the Progressive Neandertals. The problem now with these two types was how to explain this apparent paradox of more modern forms that appeared before more primitive ones.

The first specimens discovered of this older type of humans known collectively as Neandertals were those from western Europe, though we know now that they are neither typical of this type nor the oldest. Though Boule and others considered them a distinct species ancestral to *Homo sapiens* and they were initially given the name *H. neanderthalensis* by King, the evidence from numerous later finds does not support this view of a separate, distinct species. The numbers of these fossils do, however, demonstrate in the different Neandertal populations a wide range of variability that is probably no more than that found in specimens of living man. The recognition of their physical development, their variability, and careful appraisal of their cultural achievements by some workers led to the view that the Neandertals are conspecific with the Cro-Magnons and living humans, *H. sapiens*. The differences of the Neandertals, however, are sufficient to consider them a variant of our own species.

Only minor differences separate the two stocks of *H. sapiens*. Living modern humans were the first to hold the name *H. sapiens* that was assigned by Linnaeus in his *Systema Naturae* (1735). The discovery of the Neandertals in 1856, their physical differences with living humans, and the subsequent debates about the position and status of these apparently primitive remains led King in 1864 to recognize them as a species older and separate from living *H. sapiens*; he named them *H. neanderthalensis*. Shortly thereafter in 1868, however, the Cro-Magnons were discovered and quickly recognized as having no substantial differences from living man; they were identified as the earliest type of *H. sapiens*. Although the Neandertals were recognized as being different enough from *H. erectus* to merit a separate species, they were not, according to most scientists, sufficiently different from *H. sapiens* to achieve the same status. The modern reappraisals of these two types show that both are *H. sapiens* but with recognizable differences that merit separation at the subspecific level. Cro-Magnons, accordingly, because of priority, are the name bearers of the species and identified as the subspecies *H. sapiens sapiens*. The Neandertals became the subspecies *H. sapiens neanderthalensis*.

Although this action had now been accepted by scientists throughout the world, a second problem remains: how do the Neandertals fit chronologically and structurally into the lineage of modern humans so that their evolutionary pattern agrees with the evidence?

THE ARCHAIC SAPIENS

The various fossil hominids that have been identified as Neandertals all exhibit some characters so primitive that they could not be considered Cro-Magnons or equivalent to anatomically modern humans. At the same time, they had enough modern characters to preclude assignment to the even more primitive species *H. erectus*. A number of other fossils not easily assignable to Cro-Magnon or *H. erectus*, however, have specific characters that created ambiguities in making any assignments, and they have been the subject of some debate. Interestingly enough, these other skulls seem to hold important clues to understanding the evolution of modern man. One of these fossil is the **Petralona skull** (Fig. 10.11) found in a cave in Greece in 1959. Though this well-preserved skull was initially identified as a Neandertal of fairly recent age, it is now considered by most workers to be an advanced type of *H. erectus*. Trinkhaus and Howells (1979, p. 128) state

> It shows no specifically Neanderthal character, looking more like an advanced *Homo erectus*.

More recent work indicates the Petralona skull may be 300,000–400,000 years old even though it has some aspects linking it to the more recent Neandertals. Its cranial size ranges from 1,220–1,440 cc, varying with the opinion of different experts. At least one worker considers this volume too great for *H. erectus* and believes this specimen is an early type of *H. sapiens*. Howell (1973, p. 78) believes that

> . . . from its size and form it appears to rate as a moderately advanced specimen in the whole erectus spectrum, and one differing from the Olduvai and Far Eastern samples. . .

Poirier is of the opinion (1977, p. 236) that Petralona specimen with its

> . . . "advanced" features might fit well with the supposed "advanced" traits of the Vértesszöllös skull.

The latter is a skull from Vértesszöllös, Hungary and also dated about 400,000 years ago. The **Vértesszöllös skull** has several important features of the Asian types of *H. erectus*, especially being rather thick-boned with a pointed rear and having a prominent bony ridge where the neck muscles attached. Several other workers, however, believe the Vértesszöllös skull is from a more advanced population than *H. erectus* and estimate its cranial volume at 1,200–

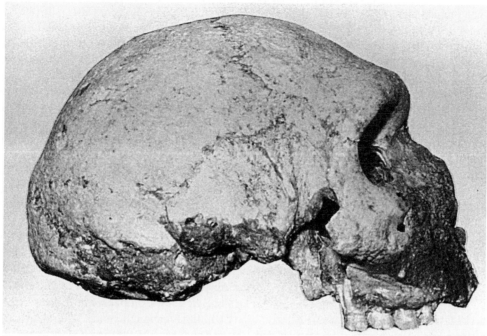
Figure 10.11 Petralona skull.

1,400 cc, much larger than the usual *H. erectus* and equivalent in size to *H. sapiens*.

Another fossil difficult to assign unequivocally as *H. erectus* or *H. sapiens* is the **Mauer** or **Heidelberg** jaw (see Fig. 9.4) from Middle Pleistocene deposits of Germany. This is a massive jaw that lacks a chin, both critical features of *H. erectus*. But the teeth are relatively small and distinctly modern in appearance. Weiss and Mann (1985, p. 346) state

> Opinion is divided as to whether the Heidelberg jaw is a European representative of *Homo erectus* or an early member of *Homo sapiens*.

Several fragmentary fossils have been excavated from the cave of Arago in the Pyrenees Mountains of southern France, along with many tools. These, too, have created problems in assignment. Both males and females are represented in this collection that consists of several lower jaws, most of the face and front part of a skull, teeth and some other bones. One jawbone and the quite massive face suggest *H. erectus*, but another jaw and some teeth suggest very primitive sapiens. These fossils, dated at about 200,000 years old, demonstrate both physical and sexual variation.

It is obvious from these problematic specimens that the evolution from *H. erectus* to *H. sapiens* was a lengthy process that involved much variability and was far from clearly defined. The reexamination of many of the older specimens considered to be Neandertals has generated new ideas regarding the status of these specimens and the likely evolutionary paths that led to Neandertals and modern humans. In general, study of all of the Neandertaloid specimens indicates that the change from *H. erectus* to modern humans began very early when some specimens developed some modern skeletal features. Although some later European specimens, the Neandertals, developed extremely conservative features, robust, sturdy builds, prominent brow ridges, low-vaulted and elongated skulls, they were not typical of the evolutionary path to modern man. These Classic Neandertals demonstrate that a conservative trend had begun in Europe during the Upper Pleistocene after a modern trend had appeared elsewhere. This conservative Neandertaloid trend may well have occurred when cold environmental conditions returned to Europe and produced the great Würm ice sheets that affected much of the Northern Hemisphere. Perhaps this conservative trend occurred in response to the chilling climate.

In general, therefore, the very gradual evolution of *H. erectus* to *H. sapiens* began about 300,000–400,000 years ago and is preserved in primitive specimens that exhibit some modern anatomical features. Such specimens are collectively termed **archaic sapiens** to indicate that they were evolving into the modern species *H. sapiens* but were still so primitive that they could not be conveniently accepted as full members of that species or retained in *H. erectus*. These specimens were very slow to develop more modern features, and this transition probably lasted for tens of thousands of years. Then, about 125,000 years ago the world climate began to chill once again, and Europe entered the Würm Glaciation and repeated the conditions of previous glacial stages. The European population did not remain static during this environmental change and apparently responded to it by becoming much more primitive and brutish in appearance. At the same time, other populations of archaic sapiens outside of Europe continued their slow modification into *H. sapiens,* but they did not develop the extremely conservative stage attained by their European cousins.

NEANDERTAL TOOLMAKING

The modifications that began to affect the physical and behavioral characters of the Neandertals were not restricted to these aspects alone. Evidence from their tools shows significant progress in their ability to make new and better stone tools, however slow it may have been in comparison to our own rapid pace of change. Their ancestors, *H. habilis,* "invented" the very simple Oldowan toolmaking industry that produced stone tools that were shaped by a few blows from other stones. Their descendants, *H. erectus,* improved upon this culture and developed the Acheulian toolmaking industry that was based on hand axes chipped from cores. Then about 125,000 years ago during the Middle Paleolithic, a new type of flaked tools appeared. These were of increased complexity that demonstrated greater skills by the craftsmen using a more sophisticated manufacturing process. The new tools are termed **Mousterian** (see Fig. 10.7) after a site in France where they were found in association with a buried Neandertal fossil. They are typical of Neandertal sites in Europe, North Africa and the Middle East where they are considered representative of these humans. Mousterian tools were made by repeatedly chipping flakes from the edges of cores, an advance over Acheulian tools which generally have bluntly shaped cores. The Mousterian toolmaking technique was able to produce a wider variety of flake shapes, principally points and scrapers, that could be useful in many different ways. With these improved tools, Neandertals were able to carry out tasks that were impossible for the earlier human ancestors and offered them the opportunity for greater success in adapting to their environments.

A possible link between the Mousterian and Acheulian tool cultures has been tentatively identified in early Mousterian tools dating from the Riss-Würm interglacial. Tool kits from this time have a high percentage of typical Acheulian hand axes and some Mousterian types and reflect an intermediate stage between the two cultures. Accordingly, these associations have been termed **Mousterian of Acheulian Tradition (MAT)** to indicate a relationship between the two cultures. Bordes (1961) suggested that the Acheulian industry gradually became the Mousterian by local advances without an introduction from outside. Such an evolutionary path is likely to have resulted from improvements in the skills of the craftsmen, and these were probably due to increases in the their brain abilities, their intelligence. Tools of this MAT type were in use between 125,000 and 75,000 years ago and then were succeeded by fully Mousterian tools. The advantages of the more advanced tool kits were obvious to these people, and they naturally adopted better tools for their needs.

Various types of Mousterian tool kits with different styles and percentages have been found, though not in an evolutionary sequence. Instead, these kits were often alternately interlayered in deposits and explained by interesting interpretations. The alternating styles could represent seasonal usage with different foods, different preferences by different groups traveling through the areas, or some other possibility. Whatever the explanation, the Neandertals became so versatile with toolmaking that many new activities were opened to them and allowed them to utilize materials and foods that were otherwise unavailable. Such expansion of uses allowed them to survive in environments that previously would have been hostile to them. With new materials, foods, tools and environments, the Neandertals may well have modified some of their behaviors as well as introduced new ones.

NEANDERTAL BEHAVIORS AND PRACTICES
BURIALS

A great many Neandertal fossils have been found since 1955. Many, if not most of these, were **buried** in shallow graves, a practice that was not initially recognized. These burials were particularly fortunate for us in reconstructing the evolutionary pathway to modern humans because many more individuals were preserved than would ordinarily have been the case. Many of these fossils themselves preserve injuries and deformities that handicapped these individuals or contributed to their deaths. In particular, some of the fossils show damage that indicates these Neandertals were capable of great violence. Some fossils indicate that these individuals had been murdered, perhaps even cannibalized. These burials also provided an insight into some of their ceremonies and practices. Data from the more recent and some older finds have enabled us to develop an even better picture of both their variability and their behaviors which are far more complex than envisioned by earlier workers. It is clear that they practiced ceremonies or rituals, including burials, though we can only speculate at their motives. They were recognizably human with many of our failings and emotions, if only because they buried some of their dead.

SHANIDAR. The Middle East Neandertals are among the most informative fossils in the search for human origins, because some specimens exhibit a blend of primitive and modern characters. Equally important is the recognition of some behaviors in the preserved evidence. Perhaps the most important specimens were those discovered by **Ralph Solecki** in the cave of **Shanidar** in the Zagros Mountains of Iraq. This very large cave was uniquely valuable to the study of human evolution because it had been inhabited almost continuously for about 60,000 years. Solecki began excavations in 1951 and continued until 1960 with great success. In 1953 he found the first skeleton, a Neandertal infant. Eventually eight other Neandertal skeletons were removed from these cave deposits that have been dated between 70,000 and 40,000 years ago. At the very least, some of these individuals demonstrate a manner of death that is particularly informative

Shanidar I is an individual who had been killed by rocks that fell from the ceiling. This fossil is the remains of a 5'3", 40-year-old male who lay where he had died. His skull is crushed on the top and left and also bears healed scars of previous trauma, perhaps also from falling rocks. This individual shows some relatively modern characters, particularly in his relatively small supraorbital torus, a condition present in all the specimens and one that contrasts markedly with the thick and heavy eyebrow ridge of the Classic Neandertals. Even more remarkable is the

evidence from the postcranial skeleton. This individual had suffered from a birth defect that had crippled his right collarbone, right shoulder blade and right upper arm. His lower right arm is missing and had been removed a considerable time before death, based on the healed nature of the bones. This was likely a deliberate amputation given the time necessary for healing and nature of the bones, but such surgery could only have been accomplished without anesthesia and using stone instruments. Other evidence showed he suffered from arthritis and was blind in one eye. Survival for this individual in a hunting and gathering society would have been virtually impossible if he had to contribute equally to hunting and other tasks within the group. Similarly, he could not have survived if he had been living alone, perhaps as an outcast. But he may have provided some skills that the group considered highly desirable. That he was seriously handicapped and survived to the great age of forty strongly suggests generosity and a cooperative social behavior in this group.

An equally surprising discovery was made on Shanidar IV, another specimen that was also found in the cave. This individual was surrounded by stones and buried with the legs drawn up to the chest. Soil samples from this burial were analyzed in 1968, long after the excavation had been completed. The soils were discovered to contain pollen from eight different colorful wild flowers that grow in profusion across the hillsides today, but certainly not in caves. The quantities of pollen in small concentrations indicate that it was not simply blown into the cave. More likely, it came from flowers that had been placed individually or in bunches around the head of the deceased. Because these flowers blossom only in late May to early June, they mark the time when the burial probably took place. Though other types of pollen grains were found in the cave, they were randomly scattered and not in the great quantities that could only have come from whole flowers concentrated around the deceased's head; these had drifted into the cave on breezes. We may easily speculate that someone deliberately picked these flowers and placed them around the deceased in some act of remembrance or honor, just as we do today. Most of these flowers are known to have herbal properties, and they may have represented an act related to healing or health. Solecki speculated that their presence might have indicated that the deceased was ". . . a very important man. . .," perhaps a leader or medicine man. Whatever the motive, the presence of the flowers included in this burial enhances the belief that Neandertals had a sense of life and death, placed great importance on a "proper" burial, and may have had an inkling of an afterlife where such flowers could be enjoyed.

RITUALS

Some of the Neandertal skulls already discussed were found with their bases around the foramen magnum area broken open, a generally simple task once the head had been severed from the body. This area is probably the weakest portion of the skull and the clear intent of such penetration is to have access to the brain. The purpose for such entry into the brain cavity, other than access to the brain, is unknown. Among living primitive peoples, access to a brain was usually performed to actually obtain the brain during a **ceremony** or **ritual** that was intended to honor the deceased. This "honor" probably involved removal of the brain to preserve it or the skull as a memorial. This honor often entailed eating the brain with the hope of gaining some of the "greatness" or "wisdom" of the deceased. Given the evidence from the Neandertals, it is likely that they, too, performed similar acts with their contemporaries. It should be noted they probably associated great significance to the brain, perhaps as the center of thought, body control or ability, and this might explain its significance.

CANNIBALISM

In 1899 a rock shelter at Krapina in Croatia was excavated and yielded more than five hundred scattered and fragmentary pieces that represented the remains of perhaps twenty Neandertaloid humans ranging from one to fifty years old. The bones of other animals, particularly the very large cave bear, *Ursus spelaeus*, were also found in this shelter along with tools and clear evidence of fire. The human skulls were broken into pieces and some of the limb bones were split lengthwise, possibly to obtain the marrow. Many of the human bones show marks and evidence of burning, and some specific human bones are missing from this accumulation. The fragmentary and scattered nature of the damaged human remains indicates that the individuals had not died in place or were buried. Instead, the entire accumulation is thought by some to represent a primitive garbage dump. The human remains have been interpreted by some workers as the leftovers from a cannibalistic feast.

Other workers, however, believe the marks are the result of ordinary cave weathering and rock falls.

Though the human skulls found at Krapina are of the right age, they are not all Classic Neandertals. Some have more modern looking features including prominent rounding of the skull base and back, and cheeks that tend to be hollowed. Though these specimens are not Classic Neandertals, they are also not fully modern humans. A similar discovery was made in 1965 at a cave in Hortus, France. There, broken and charred human bones similar to those from Krapina were found and interpreted as the remains of cannibalism. In all probability these were individuals who also had been murdered. Constable (1973, p. 104) pointed out that

> . . . the slaughters at Krapina and Hortus seem
> to have been more savage and less selective
> than any cannibalistic rite of today.

An alternate interpretation, however, suggests that the remains from Krapina and Hortus were too badly broken up to indicate cannibalism. If these fossilized remains are not the result of cannibalism, it is difficult to speculate on the nature of actions that produced such violence and destruction.

In 1938 an undisturbed skull was found in a cave at Monte Circeo in Italy near Rome. This skull was found lying within a ring constructed of large stones in a shallow trench that had been dug in the back of the cave away from general traffic. The skull was broken at the base, presumably to get at the brain. In addition to the damage at its base, the skull showed trauma from a powerful blow to the temple that had killed the individual. Three unrelated human mandibles were the only other skeletal parts found, strongly suggesting that these specific human remains had been selected and brought here for some particular purpose. There is no indication of cannibalism, however, and the entire Monte Circeo discovery undoubtedly represents some type of ritual; involving an enemy or a valued friend or ally.

The cave of La Chapelle-aux-Saints that yielded the original Old Man described by Boule was another ritual burial site, though it was not initially recognized as such. The Old Man had been placed in a shallow trench with a bison leg on his chest and surrounded by flint tools and broken animal bones. Because all these objects are associated with hunting, the Old Man is believed to have somehow been associated with hunting, perhaps as a great hunter. Similar interpretations of burial have now been made for the original Neandertal skeleton discovered in 1856 in Germany and for the two skeletons found at Spy, Belgium in 1885.

In 1924 a cave at Kiik-Koba in the Crimea was found to contain the burials of a man and a year-old child. Both skeletons were resting on their sides with their legs bent. The man was aligned east-west, similar to some other burials, and perhaps in ritualistic association with the rising and setting sun. In 1938 a cave in central Asia at Teshik-Tash near Samarkand was found to contain a burial with the body of an obvious Neandertal boy about nine years old. The remarkable aspect of this burial was that the boy had been interred in a shallow grave that was encircled by six horned ibex skulls embedded in the ground. It is unknown whether this was a routine burial or one that merited special attention.

Burials had obviously become an important and integral part of the Neandertal culture. Too many of these fossils have been found in graves to consider them as random, accidental interments. The burials indicate that these people had developed the view that human life was different from that of other animals and worthy of some measure of respect. They sought to ensure that the deceased was not simply dumped after death but was carefully placed in caves where it could be protected from weathering, decomposition and scavengers. The inclusion of specific artifacts with the deceased admits to a high level of thought in this population of ancient humans. They obviously associated particular items and activities with particular individuals. Including such items with these individuals indicated their measure of respect and honor, practices that we continue to this day.

THE EVOLUTIONARY PATH TO *HOMO SAPIENS*

The generalized geographic distribution of fossils described as Neandertals is well known. To some extent this distribution is dependent upon the definition of Neandertals; different workers have different views and criteria defining them. (Wolpoff (1989, p. 97), for example believes,

> The Neandertals themselves, however, are recognized to be a geographic variant limited to Europe and western Asia.

Their anatomical features have hinted at this evolutionary pathway leading to modern humans and we now have some understanding of how this probably happened. Until the 1950s, the numbers of Neandertal specimens discovered and studied indicated that their evolution had not progressed evenly throughout the entire body. Instead of evolving in relatively coordinated fashion as had been assumed by some earlier workers, these humans had evolved in **mosaic**, or piecemeal, fashion with different parts of the body evolving at different rates. Some body parts had become more advanced or more modern in appearance earlier than others that remained more conservative or primitive. This pattern was obvious in the ancestral stock of earlier hominids, the australopithecines. Their postcranial skeletons—pelvis, legs and feet—had evolved early and allowed them to adopt erect bipedalism while their brains and jaws were still relatively small and just beginning to change. It would seem that bipedalism and manipulative hand abilities were more critical to them initially than larger brains and intelligence. Apparently bipedalism facilitated hand advances, and these two were probably responsible for the subsequent advances in brains and intelligence.

This same pattern was also noted in *Homo erectus*, the species that immediately preceded and gave rise to the archaic sapiens and the Neandertals. *H. erectus* had essentially modern legs and feet but was very slow to evolve a modern head and brain. Again, the available evidence argues that their intelligence had improved enough to match their primitive lifestyles; they did survive, after all, for hundreds of thousands of years. Their postcranial body, however, improved much more rapidly to the modern state. We do not see a significant improvement on their postcranial skeleton in modern humans, an indication that bipedalism reached its peak early in hominid history. This, then may have been the most important aspect of hominid evolution, the transition from quadrupedal locomotion to bipedalism. Once bipedalism became established, hominid behavior patterns changed to accommodate the new lifestyle, and the other parts of the body advanced to keep pace. Teeth, jaws, skull, and brain of *H. habilis* and *H. erectus* began to modify after the feet, legs and hands had already undergone changes. Later specimens of *H. erectus* also demonstrate this trend; their heads and brains continued to evolve toward the modern condition. Even in the

Neandertals this pattern is recognizable. The early specimens of sapiens had some modern appearing features in an otherwise less advanced body and they gradually evolved into the Progressive Neandertals.

The problems associated with Neandertals, modern humans, *H. erectus* and the archaic sapiens are many. In part these are because fossil remains are found scattered in time and place across the Old World. Differences in gene pool, phenotypes derived from particular genomes in particular environments, modifications caused by adaptations (Smith and Paquette, 1989, relate changes in facial size of Neandertals to anterior dental loading, the use of their teeth for other than ordinary masticatory purposes) and many other factors contribute to differences in populations. Added to these are the differences in ages of the fossils that allow for time in which variations due to evolutionary effects can accumulate. A futher complication is the attempt to relate these fossils to modern humans which have undergone evolution from these long-dead ancestors and which are variable enough to be separable into racial strains or populations. The result is a collection of fossils of a particularly important segment of human history that are interpreted in slightly different fashions by anthropologists and paleontologists. All paleontologists do not agree which aspects of these fossils deserve emphasis in interpretations of taxonomic position and evolutionary significance. The same holds true for anthropologists; they simply do not all agree. If the positions and relationships of these fossils are not as clearly defined as we would like, there are many factors and reasons why such is the case. As with all science, each new bit of information allows for a new interpretation and for a new progress report, for that is what occupies most scientific literature—scientific reports of progress, not final statements.

SUMMARY

The appearance of *Homo sapiens* was neither sudden in time nor well defined in space. The change began with the people known as archaic sapiens who represented transitional forms that appeared almost imperceptibly in the Old World and gradually developed more modern aspects. These changes took place during the Pleistocene Epoch, when temperatures changed and increased precipitation formed glaciers in the higher latitudes and altitudes of many parts of the world. Although harsh conditions appeared in some places, notably northern Europe and its mountains, areas farther south did not suffer as much and may well have had pleasant conditions with increased precipitation. In these new environmental conditions, natural selection offered different phenotypes and genotypes the opportunity to succeed. As a result, the resident *H. erectus* populations began to slowly change into archaic sapiens and, eventually in Europe, into the Neandertals.

The transition of the Neandertals in Europe from archaic sapiens was also not an obvious one, in part because of the paucity of fossils. By 75,000 years ago, however, they were recognizably distinct though they would not be discovered by modern humans until the nineteenth century. That discovery did not bring sudden acceptance. Instead it generated much controversy and debate until enough specimens had been found to eliminate pathologic conditions as the cause for the differences with modern populations. Even so, the Neandertal population of Europe was considered distinct from other populations elsewhere. Boule described these Neandertals in 1911–13, unfortunately using the skeleton of one individual suffering from a variety of ailments, and produced a misconception of these people until relatively recent studies provided a more realistic view.

Until 1921, all Neandertals were thought to be European. At that time, a Neandertaloid skull was found in Zambia and experts began to suspect that Neandertals might not be restricted to Europe. Before long, typical specimens were found as burials in various parts of Africa and Asia. A major problem arose, however, when examples typical of the European specimens found in the Middle East also were seen to have some more modern anatomical features. This led to the categorization of the typical European Neandertals as Classic specimens, and those elsewhere with more modern aspects as Progressive. A dilemma ensued when it was discovered that the Progressive specimens were actually older than the Classic ones. Additional specimens found at Qafzeh and elsewhere quickly reinforced the concept of two types of Neandertals.

With such information in hand, scientists reexamined many of the problematic European Neandertals. Though Piltdown man was discovered to be a hoax, others were assigned to archaic sapiens. These, then, were part of the transition population that gradually evolved from *H. erectus* to *H. sapiens*. Elsewhere, these archaic sapiens did not all appear to evolve into Neandertals before becoming the modern form of *H. sapiens*. They underwent the transition directly into more modern types without the Classic Neandertal stage.

Culturally, the Neandertals were more advanced than *H. erectus*. Stone tools that they produced, Mousterian tools, exhibited improvements over the Acheulian tools of *H. erectus*. Mousterian tools are characterized by repeated removal of flakes from the edges. This technique enabled them to produce many more shapes of greater utility. And with such tools most aspects of their lives became better.

Neandertal behaviors also were advanced over those of *H. erectus*. Burials were a common practice and hundreds of specimens have been recovered from graves. Various objects included with the burials offer insights into Neandertal lifestyles. One specimen from Shanidar Cave in Iraq was surrounded by pollen from wildflowers; we continue the same practice with flowers today. Another Shanidar specimen showed evidence of physical handi-caps that made his survival impossible without the generosity of others. Certain objects from burials also demonstrate the existence of ceremonies, including burial with great honor. Yet the interpretation of murder and cannibalism from the evidence at many sites reflects the darker side of human nature.

Today we continue the trend toward modernization. Our bipedalism has not significantly changed for hundreds of thousands of years from *H. erectus*, although our teeth, jaws, skulls and brains are much modified from these ancestors. The only anomaly in this pattern is the appearance of the Classic Neandertals after the Progressive ones. Why did some later populations of sapiens became the more primitive-appearing and conservative Classic type of Neandertals even as others were continuing the trends to modernizations? Even so, we accept the Neandertals as members of our own species but classify them as a subspecies, *H. sapiens neanderthalensis*.

STUDY QUESTIONS

1. Who were the archaic sapiens?
2. Differentiate between the anatomy of the Classic Neandertals of Europe and *Homo erectus.*
3. Describe how *H. erectus* could have evolved into the Classic Neandertals.
4. Explain the Neandertal Problem.
5. Why are Progressive Neandertals earlier in time than the Classic ones?
6. How do the Classic and Progressive Neandertals differ anatomically?
7. Explain how the Mount Carmel Neandertals could be a mixed group with both Classic and Progressive features.
8. What procedure should Boule have followed in making the original description of Classic Neandertals?

9. How do Mousterian tools differ from Acheulian ones?
10. Why was it so difficult for nineteenth and twentieth century humans to accept the Neandertals as fossils of early humans?
11. Describe the anatomical differences between Classic and Progressive Neandertals.
12. Develop a model to explain how Classic Neandertals could have evolved in Europe but not elsewhere.
13. Identify some Neandertal behaviors still practiced by modern humans.
14. What is meant by mosaic evolution?
15. What evidence supports the belief that Neandertals had ceremonials or rituals?

11 The Cro-Magnons

Figure 11.1 The Dordogne River valley.

INTRODUCTION

About 25,000 years ago, a unique group of humans lived in the Dordogne District in southwestern France (Fig. 11.1). They were not the Neandertals who had inhabited this region for years but a new people with different physical features and vastly improved culture. These new-comers, originally identified as the **Cro-Magnons** (Fig. 11.2), were essentially anatomically modern humans who apparently appeared with great suddenness in the midst of a Europe that had been populated for tens of thousands of years by the simpler, more primitive appearing Classic Neandertals with their relatively unsophisticated tool in-dustry and culture.

The Cro-Magnons were abruptly different from these Neandertals in several ways. Physically they were fully as modern as people today with none of the so-called "brut-ish" Neandertal traits. They had, as we do, inconspicuous brow ridges, prominent foreheads, and high-vaulted,

rounded skulls, features not present in the Classic Neandertals. They differed intellectually and culturally, as well. The tools of Cro-Magnons are much more varied and specialized than the Neandertal tools and required much greater manipulation and care in manufacture. These better tools also reflect the greater intelligence of the Cro-Magnons and the ability to lead a more sophisticated way of life. Cro-Magnons made tools for many different purposes, virtually all of which we would recognize today, and they used these to provide a much better way of life for themselves than their predecessors. At the same time, these tools demon-strate their more complete status as modern humans. None of their numerous and versatile tools, however, demon-strated their status more than the simple sticks, twigs and brushes that they used to create magnificent works of art preserved in the caves at Lascaux, Altamira and elsewhere. Their total culture was so advanced beyond that of the

Figure 11.2 Cro-Magnon scene.

Neandertals, who apparently lacked similar art, that we have no difficulty accepting the Cro-Magnons as thoroughly modern people.

The original discovery of the Cro-Magnons in 1868 at **Les Eyzies** in southwestern France provided ample proof that essentially modern humans had lived in the distant past. These fossils verified Darwin's hint of human evolution in his *Origin of Species* published just a few years previously in 1859 and were even more important after he published the *Descent of Man* (1871). Their discovery was so important that various experts came to Les Eyzies to examine these remains with their own eyes, many undoubtedly skeptical of the idea that they could possibly have been like modern humans. These experts soon uncovered the fossils of either four or five skeletons, including two or three men, one young woman and a two- or three-week-old infant. They were found buried with flint weapons and tools and with pierced animal teeth and seashells that were probably some type of ornaments. All the experts agreed that these remains were fully modern humans who had lived long before recorded history and this view soon became common public knowledge. Comparisons with

other humanoid fossils demonstrated that the Neandertals were an earlier, less modern type of humans who had lived thousands of years before the Cro-Magnons, a daring concept for the times but one that was inevitable given the evidence. Such interpretations shocked the sensibilities of many who believed in a literal interpretation of the book of Genesis, which had long been cited as evidence that humans had lived on Earth for only about six thousand years. There was no doubt among many of the scientists and others, however, that these remains of modern humans were much older than six thousand years. In fact, these particular individuals from Les Eyzies were shown to have lived about 25,000 years ago.

THE DORDOGNE

Les Eyzies lies in the valley of the Vézère River which is cut into the bedrock of the limestone plateau of the Dordogne region. The steep cliffs along the rivers, especially the Vézère, were affected by rain, frost and groundwaters that seeped down through this soluble bedrock to form numerous caves and overhangs. These provided exceptional shelters for the early humans that inhabited the region over tens of thousands of years. These Cro-Magnon people left many traces of their existence in these caves and shelters—their bones, their tools and their art. Discovery of this evidence has allowed scientists to reconstruct their lifestyle as a complex society of intelligent hunters, complete with rituals and ceremonies.

The first people in the Dordogne region were not, however, these Cro-Magnons. Neandertals had previously inhabited this region for thousands of years during the early part of the Würm glaciation. Lartet and Christy had excavated a cave at Le Moustier in 1863 and discovered numerous flake tools in the lower of two superposed caves. These deposits provided the name for the Mousterian Industry and the **Upper Paleolithic** was established to include it. A Classic Neandertal fossil was later found buried in the lower cave and clearly demonstrated the association of these conservative people with the Mousterian tool culture.

The presence of Cro-Magnons in this area after the Neandertals emphasized how attractive the Dordogne region was for human habitation. Most of the caves and rock overhangs faced south and admitted the sun's warmth during winter, even as they offered protection from cold northern winds; they were desired shelters and prehistoric people exploited them. Grassy valleys and wooded hills stretched for miles in all directions, and large numbers of reindeer, horses, bison and other game animals roamed these peaceful settings. Water was plentiful and fish abounded in the rivers. Life in the Dordogne district might have been relatively idyllic for the prehistoric humans living there.

Many of the numerous caves in this part of France were first excavated by Lartet and Christy. Lartet had originally been attracted to the idea of ancient bones and stone tools by the discovery in 1852 of seventeen human skeletons in a cave at Aurignac in France. These skeletons had been found with stone tools and the bones of extinct animals; some of the bones were engraved. These skeletons probably represent a burial, though this was unrecognized at first. Unfortunately, the skeletons from Aurignac were also presumed to be recently deceased Christians and were given Christian burial in a nearby parish cemetery. This proved a great tragedy for science, because the location of the graves was lost and the fossils have never been recovered. Soon afterward, Neandertal man was discov-

Table 11.1 Paleolithic cultures of Europe.			
	Lartet 1875	De Mortillet 1883	Present-day Name
UPPER PALEOLITHIC	Age of Reindeer	La Madeleine	Magdalenian
		Solutré	Solutrean
			Gravettian
		Aurignac	Aurignacian
			Chatelperronian
MIDDLE PALEOLITHIC	Age of Cave Bears	Mousterian (Le Moustier)	Mousterian

ered in Germany and, with this as impetus, Lartet began his search for ancient humans by excavating caves and rock shelters in France. He explored Le Moustier, La Madeleine, Laugerie Haute, and Gorge d'Enfer, all famous habitations, and made important discoveries. He was then joined by a wealthy Englishman, Henry Christy, and together they published a monograph on ancient humans in French caves, *Reliquiae Aquitanicae* (1875), that included a classification of Cro-Magnon cultures (Table 11.1) that became the standard for stages of development of early humans.

STONE AGE TOOL CULTURES

Lartet and Christy's dual classification of the Stone Age cultures emphasized the extinct animals hunted by *H. sapiens*. The older unit was based on cave bears and named the **Age of Cave Bears**; it was assigned to the Neandertals. The younger subdivision, the **Age of Reindeer**, was based on reindeer remains and was, therefore, associated with the Cro-Magnons. These two different cultures made extensive use of distinctive stone tools that helped identify them. De Mortillet subsequently determined that this classification was too simple for convenience and revised it. He identified the Neandertal Cave Bear tool culture as Mousterian, or **Middle Paleolithic**, because of the important evidence found at Le Moustier. He also decided that the stone tools of the Cro-Magnons represented such a highly organized industry that it warranted major revision. He divided the Age of Reindeer into three units which were named for the sites where distinctive stone and bone tools were found, Aurignac, Solutré, and La Madeleine. The Age of Reindeer then became the **Upper Paleolithic**.

In 1912, the legendary **Abbé Henri Breuil** published his classification of the Upper Paleolithic and introduced new terminology, some of which is still in use. He renamed the Upper Paleolithic and subdivided it into the **Aurignacian**, **Solutrean** and **Magdalenian** cultures. With more finds, the Aurignacian was later subdivided into lower, middle and upper units. More detailed subsequent work showed that the lower and upper units of the Aurignacian overlapped and interlayered in some places creating some confusion and indicating a contemporaneous origin for the two. These two were later renamed the **Perigordian**; the middle unit, however, retained the original name Aurignacian.

Most early work indicated that the Upper Paleolithic tool cultures were produced by peoples who succeeded each other. There is no evidence, however, to indicate that the newly revised Perigordian and Aurignacian tool cultures were made by people whose industries followed each other or whether they were isolated from each other. A more likely interpretation is that these cultural units probably represented different preferences in toolmaking, perhaps by particular nomadic groups or tribes who traveled through this area of France and left tools as they migrated. In this model, the distribution of tools may have produced an apparent rather than a real succession. In contrast, the Solutrean, Magdalenian, and **Azilian** tool cultures apparently did succeed each other in time, and these cultural styles suggest that their makers had developed much better communications that allowed the quick spread of technological innovations between tribes.

The tools that provided the basis for these hunting cultures were a particularly appropriate and efficient means of identifying subtle or other changes in the human populations that could not otherwise be preserved in the bones. Though humans had spread throughout the Old World, anatomical changes in these modern humans were very slow to take place. Normal variation in human anatomy is such that there were no obvious and rapid shifts to indicate trends one way or another; some experts believe that the anatomical variations we see today in living humans was approximately the same as that in the Cro-Magnons. Though the Cro-Magnon populations were becoming imperceptibly different anatomically, their cultural changes were taking place with much greater rapidity. Tools in particular reflected cultural changes—especially their ideas and practices—and they evolved much more rapidly than the human population itself. The changes and modifications in tools and culture could be quickly spread from one region to another as individuals carried their ideas and tools with them. It was quicker and much easier for these intelligent humans to learn about new ideas, new developments and new tools from other humans than to develop or invent them independently, and with such a spread of cultures and tools, modern cultures were not far in the future.

NATURE OF THE CRO-MAGNONS

Technically, the Cro-Magnons are anatomically modern people who appeared in Europe about 35,000 years ago, although others are known earlier from different parts of the Old World. All of these modern humans are known scientifically as **H. sapiens sapiens** to distinguish them from earlier humans also included in *H. sapiens*. The earlier or older specimens of the species include those referred to as archaic sapiens and the Neandertals (**H. sapiens neanderthalensis**). The latter includes those populations that have been termed Classic and Progressive Neandertals.

Cro-Magnons are indistinguishable today from living people. The men stood about 5'8" tall and, like many species, were slightly larger than the women. The men had slightly larger heads than the women, and both men and women had high foreheads, sharp noses, forceful chins and small teeth. These people were highly variable in characteristics and differed from each other about as much as modern Europeans. But not one of their physical features or cultures suggests an origin from the Classic Neandertals of Europe. The Cro-Magnon culture was impressive and its development clearly exceeded that of the Neandertals. In these regards, they had a more advanced tool industry, greater social organization with religion, a much more sophisticated way of hunting, and art. Though some have reasoned that the Cro-Magnons were not as advanced as modern humans, the evidence does not support such an interpretation. Their art alone indicates a level of mentality that was probably equal to ours. Our achievements have been significantly greater because of our accumulated knowledge, not because we are more intelligent.

In anatomical terms, the Cro-Magnon skulls (Fig. 11.3) are recognizably different in many ways from the

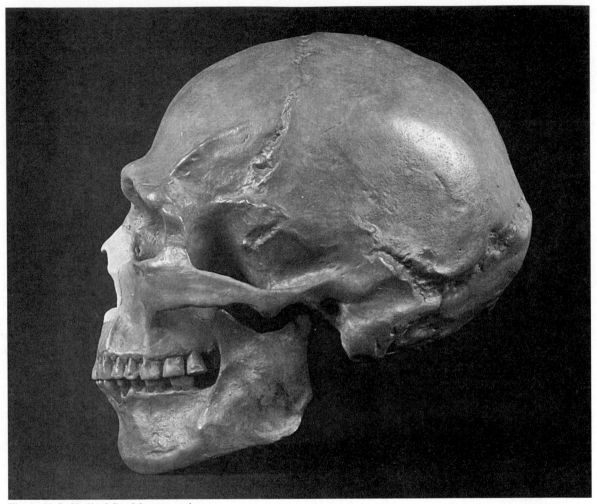

Figure 11.3 Skull of Cro-Magnon male.

Neandertals, particularly the Classic Neandertals (see Fig. 10.8). These earlier humans generally have skulls with sloping foreheads, low vaults with bulging sides and a bun-shaped rear where the skull is drawn out and often pointed. They also tend to have prominent brow ridges, receding chins, and broad noses that dominate their jutting faces. All of these features, however, may vary in size and prominence in different Neandertal skulls, and they demonstrate a wide range of variability. In the Progressive Neandertals, many of these features are less prominent and the skulls tend to be more modern in appearance. Classic Neandertals, however, obviously appear more "primitive" than the Progressive types even though the Progressive types appeared earlier in time.

Cro-Magnon skulls, in contrast to the Neandertals, exhibit features that continue the trend of modernization begun in *H. erectus* and advanced in the Progressives (see Fig. 10.8). Cro-Magnon skulls tend to be higher vaulted, relatively flattened at the sides and well-rounded at the rear. At the front, the Cro-Magnon skulls have more vertical, prominent foreheads with small or no brow ridges. Their faces tend to be smaller and flatter with eyes set somewhat closer together above a smaller nasal cavity. The jaws are smaller and feature chins that developed as the jaws became reduced in size. Such chins are lacking or very small in Neandertals. Cro-Magnon teeth are smaller than Neandertal teeth, especially the front teeth, and tended to be evenly sized. In the postcranial skeleton, the bones of Cro-Magnon tend to be more slender than Neandertal bones.

The differences between Progressive Neandertals and Cro-Magnons seem to be gradational with the Cro-Magnons simply further along the path to modernization. The Classic Neandertals, if fitted into this anatomical sequence, are far less modern and have many features that were essentially more primitive, perhaps because of their way of life in the restrictive environments of glacial-age Europe. Trinkhaus (Putman, 1988, p. 452) pointed out that the Classic Neandertals had larger and thicker bones than modern humans because of their much greater physical activity and strength. Of their thigh bones he said,

> . . . the [marrow] cavity is only a quarter of the whole, in modern man it is about half. It indicates they were hyperactive. Bone, like muscle, builds up in response to activity.

In keeping with such larger bones, Neandertals had bigger muscles to operate their skeletal structure. But what made the non-European Neandertals so different from the Classic type? Only the climate is identified thus far as having made such a difference. Why the Cro-Magnon way of life was less demanding physically than that of the Classic Neandertals is unknown, but it is likely that the greater intelligence of Cro-Magnons was one significant factor. Perhaps their new tools, new diet, and other behavioral advances—all related to intelligence—made their life less of a struggle.

In addition to all the modifications to their skeletal structure, the Cro-Magnons had brains slightly smaller than those of Neandertals, though they are equal in size to modern humans, about 1,250–1,350 cc. This slight decrease in size does not reflect a lessened intelligence in the Cro-Magnons, however, or a greater intelligence in the Neandertals, although it reverses the trend in hominids (Table 11.2). Neandertal brains apparently simply were not as advanced or capable as our modern brains. If Cro-Magnons had better brains, we should expect them to have greater intelligence, because that is a principal function of the brain. This evolutionary advance could explain the sudden improvements in cultural aspects that Cro-Magnons demonstrated over the Neandertals, especially improvements in their tools, clothing, hunting, shelters and art, particularly art which is lacking in Neandertals. With better brains, Cro-Magnons were simply better at being human. Not only were they better than Neandertals, but we note

Table 11.2 Brain sizes in hominids.	
Homo sapiens sapiens	1,350 cc
Homo sapiens neanderthalensis	1,450 cc
Homo erectus pekinensis	1,000 cc
Homo erectus erectus	900 cc
Homo habilis	700 cc
Australopithecus robustus	500 cc
Australopithecus boisei	530 cc
Australopithecus africanus	450 cc
Australopithecus afarensis	400 cc

with emphasis that the Neandertals disappeared in Europe shortly after the Cro-Magnons arrived. There is no universally accepted hypothesis to explain this disappearance; the Neandertals could have been eliminated by the Cro-Magnons through direct conflict, by absorption into the Cro-Magnon gene pool or by some other possibility. The apparent disappearance of the Neandertals may simply reflect the superiority of Cro-Magnons in one or more aspects that made them much better humans who displaced the Neandertals.

History records numerous cases of superior cultures displacing inferior ones as the less-advanced populations recognized and adopted better ways. Given the explanation by Trinkhaus that Neandertals had sturdier and more robust body builds and skeletons because of their greater physical activity, we must also acknowledge that the Cro-Magnons with less robust and sturdy body builds were not as strong and, perhaps, led less strenuous lives. Such a lifestyle, therefore, could have been based on their abilities to obtain better shelter and better clothing, and to utilize better hunting techniques, all of which required less strength. It seems logical, therefore, to believe that the Cro-Magnons, modern humans, depended on greater intelligence, more and better tools, and more efficient social organizations for their success and that this may have been the basis for their superiority and the decline of the Neandertals.

THE MEANS FOR CULTURAL CHANGE

The Cro-Magnon people of Europe survived during the last millennia of the Würm Glaciation by capitalizing on their intelligence and inventiveness, not their limitations. Like the earlier humans they knew that their own physical strength was barely adequate for survival under the ice-age

Figure 11.4 Cro-Magnon stone tools.

conditions of Europe without tools. Fortunately, they were smarter than their hominid ancestors. Though they did not need tools to gather plant food in growth seasons, they may not have been able to collect adequate quantities to last through the cold seasons. They would naturally have relied heavily on animal meat for sustenance during these other times of the year. But their physical ability alone was probably inadequate to obtain the quantities of meats adequate for their needs. They were neither as swift as most animals nor as strong as others. They did, however, have important advantages. They had learned from their ancestors and predecessors that tools could extend their natural abilities, and they used these tools to obtain more meat than was possible using their bare hands. They learned that with appropriate weapons they could kill any animal within their hunting range, including giant bears, woolly mammoths, woolly rhinos and more. The use and manufacture of these tools (Fig. 11.4) were, at the same time, a product of their increased intelligence and a stimulus to it.

The Cro-Magnons were much better at making and using tools essential for survival than the Neandertals. The Cro-Magnons became highly skilled and efficient toolmakers, producing a wide variety of weapons, utensils and

implements, including some highly specialized and beautiful tools (Fig. 11.5). Such delicate tools were probably intended as works of art or for use in ceremonies rather than for actual hunting purposes. In contrast to these more improved tools of Cro-Magnons, the tools of the older species *H. erectus* did not differ greatly from place to place or continent to continent; they looked much alike through time and were recognizably similar. Apparently the craftsmen in the *H. erectus* populations did not have the intellectual capacity to do much more than continue making the simplest of stone tools once invented.

The Neandertal craftsmen were not particularly advanced over the *H. erectus* craftsmen, though they produced better tools. The Neandertals' tools were generally similar in shape and function from place to place; apparently they rarely modified any tools for their own specific local needs. Their tools do, however, reflect a slightly greater intellectual ability than *H. erectus*' tools because they show more preparation and much better edges. Occasionally Neandertal craftsmen made some new types of tools and eventually they developed tool kits with up to seventy different tool types.

Figure 11.5 Cro-Magnon laurel-leaf blade.

It was among the Cro-Magnons, however, that tool-making became a highly developed industry. They produced more than one hundred different types of tools and instruments from a variety of common and exotic materials. In fact, stone tools made by the Cro-Magnons were discovered much earlier than their skeletal remains, though no one really was able to connect these artifacts with truly ancient humans at the time. Fagan (1974, p. 4) reported that

> The tools of prehistoric man were familiar sights in private collections and museums three hundred years ago, including crudely flaked stone axes found in river gravels and plowed fields all through Europe.

He also noted that hand axes were commonly found in ancient river gravels and lake beds along with the bones of extinct animals including saber-toothed tigers and long-tusked elephants. These reports, though exciting to us today, did not really capture the imagination of eighteenth century Europe and most people then failed to appreciate their significance. A few far-sighted other individuals, fortunately, were interested and they collected the fossils of humans and extinct animals that had been found together with stone tools.

In 1715 an apothecary named **Conyers** found a stone axe with elephant bones in an excavation in Gray's Inn Lane, London, though many people scoffed at the possibility that ancient humans hunted elephants in England. In 1771, a German clergyman, **Johann Esper**, found human bones in a cave soil layer that also contained bones of cave bears and other extinct animals. He questioned this association and origin and suggested that they were perhaps indicative of ancient humans who hunted once-living species, though ultimately he ascribed this relationship to chance. Scientific discoveries were not always accepted by the masses or the knowledgeable as factual, a situation that is still repeated today. In 1797, however, **John Frere** found and illustrated flint hand axes from Suffolk that were associated with the bones of extinct animals at a depth of thirteen feet. These dressed tools clearly identified and established for the first time that such artifacts of antiquity were made by unknown humans.

In 1820, the German paleontologist **Baron von Schlotheim** reported that he had found human teeth with remains of a mammoth at Köstritz. He immediately accepted the humans as contemporaries of the mammoth, now long extinct. Cuvier, however, interpreted the teeth as being from a very deep burial that had emplaced the deceased human into the layers that contained the mammoth, and his stature prevailed. Two years later, Schlotheim found additional human remains with extinct animal fossils in Bilzingsleben and another year later found more human fossils at Lahr in Baden. This time he accepted that humans had perhaps lived at the same time as these extinct animals. Once again the conservative view of others convinced him that he was wrong in this interpretation and he disposed of all the human fossils and turned to other pursuits.

In 1823, a geologist and clergyman of distinction, **William Buckland**, excavated the cave of Paviland in Wales and found a human skeleton with stone tools and the bones of the woolly mammoth and woolly rhinoceros. This human skeleton, later determined to be a male, became known as the Red Lady because of the red staining of the bones by red ocher and is dated today at about 18,500 years. He believed, however, that this association was a chance occurrence of no particular significance for prehistory. He explained that modern men accidentally introduced the human remains into the cave during a commercial operation and mixed them with the fossils of extinct animals already present. There was, however, no trace of any commercial operations and we must wonder at this marvelous bit of speculation without supporting evidence. Although a number of such sites cried out for recognition of ancient humans, the cavalier attitude prevailed.

In 1825 the English priest, **John MacEnery**, excavated Kent Caverns and found numerous stone tools associated with the bones of many extinct animals, including woolly mammoths, woolly rhinos, horses and cave bears. MacEnery suggested that the animals and humans had once lived together, but this idea was passed over by experts, including Buckland, as being due to accidental losses of tools by ancient Britons who had dug ovens in the cave floor. Despite the evidence, especially the failure to find any holes in the cave floor, the influence of experts, including Cuvier, was too strong to overcome reality. Although English geologists had visited Kent Caverns and agreed that humans had lived in association with mammoths and other, now extinct animals, this cave collection and that from other caves could not be dated with accuracy. Accordingly, these fossil associations were not accepted for what they really were.

What finally convinced the British of the existence of early humans was the evidence found at Brixton Cave. There, **William Pengelly** found stone artifacts with extinct animals in an excavation. Because of the earlier finds, the idea had been growing that perhaps humans had indeed lived at the same time as these now extinct animals many thousands of years ago, a different situation than the popular and prevailing view. It was decided that further excavations were to be carefully overseen by a committee from the Royal Society and the Geological Society of London. Examination by appointed experts of the remains found during these operations resulted in the acceptance of this important evidence that introduced modern humans to ancient humans.

In continental Europe, the French customs officer, **Boucher de Perthes** finally and firmly established in an epic work that ancient people made and used stone tools and hunted animals now long extinct. An amateur naturalist, he had watched the commercial excavation of river gravels along the Somme River near Abbeville, France and collected hundreds of fossil bones and stone tools (Acheulian Culture) found together, similar to those found by Frere. The implication was that the tools were deposited simultaneously with the extinct animals in the gravel deposits, and that humans had lived at the same time as the animals and probably hunted them. He published his results in several volumes in 1847 and created a stir in scientific circles. Like other workers, he was derided at first but was vindicated after geologists studied his sites two years later. Several prominent though skeptical British geologists, Prestwich, Lyell, Evens and Flower, insisted on examining these

Figure 11.6 Engraved mammoth tusk.

fossil materials and the deposits. Their survey unequivocally supported de Perthes' claims. It was obvious, based on all the evidence that was being amassed from many sites, that ancient humans had lived in the distant past, had made tools of stone and had successfully hunted large and small animals long extinct. The evidence was also clear that these humans had apparently been in Europe for a very long period of time. Unfortunately, this idea conflicted with the generally accepted date that placed the origins of Earth and mankind at about 6000 years B.C. This produced resistance in many people.

By 1865 Lartet and Christy had found an engraved piece of ivory, a mammoth tusk upon which a woolly mammoth had been carved (Fig. 11.6). It was then impossible to deny that humans had lived with these extinct animals from the past. Clearly, some artistic individual had been at work to create this beautiful carving. But who? The answer was apparent later when railroad builders excavated the rock shelter at Les Eyzies and discovered the remains of the original Cro-Magnon man.

THE GREAT VARIETY OF TOOLS

Tools were as important to the Cro-Magnon people as they are to us and appear to be the principal means by which they advanced their culture. Simple tools enabled the earliest humans, *Homo habilis*, to perform work that was otherwise difficult or impossible with bare hands and greatly enhanced their abilities to perform many tasks. Tools can be shown to have evolved through time and reflect the advances in cultures that demonstrate the abilities of the craftsmen. Although the early humans probably used any materials they could opportunistically find, stone tools are the only ones likely to have been preserved and found. Recognizing the importance of tools, Raymond Dart had proposed that the gracile *Australopithecus africanus* developed a culture using stone, bone, and wood tools, his osteodontokeratic culture. There is little evidence to support this culture, however, and it must be discounted without some new substantial proof. There is, however, ample evidence to support toolmaking in the descendant human species, *H. habilis*.

Our predecessors, *H. habilis* and *H. erectus*, are well known to have used stone for their tools and probably wood as well, but there is little evidence that they made and used wooden tools and none that bones or other materials were used. The Neandertals and the Cro-Magnons, however, used a great variety of confirmed tool materials for their implements and weapons, including ivory, antlers and bones. Given their ability to utilize these various materials, we should undoubtedly also include sea shells, amber, animal hides, grasses and numerous other materials that are not often found because they are too easily destroyed by the forces of weathering. Some of these materials were made into tools that were obviously basic utensils, but some others were probably simple decorations that the Cro-Magnons added to their wardrobes. Still other items became tools that were used to make and prepare their clothing, tents, basketry, ornamentation and even other tools. In this way, Cro-Magnons became the ultimate efficiency experts using all manner of natural materials and shaping them for their needs; like the Indians of North America, they wasted little. They used animal sinews for thread, animal bones for needles, animal antlers for picks, and much more. For the first time, humans were making tools and using materials for something more than the basic requirements of survival, they were indulging themselves with luxury items.

HUNTING WEAPONS

The Cro-Magnons learned to be much better at hunting than their predecessors, perhaps the most important step forward for these people. They made better tools and weapons and developed better hunting techniques. They were so efficient at hunting that many species became extinct in Europe, including woolly mammoths, woolly rhinos, saber-toothed tigers, cave bears and cave lions. Other species disappeared in Asia and still others disappeared later in North America after the emigration of the Indians from Asia. The disappearance of the very large and dangerous species, however, did not eliminate hunting by the Cro-Magnons. They had to eat to survive and continued to hunt other animals ranging from the large ones, like the horse, bison, antelope and wolf, to smaller ones like fox, rabbits, fish, seals and birds.

As the Cro-Magnons invented new weapons, tools and hunting techniques that increased their abilities, they became more efficient at life in general. Their stone tools were not radically different in design from the their predecessors' tools, but they were different in quality. They worked or shaped these tools to a much finer degree with greater care and precision. They removed more and smaller flakes to produce a delicacy of shape that served for various functions. Prideaux reported (1973, p. 66) that Cro-Magnon man made

> . . . knives for cutting meat, knives for whittling wood, scrapers for bone, scrapers for skin, perforators for making holes, stone saws, chisels, pounding slabs and countless others.

The Cro-Magnons made a major discovery when they learned how to increase an individual's strength with tools. Prideaux reported (1973, p. 66) that Cro-Magnon man

> . . . is believed to have taken to putting bone and antler handles on many of his stone tools, such as axes and knives. The handles increased his application of energy to the tools by as much as two to three times, by providing him with a firmer grasp and enabling him to use to a much greater extent the muscle power of his arm and shoulder.

This innovation allowed the Cro-Magnons to significantly advance their abilities in hunting much beyond that of the Neandertals, although there is some evidence to suggest that Neandertals may have made spears tipped with stone flakes. For the first time since the invention of tools, humans were able to exert much greater effort and strength than possible with their own raw strength and in other ways.

Though the Cro-Magnons invented many different tools which have come down to us through the years, probably the most important invention was the **atlatl**. This is a short stick about two to three feet long that was used to assist the throwing of a spear (Fig. 11.7). This device increases the leverage and force applied to the spear by means of a simple hook at one end that engages the spear shaft and a handle at the other. This atlatl, when thrown forward, extends the length of the throwing arm. Using such a tool, Cro-Magnons were able to throw spears more than twice as far as with the arm alone. The effective killing distance of such an atlatl-thrown spear was perhaps thirty yards and, with it, hunters were able to get within killing range of more animals than ever before, and this increased their kill rate. Because they did not have to get as close to animals, that is, within spear-stabbing range, they were at less risk themselves than ever before when hunting potentially dangerous large animals.

The hunting techniques of Cro-Magnons contrasted with the early techniques developed by *H. erectus*. *H. erectus* was able to drive large animals into areas of soft ground where they would become mired and killed. These people may have simply thrown stone missiles at these handicapped animals from a safe distance or perhaps closed to within arms length to stab or strike them with stone points or stone axes. Perhaps they simply waited until the animals died, then carved the meat at their leisure. The Neandertals improved upon such techniques by developing some better weapons and better hunting methods. Of great value to them would have been spears tipped with stones which enabled them to stab the prey at close range. Throwing spears are not particularly efficient against wary animals that can quickly outrun the hunters, but stabbing spears are excellent for use against animals that are mired in soft ground and cannot flee. The range of thrown spears is obviously limited and hunters must get very close to have any chance of success. Stabbing spears, on the other hand, require the hunter to move close to the prey. There the hunter is exposed to injury, especially from large animals and wounded animals that are fighting for their lives.

The development by the Cro-Magnons of the atlatl changed much of these hunting techniques and thereby

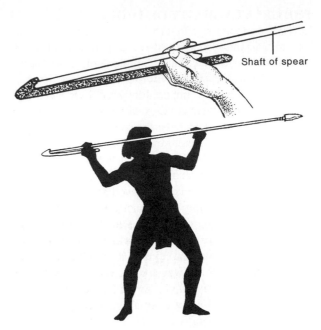

Figure 11.7 Hunter using an atlatl, a spear-throwing stick.

radically changed hunting. The Cro-Magnon hunters soon discovered that they could kill using atlatls from a relatively safe distance. Further, they learned that they could also hunt alone and still be successful. No longer was it necessary for them to close in on the animals and stab them with a spear nor was it necessary to hunt in groups. Using this new weapon, a solitary hunter could increase his kill rates with relative safety and provide a much better life for himself and his family than that enjoyed by earlier men. Different animals could now be hunted, and different techniques were developed to cope with these new prey. With each successful change or new development in the tools and hunting techniques, feedback provided positive reinforcements that led to other new developments and ideas that were of great value to them and their culture.

NONSTONE TOOLS

Although we dramatically visualize the Cro-Magnons hunting very large animals for their meat and skins, we rarely think of them as the hunters of smaller animals. But they hunted small animals of many different types, prey such as rabbits and squirrels. They also fished and even made nets to catch birds. Both fish and birds are much more difficult to catch than small animals that can be stoned or snared in traps. Fish live in a foreign environment and catching them required intelligence and skill. These hunters had to learn about fish behavior and feeding practices, just as modern anglers must do today. Cro-Magnons probably caught fish initially using their bare hands, then progressed to spears, to barbed sticks and small harpoons, then to nets, traps and hooks. Many of these tools or implements have been found. They made hooks and harpoons from bones, ivory, wood and soft stone. They even learned to make detachable harpoon heads from similar materials and learned to attached them, like hooks, to lines or strings to keep control of the quarry once speared. These harpooning and fishing lines were rarely preserved, but the presence of eyes on the hooks and harpoons clearly demonstrates their function and the presence of lines.

The use of strings, lines and cords demonstrated another major cultural advance for the Cro-Magnons, although it is possible that their predecessors also used these. The lines at first were likely vines or stems of tall grasses twisted together. The versatility and ubiquitousness of these materials would have encouraged them to continue and expand the practice, even making better examples and having them on hand beforehand rather than simply making them as needed. Once they had achieved mastery of the technique for making such lines or cords, it was a simple task to plait or braid strands of plant fibers or vines into thread, string and rope. More importantly, the manufacture of such plaited lines and threads was just a short step away from the weaving and tying of nets used in both fishing and fowling. These nets probably originated by modification from woven baskets and bags of skin that were intended initially to carry various objects, itself a major step forward in food gathering abilities. With such bags they could increase their carrying capacity well beyond that possible with their bare hands. This allowed them to take advantage of times of plenty, which in turn may have led them to stockpile items for times of shortage. Furthermore, the invention of such carrying bags provided an significant added benefit, because it freed their hands for weapons or other items.

CRO-MAGNON LIFESTYLES

Although hunting was a principal activity in which new tools played a major role, such tool development and invention carried over into many other routine activities beyond the basic demands of hunger. The Cro-Magnon's

lives were made substantially more pleasant after they had learned how to provide themselves with enough food for their needs and still had time for more than such necessities. With leisure time they were able to make and use a variety of tools that assisted them at almost every task, including preparing foods, making skin clothing, building shelters and creating art. Not only did they learn to use various materials to their advantage, but from these they made a multitude of useful objects, expanding and improving on them with time.

They made clothing and shelters from furs, skins and other materials, probably continuing the practices initiated by their predecessors. With much better skin and fur clothing, they not only survived during cold weather but were also able to migrate to less favorable climates, sometimes great distances to unknown lands. They obviously used sacks of skin for storage and were able to carry food and supplies long distances on their journeys. With such supplies they could risk travel into relatively inhospitable regions, something they were not able to do previously. They became highly adaptable, using every imaginable technique to survive and learned to live virtually everywhere. These techniques are, of course, the same types of activities that we practice today.

In many parts of the world, Cro-Magnon, modern human life was simple because the climate was mild and the resident humans had little need for shelter from the elements. In some parts of Europe, Africa and Asia, however, the climate varied with the seasons and the modern humans were forced to find protection from harsh and inclement weather. Like the Neandertals, they probably lived outdoors during the warmer parts of the year. When cold weather set in, they either erected tent shelters of furs and skins or sought caves and rock shelters to be out of the rain and snow. Tents can provide some protection during winters, but they are temporary shelters at best. Caves and shelters under rock overhangs, however, can be substantial habitations when warmed by fires, and they can provide security from hungry wild animals. One problem with caves is that they frequently were the dens of the giant cave bear, *Ursus spelaeus*. Obviously these fearsome animals had to be killed or driven out if the modern humans were to lay claim to the caves. Probably these attacks took place when the bears were asleep during their winter hibernations and could be approached with relative ease. With all the advantages afforded by caves, therefore, it is no surprise that Cro-Magnons took to living in them. After all, they continued the practices that the Neandertals had used before them in the same regions.

When the Cro-Magnons developed new hunting techniques and invented new tools and weapons, they began to move into environments that previously had been inaccessible to them because of climate, scarceness of animals and plants, or other reasons. Their new and better tools and techniques enabled them to survive in many environments that earlier peoples avoided, because they were marginal or even hostile to human survival. As a result, the modern humans migrated from place to place with greater ease than ever before, eventually to all continents except Antarctica.

Today we continue all the initiatives began by our predecessors and much more. We have improved upon the tools, techniques and practices of our ancestors, recent and ancient; we invent and manufacture a great variety of tools for our convenience and safety, for our needs and for our pleasures. We use these tools to grow foods, travel great distances, and communicate with each other almost instantly. This manuscript, for example, is being written on a computer that is a significant advance over a modern electric typewriter that was better than the old method of writing with pen on paper, and all of these are much better, quicker and more efficient than the ancient technique of using a wooden stylus to produce cuneiform writing on clay tablets. We have built upon experiences of earlier people and have learned how to make tools to perform almost any task and do it faster and better then ever. And each year we see the development of new tools and new ways of doing things.

CRO-MAGNON DISTRIBUTIONS

By the height of the last Pleistocene glaciation, modern humans are known across most of the Old World. Their range and distribution was obviously influenced by the plants and animals upon which they depended for survival and by the presence of the great continental glaciers that covered so much of northern Europe and mountains elsewhere. In those days Europe was very different from the landscape seen today. Most of the countryside was wooded, probably like the present-day evergreen-forested lands of northern Europe. In western Europe, hilly woodlands dominated, and the principal prey animals were reindeer. In eastern Europe the landscape was much flatter with miles of grassy plains. Here, the principal food animals included

the woolly mammoth which thrived on the abundant grasses and were not deterred by the cold and snow. In northern Spain, where the landscape and topography of the Pyrenees Mountains were quite rugged, the game animals did not form large herds and the meat supply was more varied.

Wherever the animals lived, in th forests, the mountains or the plains, they were a significant source of food for humans long before the appearance of the modern humans. When these animals migrated because of shifts in the changing climatic conditions during the last years of the Pleistocene, the hunters followed. As modern humans appeared, they did not suddenly change lifestyles, they simply continued the hunting practices of their predecessors with improvements. Because they were developing specialized tools and skills, few climates and environments were closed to them; they were a highly capable people able to find foods and survive almost everywhere.

THE SPREAD OF MODERN HUMANS

By all evidence modern humans should have been able to appear with relative ease through the Middle East, eastern Europe, western Russia and Asia. The few remains that have been found in these areas, however, tend to be younger than expected and are roughly equivalent with the dates of their remains in western Europe. These dates suggest that modern humans were somehow inhibited or prevented from traveling during the period from perhaps 75,000 to 35,000 years ago. A logical explanation, if this is a valid pattern, is that climatic conditions blocked or restricted their migrations. The Würm glaciation was in full swing in Europe about 75,000 years ago and then slowly declined into a warming trend. When conditions ameliorated enough, the modern human people could have easily penetrated into Europe, presumably around 35,000–40,000 years ago, producing the pattern now observed.

In central Europe and western Russia, the few fossil remains that have been found may indicate a relatively small population of modern human inhabitants. These regions were largely grassy plains with few trees and rocky hills. Caves are rare in this region, and this lack of natural shelters may have restricted their population. Humans settling there would have been severely handicapped during the harsh winters, especially if they could not find caves for shelters. Remarkably enough, the people in eastern Europe and western Russia found substitute habitations for caves and were able to endure and survive. The game animals of western Europe were not common in these eastern plains. Instead, woolly mammoths were abundant herd animals that thrived on the lush grasses. There were some advantages to hunting such large animals, and these people learned to hunt them despite the dangers. Each animal provided a great quantity of meat, and a single kill could feed an entire hunting band for some time. Though the meat was valuable to these hunting people for their very sustenance, other parts were also of great value. The mammoth's great shaggy hides could be used as effective protection against the bitter cold, either as clothing, blankets or coverings for shelters. Their bones were also sufficiently large to be of use in the construction of shelters. Though seemingly far-fetched, the remains of mammoth-bone-supported shelters have been found at Dolní Vestonice in Czechoslovakia, and at Kostenki and Gagarino and elsewhere in Russia. Dated at about 25,000 years old, these structures were commonly built in slightly excavated depressions about fifteen feet in diameter. The very large mammoth bones were apparently used for walls and as supports for roofs of mammoth skins. In some cases, several of these structures were linked to produce a long dwelling that could have sheltered several families. With campfires available for cooking and heating, such dwellings would have been more than adequate shelters during the long, cold winters. They are not radically different from shelters used by American Indians and Eskimos.

The dependence on these animals for such foods undoubtedly demanded a nomadic lifestyle as these hunters followed the mammoth herds on their migrations across the plains. Not only could these mammoths provide huge amounts of fresh meat when killed, but they also could have provided abundant supplies for the future. These hunters certainly must have been aware that sufficiently cold temperatures would freeze the meats and keep them from spoiling. Humans are adaptable if nothing else, and these people would have quickly learned to freeze meat for their future needs. Then, when winter hunting was poor, they would have been able to eat some of their preserved meats as needed, just as we do today with our mechanically frozen meats and other foods.

The mammoths were to these modern human hunters as the buffalo were to the nomadic American Plains Indians. They provided much more than just meat and hides. Most peoples living today and dependent upon hunting will utilize as much of their kills as possible within the limits of effort. They use the bones, sinews, tendons, intestines, various internal organs and much more for food and other purposes. The mammoth hunters would have been equally opportunistic. Evidence from the Kostenki site indicates they used the bones for many different purposes. They raised them as the frameworks of their dwellings, carved them into pins and needles, modified them into shovels, shaped them into numerous small tools, and drew artwork on them. In their wood-starved environment they even burned bones in fires as other peoples used wood. Like modern hunting people, they had to be clever and inventive to survive, and they were both of these. When the mammoths were finally killed off, these people had learned more than enough about survival to find alternate foods.

CAVE ART

The ultimate achievement of the Cro-Magnons was their development and use of **art** that has been found on cave walls (Fig. 11.8) and on various objects. Some Neandertals have been found with ground pigments that they probably used to paint colors on hides or their own bodies, but there is no direct evidence of any art. Cro-Magnon artists, however, were prolific in producing quantities of large and small sizes of art and in many different media. They engraved bones, sculpted statuettes and created magnificent drawings and paintings that have been preserved in more than one hundred caves and rock shelters in France and Spain. Many of this art has been interpreted as having religious, ceremonial or ritualistic significance. In general, however, their art was commonly associated with animals, particularly those that were hunted for meat, hides, and other body parts. The Dordogne region of France, for example, has provided numerous examples of cave art. Few examples of this art have been found in the rock shelters under cliff overhangs, but these habitations are somewhat exposed to the weather and any art was likely destroyed by its effects over the years. The caves, fortunately, provided numerous and spectacular examples of their artistic abilities and skills.

Although much is known of Cro-Magnon cave art, its significance is not yet fully understood. The caves that contain the artwork are typically deep, dark caverns with large and small passages and rooms. These works of art were not casual graffiti painted at random, but appear to be carefully planned creations that required much preparation. The paintings are generally found deep within these

Figure 11.8 Drawing of a small horse in Lascaux Cave.

caves, places that are distant from the entrances and pitch dark. The artists then required torches or lamps for illumination, scaffolds to reach high on the walls and ceilings, and materials for the actual painting. They had to understand their pigments, how to make them and how to use them. They had to have explored these caves previously to find surfaces suitable for the project, then decided on a course of action for these endeavors which included the gathering of the minerals for pigments, brushes for the actual painting, tree saplings for the scaffolding, cords to lash the scaffolding, and lamps and fuel for lighting. Then they had to carry all this deep into the caves. Apparently these caves were never used as habitations but were restricted for use as art galleries. The reason for this is not known but a reasonable explanation is that this art was used in ceremonies or for shrines.

Virtually all of the Cro-Magnon cave artwork is of animals that have been painted on the walls and ceiling. In some cases, the contours of the walls have been used to give realistic relief to the animals. Some animals have even been carved as statuettes on the cave floors from clay. The animals depicted in the cave art include horses, bulls, deer, mammoths, bears, cattle, bison, wolves, wild boars and others. These are commonly portrayed in great profusion in some caves in realistic colors with shape and grace that would impress a modern art critic (Fig. 11.9). The animals painted are thought to be the principal meat animals hunted by these people. Occasionally both male and female humans are portrayed. For the paint media they used various colored, ground, mineral pigments mixed with a grease vehicle. Some of the animal paintings and some of the sculpted clay figures have been marred by spears, stones or sticks. It is speculated that this marking occurred in ceremonies as if the animal paintings and figures were actually being hunted. Other animals are shown pregnant, possibly suggesting a desire for these animals to reproduce and replenish the stock lost to hunting.

Some cave art was painted in places where viewing is inconvenient or difficult. Such placement strongly argues that this particular artwork was not intended for casual viewing but was reserved for special occasions, perhaps for ceremonies or rituals. The purpose of these ceremonials is unknown, but some workers have suggested that they may have included rites of passage ceremonies in which adolescent boys were accepted into manhood, much like those practiced by hunting peoples around the world today. Others think these ceremonies were deliberately practiced in hidden places to avoid public scrutiny. In such a case they may have been religious in nature and restricted to selected members of the tribes. Perhaps these ceremonies included prayers to supernatural forces requesting the abundance of game animals and successful hunting.

Altamira in Spain and Lascaux in France are the premier galleries in which cave art has been preserved. Altamira was discovered in 1868; its ceiling is covered with magnificently painted and preserved animals, many life size, in shades of red, brown, yellow and black. Many of these figures, mostly bison (Fig. 11.10), were painted to take advantage of the rock irregularities and produce a realistic three-dimensional effect. Unfortunately, many scholars initially rejected this artwork and insisted that it was the work of a contemporary artist; they argued that the colors were too rich and bright to be more than a few years old. It was 1895 before this cave and its art began to be accepted as the work of ancient people. Then, another cave filled with very similar art was discovered in France at La Mouthe. Subsequent discoveries shortly thereafter in France at Font-de-Gaume and Les Combarelles finally established the truth of Altamira art discoveries. Today Altamira is known as the Sistine Chapel of the Ice Age.

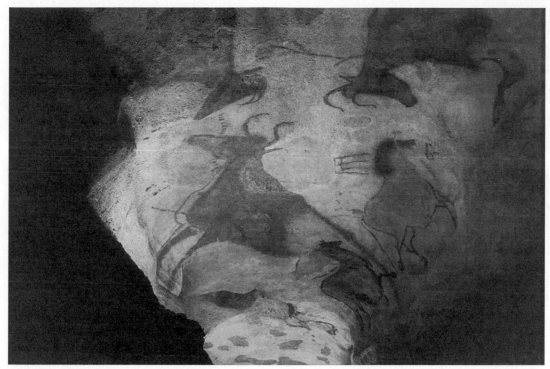

Figure 11.9 Ceiling of Lascaux Cave showing various animals.

Possibly the most spectacular example of cave art was found in 1940 when some boys followed a dog down a hole in the ground at Lascaux in the Dordogne. They discovered amazing paintings of bulls, stags and horses in brilliant color that had been created some 17,000 years ago (Fig. 11.11). These animals, in more than 600 paintings and nearly 1,500 engravings, were depicted in motion, often in wild run, in direct contrast to the stately repose of the Altamira animals. Prideaux (1973, p. 108) noted

> Where the Altamira artists had a firm control of color and movement, the Lascaux painters applied pigments loosely and used wavy lines that are almost baroque in their swirl and dash. Where Altamira appears classic and orthodox, Lascaux is freewheeling and, to modern viewers, exotic.

Unfortunately, the opening of Lascaux to the public has been its downfall. Visitors have introduced great quantities of carbon dioxide just by breathing, and their warm breath caused condensation on the walls. In 1958 an aeration system was built to filter the air, cool it, and remove humidity and carbon dioxide. By 1962, however, green spots of algae, energized by the lighting, were discovered growing on the walls and these rapidly growing plants threatened to destroy the very art work that vistors came to see. In 1963 the cave was temporarily closed to the public so that experts could spray chemicals to stop the algal and bacterial growth. By 1965 the cave pollution was arrested, but the cave was closed permanently to the public to prevent future destruction. Today, the government of France has erected a duplicate of Lascaux Cave nearby so that tourists can see the wonders created by Cro-Magnons so long ago.

Besides the cave paintings, Cro-Magnons created many other types of artwork. They carved sculptures from stone, bone, ivory, and clay mixed with bone powder. The latter sculptures were fired, perhaps in a kiln. This heating turned the mixture of clay and bone powder into a hard ceramic. Evidence for this was recovered at Dolní Vestonice, Czecho-slovakia that indicates a workshop to produce such objects was in operation there. Numerous fragments of animal heads, animal figures and human figures were scattered across the shop floor, including many broken examples of this artwork along with scraps of waste clay-bone material.

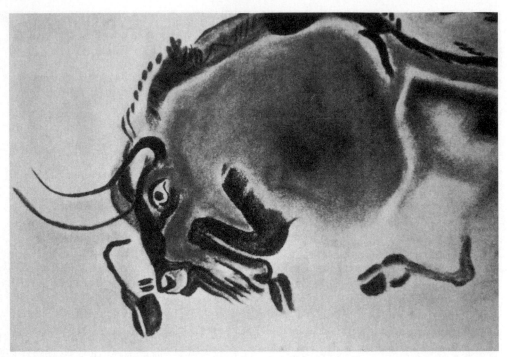

Figure 11.10 Painting of bison in Altamira Cave.

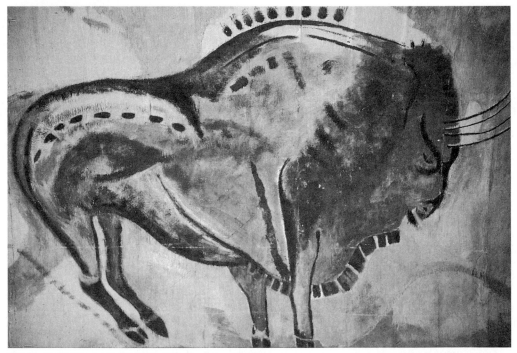

Figure 11.11 Figure in the Hall of Bulls, Lascaux Cave.

Figure 11.12 Willendorf Venus.

This workshop is dated at about 27,000 years ago and predates by some 15,000 years the next oldest use of fired clay being made into pots. It also represents the oldest known use of a fired technology that is common today.

More than one hundred statuettes carved in stone and ivory or shaped of clay have been found from France to Siberia. These are generally small in size and range in age from 27,000 to 20,000 years. They are commonly known as **Venus Figurines** and typically are voluptuous caricatures of the female body, commonly swollen with pregnancy, with enlarged breasts, buttocks and bellies, but with little detail. One example, the **Willendorf Venus** (Fig. 11.12), was found along the Danube River near Vienna, Austria. They are thought by most experts to be fertility figures representing the "mother-goddess revered by the Cro-Magnons as the source and protector of all things good." Their widespread distribution across Europe suggests that they represented a theme common to many different groups of modern humans, perhaps as objects used in religious practices.

Other objects were also turned into works of art by these modern human artisans who engraved various designs into their surfaces. Sometimes these engravings were geometric patterns without apparent function except beauty, at other times they were realistic carvings of animals. Gracefully engraved spatulas have been found, as have atlatls, buttons and more. Their artistic bent was unrestricted. Of particular importance were engravings found on bone and other materials that appear to be numerical counts. In one case (Marshack, 1972, 1991, p. 193), the counts are associated with an obviously pregnant horse and seem to be related to the gestation period. Another carving on a piece of reindeer antler (Fig. 11.13) seems to mark phases of the moon (Marshack, 1972, 1991, p. 193). If correct, these two pieces indicate that humans had developed a method of making written notations some 30,000 years ago, about 25,000 years before the earliest appearance of cuneiform writing by the Sumerians.

Though the Cro-Magnons produced some magnificent works of art by any standards, they did not totally neglect their own images. One painting on a wall at Lascaux depicts a male as little more than a stick figure. Other drawings and sculptures depict human bodies with animal heads and the outlines of hands. These give rise to speculations that these hunters sought power over these animals, perhaps in religious or other kinds of ceremonies. Howells (1958, p. 208) reports

Figure 11.13 Reindeer antler with counting marks.

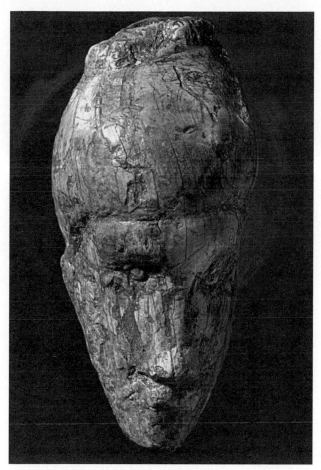

Figure 11.14 Human head carved from ivory.

The one actual painting of a man so far known, a small bas-relief at Angles-sur-Anglin, gives us a White man (actually a lightish purple, but the effect is white), with black hair and a black beard.

The paintings of animals are often brilliantly created, many in color, but most human figures are very poorly executed. They typically lack depth, detail and emotions. A notable exception, the "Sorcerer," is a vivid dancing figure. One marvelous sculpture of a human head (Fig. 11.14) was found in Czechoslovakia near Dolní Vestonice in the 1890s. This head, carved from ivory, is dated at about 26,000 years ago and could easily be of a human from today. Remarkably, this head had eye sockets, eyeballs, and eyelids.

ORIGIN OF THE CRO-MAGNONS

The origin of the Cro-Magnons, the first identified fossils of modern humans, has long been a muddled picture confused by the distribution in time of the relatively few specimens. At least it was until relatively recently when the sheer mass of accumulated fossils of these and earlier forms allowed for acceptable interpretations of their positions within the family tree and their origins. Even these views, however, are not without their difficulties. Originally, many of the various fossil specimens of early humans were considered sufficiently different from each other that they represented distinct species; they were named accordingly as different forms somehow related. As more fossils were found, these new specimens were grouped into fewer species and better relationships were plotted. Various evolutionary pathways were proposed to account for the origin of modern humans. Today, however, these paths have been narrowed to one.

The evolutionary origin of the Cro-Magnons from the earlier species *H. erectus* is generally accepted by most scientists today, but there are diverging opinions among these experts as to exactly how, when, where and why this evolution took place. The basic pattern involves *Homo erectus*, the archaic sapiens, the Neandertals and the modern humans. Many of the details of the pathway remain unclear in the absence of sufficient fossils distributed at critical points in time. The various hypotheses proposed to explain this pattern have considered the unusual distributions in time of the different specimens of archaic sapiens, the two types of Neandertals, Classic and Progressive, and other problematic specimens with mixed modern and primitive characters, and the role played by the ancestral stock, *H. erectus*. Especially crucial to these hypotheses are explanations for how these various forms arose, their status as humans, and their relationships to modern humans and to their ancestral stock.

Complicating this issue is the factor of climate; the important phases of this evolution into modern humans took place during the Pleistocene Epoch when intervals of significant cooling resulted in the development of great ice sheets that alternated with warming trends and melting. Such climatic conditions obviously exerted great influence on the plant and animal life of many areas, particularly Europe where the Classic Neandertals were first discovered and where they reached their greatest expansion. But how much influence did these conditions exert?

It is clear, therefore, that pathway to modern humans is generally understood. A new level of inquiry is now in full bloom as more researchers are attempting to find the answers to these questions. Obviously, more fossils must be found to establish the living range of Neandertals, to understand where, when and how the ancestral stock began to develop the modernizing trend. More specimens must be found to satisfy the questions of the disappearance in Europe of the Neandertals after the Cro-Magnons migrated there, and so much more. One of the infuriating aspects of science is that finding answers to questions inevitably generates even more questions. That is certainly the case with human origins. Somehow we must satisfy some of the questions that abound in this intriguing mystery. After all, these are questions about our ancestry, about where we came from and how we got here.

SUMMARY

The Cro-Magnon people were first discovered in the caves and rock shelters of the Dordogne region of France shortly after the Neandertals had been found. These people were clearly different from the Classic Neandertals both physically and culturally. They were taller, more slender, and had more delicate features. They were more similar, moreover, to the Progressive Neandertals of the Middle East. Their stone tools were sophisticated and highly variable to suit the myriad tasks they performed, often being very delicate. They were so abundant and obviously important that Lartet and Christy referred to their period as the Stone Age. Their presence in great numbers and variety led to the recognition of several cultural subdivisions; even these subdivisions were further divided.

The Cro-Magnons appeared fully developed and suddenly in Europe only about 35,000 years ago. Once present, they quickly became the resident population, displacing or somehow superseding the Classic Neandertals. No single explanation for this sudden population change has satisfied all researchers.

The culture of the Cro-Magnons was clearly superior to that of the Classic Neandertals even though they apparently had less physical strength. They relied instead on their greater intelligence to compensate for this difference. Their tools, more than one hundred different kinds, and the ways they used them all reflect this asset. What they learned was taught to their offspring, and this trait quickly increased their storehouse of knowledge in their improved brains.

Though stone tools were known and documented in Britain from 1715, their precise nature had not been determined at that early time. A number of curious individuals explored caves and other sites to recover these artifacts, and they developed a picture of the past that was not accepted until the mid-nineteenth century. The Cro-Magnon people were discovered shortly after this and were confirmed as the toolmakers. Eventually these tools allowed recognition of their lifestyle that was essentially modern. They hunted for all manner of large and small game, fished with nets, hooks, harpoons and spears, made baskets and bags to carry various materials and used their tools to expand and exploit their environments. They became adept at making clothing, tent shelters, ropes and cords, sacks and knapsacks, and they occupied caves for shelter against winter cold. They did not hesitate to attack and kill cave bears to preempt their winter dens. All these assets, especially the tools, allowed these modern humans to spread into all variety of environments, including those that had been previously uninhabitable because of severe conditions. If they found a local environment hostile, their intelligence allowed them to adapt to it.

The modern humans appeared in Europe fully developed only about 35,000–40,000 years ago. Clearly they had been evolving for some time, but not in Europe. Dates of modern human finds in the Middle East, eastern Europe and western Russia are all much younger than expected, leading researchers to seek explanations. Climatic conditions of the Würm Glaciation may have been the barrier that kept these areas free of modern humans. When the Würm began its slow decline they moved into the region to exploit its potential. The region still must have had harsh winters because they constructed shelters of mammoth bones covered with mammoth skins. These modern humans had been evolving elsewhere for some time. But where?

The caves of Europe were more than just habitations for the Cro-Magnons; uninhabited caves were sites for ceremonies. These ceremonial caves are covered with art of great quality and skill. Such art gallery caves are typically deep and dark, and the art was the result of sophisticated efforts involving exploration, lighting and scaffolding. Almost exclusively of animals, the art was likely not decorative in such remote, dark places of difficult access. Instead, this art may well have been used in ceremonies; support comes from various lines of evidence. Other Cro-Magnon art takes the form of sculptures and carvings. Bones, teeth, antlers, tusks and other materials have been utilized for both decorative and utilitarian purposes. Some even record counts and possibly astronomical information.

These modern people, our ancestors, were highly successful in Europe as elsewhere. They utilized behaviors, ways of surviving, and lived in cultures not too different from that of their ancestors from whom they had learned much, modifying it as needed, improving it slowly. Eventually, as they gradually accumulated knowledge, these people developed cultures that became our cultures and the techniques that led to our technology. It is not difficult to imagine Cro-Magnon people clustered around a back-yard barbecue grill, fixing dinner on a warm summer evening just as we do. Such a situation probably has not changed radically over time. Though our setting and our devices and our foods are a bit more sophisticated, the basic pattern is the same. The difference between our backyard grilling of foods and their cooking is not much more than the time in which to accumulate additional knowledge and experience and put them to use. It is more difficult to visualize a Cro-Magnon family grouped around a computer, yet again the difference may be nothing more than the time and experience to develop the computer technology that we as intelligent beings have produced. Certainly their art is fully as emotionally and intellectually satisfying as our art. The obvious explantion is that they are essentially human beings like us.

STUDY QUESTIONS

1. Describe the environment of western Europe inhabited by the earliest *Homo sapiens* people.
2. How do the brains of modern humans differ from those of Neandertals?
3. How do the skulls of modern humans differ from those of European Classic Neandertals?
4. How do the stone tools of Cro-Magnons differ from those of Neandertals?
5. How do the Progressive Neandertals relate to the modern humans?
6. Describe the criteria upon which the first subdivisions of *Homo sapiens* culture were based.
7. Explain the cultural differences between the Neandertals and the Cro-Magnons.

8. Explain the sudden appearance of the modern humans in western Europe.
9. Explain the disappearance of the Neandertals in Europe shortly after the modern humans arrived.
10. How did the hunting techniques of the modern humans differ from those of the Classic Neandertals?
11. Why does the Cro-Magnon cave art support the view that these people had ceremonials?
12. Cite any evidence for unique technologies of the Cro-Magnons.
13. Differentiate between the Cro-Magnon people and modern humans.

12 The Rise of Modern Humans

INTRODUCTION

The basic pathway of human evolution leading to the Cro-Magnons, the first identified modern humans, is reasonably well known in broad terms, but many of the details are unclear in the absence of sufficient fossils at crucial intervals in time. In recent years the sheer mass of accumulated fossils has allowed for acceptable hypotheses to explain this evolutionary pathway. There is no reason to believe that the australopithecines were not direct ancestors of the genus *Homo*, nor any reason to believe that the hominids did not evolve from the apes. The fossil evidence combined with biochemical studies amply demonstrate these evolutionary relationships.

With each discovery of a fossil hominoid and hominid specimen a little more information was added to the collective knowledge of human evolution. Until a sufficiently large number of these discoveries were made to allow for recognition of a pattern, the lineage remained enigmatic. Each specimen was given its own name and assigned to a position relative to the other few known fossils. Some were included in other species but many were considered sufficiently distinct to warrant their own taxon. Today, however, the numbers of specimens available for study has increased enough to revise these early thoughts and patterns. Today, this early view of human evolution has given way to the idea that these types were not so separate and distinct after all and that a more coherent picture of the lineage can be offered. Howell (1984, p. xiii) said,

> There is now a near consensus among students of human evolutionary biology that the origins of our species, *Homo sapiens*, is somehow intimately linked with the first intercontinental ancient hominid, *Homo erectus*. However, neither the transformation of *erectus* to *sapiens* nor the transformation of ancient (archaic) populations of *Homo sapiens* to their anatomically modern succedents (*H s sapiens*) are matters of agreement in this scientific fraternity.

There is a strong case that *H. erectus* had evolved from *Australopithecus* through *H. habilis,* and then into *H. sapiens* over a span of perhaps 200,000 or more years (Fig. 12.1). This transition of *erectus* into *sapiens* was a

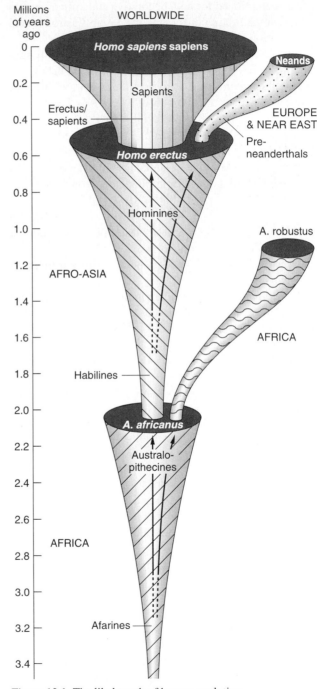

Figure 12.1 The likely path of human evolution.

geologically quick process that began about 300,000–400,000 years ago when some specimens of the existing human population, *H. erectus*, that had spread around much of the Old World, began to change. These new types, collectively termed archaic sapiens for those specimens that fit between *H. erectus* and modern humans, did not evolve rapidly from the general *erectus* pattern but developed very slight changes over time that eventually resulted in their evolution into the new species, *H. sapiens*. The dividing line between these two species, *H. erectus* and *H. sapiens*, cannot be sharply drawn using specimens from this time interval, suggesting that this change is probably a gradual one. And this well illustrates Darwin's fundamental view that changes in species take place gradually over time. But, however slowly these changes took place within the genus *Homo*, there is little doubt that *H. erectus* evolved into *H. sapiens* somewhere in the Old World as these people gradually became more modern.

Various hypotheses have been proposed to explain the distributions of humans in time, particularly the patterns shown by the different populations of archaic sapiens, the Classic and Progressive types of Neandertal, and other specimens that exhibit a blend of modern and primitive characters. Reasonable models offer the view that anatomically modern *H. sapiens* could have evolved in several different ways from *H. erectus*. Especially important are the explanations for the Classic and Progressive types of Neandertals, how they arose, their status, and how they relate to modern man and to the ancestral stock. Furthermore, these important phases of the evolution of modern humans took place during the Pleistocene Epoch, the time of the great Ice Age, when the climate cooled and formed great glaciers that alternated with warming trends and melting of this ice. Such conditions undoubtedly exerted great influence on the plant and animal life in many regions, particularly Europe where the Classic Neandertals were first discovered and where they reached their greatest development. Constable (1973, p. 26) pointed out that the extremely primitive characters exhibited by Classic Neandertals could have resulted from changes in the European environments due to the Pleistocene glaciations. He maintained that

> . . . the Neanderthals were a chill-resistant breed who arose in eastern Asia and migrated to Europe during a bitterly cold glacial period; meanwhile the progressive Swanscombe

people had the good sense to get out of Europe and go to Africa or tropical Asia. Later, when the weather improved, they snatched back their old territory from the Neanderthals.

Many workers accept that the general pattern of Pleistocene climatic reversals influenced this portion of hominid evolution, though exactly what effects it had on this process and the evolved individuals is largely speculative. Other workers tend to reject this hypothesis in favor of other views, but the very idea that climate could have been an important factor in this evolutionary process leading to modern humans should be self-evident. It is, after all, a major aspect of natural selection, a fundamental part of Darwin's original concept of evolution.

THE ANCESTRAL SPECIES— *HOMO ERECTUS*

The ancestral stock that gave rise to modern man, *H. sapiens*, was undoubtedly *H. erectus*. There is no other viable candidate for this role and *H. erectus* fossils show structural changes at the appropriate time in the right direction for such a relationship. Populations of the stock had likely originated from *H. habilis* in East Africa about 1.5–2 million years ago and lived until about 300,000–400,000 years ago. These oldest unquestioned fossils of *H. erectus* have been found along the Rift Valley at Olduvai Gorge (Upper Bed II), at Koobi Fora (KNM-ER 3733, see Fig. 9.8), at the west side of Lake Turkana, and at Olorgesailie. These people could easily have spread along the abundant and food-rich watercourses of the Rift Valley that extend from South Africa into the Middle East. Their fossils have been found at many places along this relatively hospitable landscape all the way north to Latamne in Syria.

Both the appearance and disappearance of the *H. habilis* stock were very gradual and took many hundreds of thousands of years. The record of this long history is, unfortunately, incomplete at this point in time; not enough fossils have been found to fill in all the blanks. The hominid descendants of this stock, *H. erectus* and *H. sapiens*, are known from fossils also distributed unevenly through the geologic record. It is unknown if significant hominid evolution took place in these gaps in the fossil record, and this irregular pattern has complicated our understanding of possible pathways from *H. habilis* to *H. erectus* to *H. sapiens*. There are, however, no major structural changes in this lineage and the fossil evidence, thus, records a slow, general trend toward modernization

Figure 12.2 The spread of *Homo erectus* from Africa.

in which many individuals of the late *H. erectus* and early *H. sapiens* populations developed some modern features relatively soon in the evolutionary process.

H. erectus migrated from Africa throughout the Old World (Fig. 12.2) seeking suitable locales for their needs, perhaps by following migratory game animals. These sites, which may be characterized as having acceptable climates, ample water, and adequate supplies of both edible plants and game animals, were not too difficult to find. The far distant island of Java, where the first specimens of *H. erectus* were found, was just such an ideal habitat with all of these criteria, just as it is today. Recent work indicates that Java man may be as old as 1.8 million years.

At this time, about 1.8–1.5 million years ago, much of the world enjoyed a relatively mild climate when the Pleistocene ice age began (see Table 10.1). Suitable vegetation could be found in many places and such plant foods could be had for the picking. The early *H. erectus* people are not characterized as exceptional hunters like their descendants and they probably subsisted mostly on plant foods that they gathered and meats they scavenged. There is, nonetheless, some evidence that suggests that the earlier robust australopithecines may have eaten more meat than previously believed (see Chapter 7) and that *H. erectus*

could have learned its meat eating ways from *H. habilis*, a contemporary of the robust types. Eventually, however, the *H. erectus* people began to include more hunting in their way of life as shown by the increased number of animal bones found in their caves. Perhaps this emphasis took place in temperate regions like Zhoukoudian where Peking man was first discovered. In such regions plant foods are readily available only during the warmer seasons and have very limited potential for supporting a population or band of any size throughout the entire year. During the winter season in such areas the local bands of these people would have probably turned to hunting to provide more of their needed food. When this happened they would also soon have discovered that meats have much more caloric value than plant foods and provided more energy. Perhaps after they had learned to use fire, they also discovered the delights of roasted or barbecued meats; there is ample evidence of campfires and charred bones at Zoukoudian. Whatever the reason, these people began to hunt more than earlier peoples and became proficient at it.

The trend to hunting behavior in *H. erectus* was natural and in time these hunters learned to successfully exploit game animals and obtain adequate quantities of meat. More meat meant more food, and the human populations probably increased in proportion. Because an increase in population required more foods, the supply of plants and animals in any area was depleted faster than if the human population was small. Larger bands or tribes would have required the hunters and gatherers to travel farther afield to find adequate foods, perhaps even migrating with the animals during their seasonal wanderings. Greater travel meant much more difficulties and greater effort from the hunters and gatherers. Increased hunting, for example, tends to increase the wariness of the prey animals. As these animals became more cautious, hunting became more difficult, and the hunters had to develop greater skills for continued success.

The natural solution, of course, was for an inflated population to separate into groups small enough to be supported by a given area. Obviously, fewer people locally could survive without overharvesting the plants and the animals. Small groups could have found a better life by simply leaving a main tribe for a productive area some distance away. This process, termed **budding**, results in the development of small subgroup populations. In effect, because these people were successful and increased in numbers, they had to separate and disperse to survive. That they were successful can be seen in the fact that they had spread from Africa across Asia to Java by about 1.8 million years ago.

About 600,000 years ago the inhospitable cold of the Mindel Glaciation covered the mountains of central and northern Europe and may have restricted or prevented the influx of *H. erectus*. Conditions along the Mediterranean Sea, however, were probably very pleasant and much more suitable for life. These people had already spread into the relatively hospitable Middle East with its climates and varied foods, and their migration along the coasts of the Mediterranean was to be expected. Their remains, dated around 600,000 to 400,000 years ago, have been found along the Mediterranean coast as far west as Algeria at Ternifine, at Salé, Rabat and Sidi Abderrahman in Morocco and Terra Amata near Nice in France. Evidence from their camps at these sites suggests they were successful in obtaining a wide variety of foods from these areas. They ate fish and shellfish from streams and the nearby shallow sea and supplemented their diets with fruits, seeds and berries picked in season and with roots, tubers, leaves and other plant materials that were collected as available. It is clear from this abundant evidence

that they were capable hunters and gatherers able to migrate along the sea coasts of the Mediterranean where they survived on foods readily available.

Once the *H. erectus* people became established along the coast, they would likely have spread out as roving groups and occupied all of the available sites, either for permanent or temporary encampments. As their numbers increased and they began to deplete the available local food supply, the bands began to split up to find their own hunting territories; some individuals and groups undoubtedly moved inland. The retreat of the great ice sheets had left behind vast areas of exposed fertile ground that consisted of clay, sand, gravel and boulder fields deposited from the melting glaciers. This rich surface quickly sprouted a cover of vegetation that attracted migrating herd animals, and these animals attracted the hunting peoples who followed them away from the coast to the continental interior and northern Europe. The major rivers, in particular, were important migration routes to these animals and became so to these nomadic people.

The presence of these people at Vértesszöllös in Hungary, at Heidelberg in Germany, and elsewhere demonstrate that they found suitable living conditions well inland and far from the coasts. Fossils of butchered elephants found in Spain show that these people had spread inland there where they effectively exploited the available resources of the countryside. They probably also migrated all the way to the Atlantic coast at this time. Their adaptability to survive in so many different environments and to eat a great variety of different types of foods contributed significantly to their successes in so many places. This ability was emphasized even more when cooking was discovered and enabled them to convert many previously inedible foodstuffs into tasty and desirable foods.

From the Middle East these people spread eastward into Asia, reaching far-off Java by perhaps 1,800,000 years ago and China thereafter. The earliest discovered specimens of *H. erectus*, Java man, were typical of that species in that they did not evolve uniformly and totally toward modern man. They had evolved an essentially modern pelvis with modern legs and feet even as they retained other more primitive characters. This blend of primitive and evolved features is well seen in their skulls which exhibited typical primitive features including prominent brow ridges and a prognathous snout. At the same time, their skulls contained evolved brains of about 900 cc volume, about 200 cc larger than those of the first humans, *H. habilis*. This form of *H. erectus*, Java man, then evolved very slowly and with little change of features presumably into Peking man, *H. erectus pekinensis* (see Fig. 9.5). This latter stock differed essentially in having a slightly more modern appearing skull, a slightly larger brain of about 1,000 cc and more advanced behaviors. By this time, about 600,000 to 500,000 years ago, they depended on caves for shelter during harsh seasons and had learned to use fire. Evidence from southern France shows also that they constructed temporary shelters and were able to carry objects in bags or sacks.

H. erectus pekinensis was originally discovered in cave deposits at Zhoukoudian, just outside of Beijing (Peking), China. These deposits indicate that they had inhabited this site for several hundred thousand years, from about 600,000 to about 250,000 years ago. They were probably able hunters who took a wide variety of large and small animals for food. Some of these animals were swift, others were large, and both probably required sophisticated hunting techniques. The hunting of large and swift animals was undoubtedly a cooperative activity similar to that practiced at Ambrona and Torralba by their European cousins. There, in Spain, the remains of elephants suggest they had been trapped in boggy ground by the hunters and had been systematically butchered some 400,000 to 300,000 years ago. The remains of numerous animals at Zoukoudian, Ambrona and Torralba provide adequate testimony of the ability of *H. erectus* to obtain large quantities of meat.

Recent discoveries (1988) of bone fragments of various animals from the cave at Swartkrans in South Africa indicates that they had been burned, probably in campfires. These bones can be anywhere from 1 to 1.5 million years old and could represent the earliest known use of human-controlled fire. It is unknown whether these bones were from meat that was being cooked or whether they were simply refuse thrown into the fire. Though bones of both *H. erectus* and *Australopithecus robustus* have been found in the Swartkrans cave, neither species can be positively identified with the fire-burned bones. However, in view of the fact that younger *H. erectus* is known to have used fire, that species must be the logical choice as the user. Other evidence from Kenya suggests fire was being used there by *H. erectus* sometime after 1.4 million years ago. The quantities of charcoal and ash at three younger sites, Zoukoudian, Ambrona and Torralba, amply record the use of fire.

Fire provided *H. erectus* with several important new aspects of life that gave them significant advantages. It provided warmth, protection against large predators, light at night, and the source of heat that allowed them to develop cooking. Cooking was a major step forward because it allowed the exploitation of new foods that were inedible as raw materials. Once they learned how to make fire and cook foods, they were able to move into some environments that were previously inhospitable. Once they made such a step, their intellectual abilities enabled them to increase their numbers, expand their potential, and eventually become *H. sapiens*.

In recent years debate regarding the position of *H. erectus* as a legitimate species has arisen (Howells, 1980). Day (1986, p. 409) states

> No longer is *Homo erectus* a clearly defined taxon temporally, morphologically or even geographically.

Some workers have questioned its position with regard to *H. sapiens*, whether the European specimens are truly *H. erectus*, and if the African specimens are the same taxon as the Asian ones. At the same time, other workers have supported one view or another, or the original validity of the species. Part of the problem arises from the philosophical viewpoints of the different workers. Another difficulty is the lack of a continuum of specimens that are, instead, scattered in time and place. These have combined to allow the assignment of specimens to specific time intervals separated by gaps, and this has presented workers with the opportunity to speculate which, if any or all of these specimens, actually gave rise to modern humans. Whatever the outcome, we must remember that the establishment of such a species is a human endeavor dependent upon the available fossil materials. Despite these questions of *H. erectus* and whatever the assignment of specimens and their names, some or many or all of these played an important role in the evolutionary pathway to modern humans

ARCHAIC SAPIENS IN EUROPE

About 400,000 years ago, during the middle Pleistocene, the climate began to cool and glacial conditions of the Riss Glacial Stage returned to Europe. With the seasonal increase of snow and ice, life was more difficult for the local *H. erectus* population, except for those who had the ability and the adaptations suitable for survival under such harsh conditions. By about 300,000 years ago the Riss Glaciation had given way to the Riss-Würm Interglacial Stage. The return to a long interval of milder conditions saw the appearance about 200,000–250,000 years ago of a human population that had begun to evolve into somewhat more modern looking individuals. The fossils of these latter types are considered by some experts to be early versions of the Neandertals, *H. sapiens neanderthalensis* (see Fig. 10.2) though others believe them to be very late *H. erectus*. There is no sharp dividing line between these two species, and the available evidence supports a gradual change from *H. erectus* into *H. sapiens* by the slow development of some more modern features and characters. The spread of these newcomers throughout central Europe is demonstrated by their remains at Ehringsdorf, Saccopastore, Steinheim, Swanscombe and elsewhere. Today, many workers commonly refer to these people as archaic sapiens in recognition of their status as the earliest and not fully developed specimens of *H. sapiens*. These fossils, from widely separated European sites, probably represent a general trend of the population as it was gradually evolving into a new species. At the same time, the populations of *H. erectus* that were living elsewhere, the Middle East, perhaps in South and East Africa, and southern Asia, were also evolving some more modern aspects, and it is possible that individuals throughout the entire *H. erectus* population were experiencing a slow evolution.

Conditions were much more pleasant in Europe and elsewhere during the Riss-Würm interglacial as the archaic sapiens populations were evolving into more modern humans. About 125,000 years ago in Europe, however, the climate cooled into the Würm Glacial Stage and conditions once again became harsh. For perhaps 100,000 years severe glacial conditions continued, though there were minor reversals that resulted in brief warming trends. The archaic sapiens population of Europe that entered this episode of harsh conditions did not continue the modernization trend like their relatives living elsewhere under more hospitable conditions. Instead, about 100,000–120,000 years ago, the European archaic sapiens responded to these conditions by developing more primitive characteristics and features; they evolved into the Classic Neandertals.

These Classic Neandertals had the typical robust body build, thickened bone structures, powerful muscles and culture that they had inherited from *H. erectus* with slight modifications. This trend spread quickly through the gene pool and soon they were the only human inhabitants of Europe until the arrival of the Cro-Magnon people about 35,000–40,000 years ago. We often speak of the Neandertal inhabitants of Europe as being primitive, but this is unfair; they were primitive only in the sense of being nonmodern like the Cro-Magnons. The Neandertals were, after all, more advanced than their *H. erectus* ancestors. As the products of evolution, they were well adapted to inhospitable environments of Ice Age Europe in which they thrived for more than 50,000 years. These people, the Classic Neandertals, had appeared and spread through Europe at approximately the same time or slightly after that when the archaic sapiens populations in Africa were continuing their evolution into anatomically modern humans.

Though the European Neandertal specimens comprised only one portion of the world's population of *H. sapiens*, they were accorded a position of great importance in the study of human evolution. Probably much of this importance was simply because the principal scientists studying and writing about human origins were European workers. In addition, the sites where these fossils were found were immediately at hand and did not require travel to other continents. Other populations of archaic sapiens living elsewhere were not as well represented in collections because of great distances and lack of ready sites like the European caves. The specimens that were found, however, were enlightening. They did not exhibit the same changes seen in the European Neandertals. These non-European populations were living under different environmental conditions with different stresses, and they adapted to their environments by producing structures, features and behaviors that were as conservative. And because these other populations were not as well studied or as well known as the European stock, they were not originally considered as important to human evolution.

During the last glaciation, the Würm, significant portions of Europe were covered by great ice sheets that undoubtedly isolated directly or indirectly some stocks of the resident archaic sapiens population. As this happened, the result of the effects of glaciation was to limit their access to the larger gene pools available in Africa and Asia.

With limited gene pools and under physical and other stress, the archaic sapiens of western Europe simply could not or did not evolve with the full range of genomes as did their archaic sapiens population relatives in more hospitable parts of the Old World. As a result, the European stock evolved from their available gene stock into phenotypes that were more suitable for their environments. These were the sturdy, rugged phenotypes of their ancestors and these local European populations became the Classic Neandertals. Elsewhere, however, the archaic sapiens were evolving slowly under much less harsh conditions and could be more experimental with their larger genomes and more experimental phenotypes. They tended to lose many of the typical sturdy *H. erectus* aspects and, in Africa and Asia, gradually became more modern. In the Middle East, however, the evolution of these archaic sapiens into more modern types was not clearly defined.

The robust and sturdy nature of the Classic Neandertals, which provided such a contrast to the modern Cro-Magnons, has been ascribed by many scientists to their life in Europe during the restrictive times of the Pleistocene glacial and interglacial stages. Their sturdier bone structure clearly indicates much greater physical activity than in modern humans. Obviously, these people lived a strenuous and difficult life that depended on their physical ability and strength to endure the inhospitable cold of the Pleistocene winters. They survived despite the relatively limited supply of food available to them during the European winters. During the warmer seasons they hunted animals and gathered plant foods, and followed the animals on their migrations. The record is clear; they survived for many thousands of years.

Communication by these people with other archaic sapiens outside of Europe may have been impossible because of climatic circumstances, cultural differences or some other mechanism, and they may have been denied the advantages of genetic and cultural developments that other peoples elsewhere were making and freely exchanging. Under such circumstances, the Classic Neandertals developed a culture based on their genomes and phenotypes that enabled them to successfully exploit their particular environments. Though they used their intelligence, abilities and strength to survive and multiply, we cannot identify and ascribe to them any level of intellectual achievement significantly beyond *H. erectus* except for some minor advances in toolmaking, certainly nothing comparable to the Cro-Magnons. It is likely, therefore, that they suffered restrictions in culture that emphasized those physical aspects that enabled their survival in Ice Age Europe.

While the Classic Neandertals were struggling to survive under the difficult conditions of Ice Age Europe, the other populations of archaic sapiens were finding life much better in Africa, the Middle East and parts of Asia. Those regions were essentially free of great ice sheet glaciers, and the evolution of the resident human populations contrasted with that in Europe. The Asian and African populations of archaic sapiens lived under far less restrictive conditions and enjoyed access to a much larger gene pool and greater cultural developments than the European people. They were able to migrate through the Middle East and exchange genomes freely. They were able to live without the constraints of survival in a harsh glacial region and, as a result, continued the early trend of evolution into more modern humans; they became the Progressive Neandertals (see Fig. 10.7). In fact, much of Africa probably experienced better conditions for human survival than Europe at this time because of increased precipitation and ameliorating temperatures. That the humans in Africa and Asia did not evolve the same conservative phenotypes as the Classic European Neandertals must be considered in light of these less-restrictive climatic conditions and the larger, more accessible gene pools.

Somewhere outside of Europe, in the Middle East, western Asia, or Africa, these Progressive Neandertals continued to evolve and eventually became modern humans. Stringer, et al., (1984, p. 115) discounted western Europe as the birthplace of Cro-Magnons because,

> . . . western Europe has no good evidence for the actual origins of a.m. [anatomically modern] *H. sapiens.*

They argue that no western European specimens exhibit any morphology intermediate between Neandertals and modern sapiens and that modern humans appeared too quickly in Europe and with great contrast to the resident Neandertals. Therefore, no evolutionary transition between the two types took place in Europe. But if not in Europe, where did it occur? Africa is the likely possibility but the Middle East must also be considered. Both are potential sites for the origin of modern humans because of the fossil evidence and archeological evidence found there. However, some of these Middle East specimens had crude features typical of the Classic Neandertals yet others had features that were not as conservative but were more modern or more advanced; in fact, one specimen showed great similarity to modern humans. This mixture of individuals with different features suggests that some individuals had the genes of both Classic Neandertals and their more modern cousins. How they obtained these genes could be a key question to their origin.

The discovery in the Middle East of human skeletons with mixed Classic and Progressive Neandertal characters initially raised difficulties in interpreting how these two sets of humans could have evolved and been present together. The age of these specimens, probably in the range of about 35,000–70,000 years ago, indicates that they were very close in time to the fully modern type of fossil humans, Cro-Magnons. Specimens from the caves at Shanidar in Iraq, and from Tabun, Skhül, Wadi Amud and Jebel Qafzeh in Israel demonstrate some aspects of both types of Neandertals. The appearance of these dual aspects in so many specimens of these populations suggests one hypothesis for human evolution. The Middle East, therefore, must be considered a crucial area in the evolution of modern humans.

The Middle East had been an important natural crossroads between Europe, Asia and Africa for tens of thousands of years. Peoples have migrated through this region individually and in groups, carrying and spreading their social systems, languages, cultural developments and their genes, sometimes residing, sometimes continuing on their journeys. These migrating peoples exchanged cultural ideas and technology, commonly adopting the best of whatever they saw. In relatively modern times mathematics, medicine, science, philosophy, art, architecture and other disciplines flourished under the Arabs in the Middle East as Europe slid into the decline of the Dark Ages. These ideas and disciplines all experienced growth and development in the Arab world that recognized and adopted them. It is no accident, for example, that three of the world's great religions also arose in this region. Eventually these ideas and technology were carried from the Middle East into Europe by travelers. In similar fashion, the archaic sapiens must have spread their culture and technology wherever they traveled. Some Classic Neandertals could easily have traveled from Europe into the Middle East, spreading their more specialized conservative aspects, especially genes and culture. Any contact between Classic Neandertals and the local populations of Progressive Neandertals could

have allowed the exchange of genetic materials and resulting adaptations that led to slightly different phenotypes. The offspring of such potential Classic and Progressive Neandertal mating could easily have had mixed characters, just as the specimens at Shanidar, Tabun, Skhül, Wadi Amud and Jebel Qafzeh demonstrate.

BIOCHEMICAL EVIDENCE FOR HUMAN ORIGINS

New evidence for the origin of modern humans has been proposed through innovative methods of investigation that have been developed in recent years. Paleontological evidence today supports the belief that Africa was the birthplace of modern humans. Fossils of early modern humans have been found at several places, though the status and dates of some have been challenged or doubted. In 1963, Morris Goodman provided a new line of testing that did not rely on fossils. He made immunological comparisons of proteins in blood from living apes and humans. Other workers (including Wilson and Sarich, 1967, 1969; Sarich, 1968, 1971; Goodman and Tashian, eds, 1976; Cronin, 1983; Britten, 1986; Wainscoat, et al., 1986; Cann, et al., 1987; Wainscoat, et al., 1989; Vigilant, et al., 1991; Wilson and Cann, 1992; Satoshi, et al., 1992; Thorne and Wolpoff, 1992) soon followed and used various biochemical techniques to analyze the DNA, blood proteins, amino acid sequences and more to identify and measure relationships between living humans and various living hominoids. From this accumulated data of living hominoids, some workers were able to project these to fossil forms and to propose hereditary patterns for their evolution.

Wainscoat, et al., (1986, 1989), for example, used an analytical technique involving the DNA from a gene that produced part of the hemoglobin molecule of red blood cells. In their first test, they were able to isolate five DNA enzyme sites in the ß-globin gene cluster (the combination of sites on one chromosome is identified as a haplotype). They consequently examined these haplotypes on 601 chromosomes from people in eight populations groups, only one of which was African. They noted that Africans had two haplotypes in common that were unique to them and not present in non-Africans. The non-Africans, however, had a limited number of combinations in common. These researchers concluded that the patterns indicated that Africans had evolved from a small, inbred stock and spread out to give rise to other peoples. After some time away from Africa, these now non-Africans experienced evolutionary changes that modified and gave rise to their own DNA patterns.

Later work by Cann, et al. (1987) offered another study of human inheritance, though the possibilities are still to be fully understood, and differences of opinion among experts have produced alternate and contrary interpretations. These researchers examined the genetic composition of **mitochondrial DNA** (mtDNA) in human cells. Mitochondria are cell-nuclei structures whose DNA mutates "several times faster than in the nucleus." Further, they are derived only from females and thus present a record of maternal inheritance. Testing of the mtDNA from 147 people of five geographic populations indicated a pattern that allowed for several hypotheses, including the one that has sparked great interest.

Cann, et al. proposed in this particular version of inheritance (p. 33) ". . . that Africa is a likely source of the human mitochondrial gene pool." They constructed a genealogical tree in which modern population maternal lineages could be linked ". . . to a common ancestral female." In effect, their view allowed modern humans to trace their origin to a single female, a "**mitochondrial Eve**," who lived in Africa between 140,000 and 290,000 years ago and whose descendants now cover the globe. She presumably provided the source of the mtDNA that is now widespread among humans.

Other workers since (including Saitou and Omoto, 1987; Darlu and Tassy, 1987; Maddison, 1992; Templeton, 1992; Hedges, et al., 1992; Maddison, et al., 1992; Goldman and Barton, 1992, and others) have raised objections of various types to dispute these interpretations. Saito and Omoto (1987, p. 288) pointed out that the rate of mitochondrial mutations, which is a key in this analysis, is probably half or less than that used by Cann, et al. If the mutation rate used by Cann, et al. is in error, it could indicate that the origin of modern humans from the archaic sapiens could have taken place farther back in time, perhaps as much as 400,000 or 500,000 years ago. This would necessitate a significant modification of the Cann, et al. hypothesis.

Avise (1991) reported research indicating that mtDNA may not be exclusively maternal in inheritance; some of it may derive paternally. The significance of these biochemical studies is hardly finalized; they have only just begun

this complicated research program. Experts may eventually refine the analytical techniques and provide a more definitive interpretation of human origins.

What is remarkable about these types of biochemical analyses is how different workers view the same evidence and how different combinations of evidence support their favored model of the origin of modern humans. Stringer and Andrews (1988), for example, support a recent African origin for modern humans and cite both genetic and fossil evidence in their model. Their basic data is from fossils and is supported by the genetic data including mitochondrial DNA. They argue that these are ". . . strong evidence against the multiregional model."

THE TRANSITION TO MODERN MAN

Archaic sapiens specimens ranging from about 400,000 to perhaps 150,000 years ago are known only from the Old World—Africa, Europe and Asia—although younger Cro-Magnons (anatomically modern humans) are common in these areas. Australia and the New World remained barren of the archaic sapiens because there were no effective land bridges to connect with the Old World. Modern humans were able to invade Australia and the New World only during relatively recent times. They were able to reach these hitherto remote areas only because of the consequences of Pleistocene glaciation.

During the Pleistocene, vast quantities of water were removed from the oceans to form the massive ice sheets, and sea level was lowered by as much as three hundred feet. This lowering of sea level exposed much of the sea floor on the continental shelves and produced connections between some otherwise isolated landmasses. These land bridges allowed modern humans to emigrate into Australia from Southeast Asia by simply walking across the land of the former sea floor exposed around the islands of the Indonesian Archipelago and sailing the short distance between New Guinea and Australia, presumably on simple log rafts. Many workers believe these immigrants arrived in northern Australia by about 60,000 years ago and they probably reached southern Australia by 40,000 years ago. In the north Pacific, Asians were able to travel from Siberia into the New World across the newly exposed land bridge in the Bering Sea area between Siberia and Alaska. This travel took place between 20,000–40,000 years ago and these immigrants were able to spread throughout the New World, first into North America and eventually to South America.

Despite all the information and interpretations, the specific transition from archaic sapiens to modern humans is still not clearly defined. Although fossils at a great many sites provide much evidence, different workers have placed different emphases on the various paleontological, biochemical and archeological evidence and how to interpret it. Their interpretations have led to a number of hypotheses that explain the relationships between archaic sapiens, the two types of Neandertals and modern man. None, however, is supported conclusively by the evidence, though each enjoys or has enjoyed a measure of popularity.

Today, the various hypotheses have been critically screened by proponents and opponents, and two seem to be the favored ones. These can be characterized as a single-origin model and a multiple-origin model. Stringer and Andrews (1988, p. 1263) discuss these and are of the opinion that

> These two models are not the only ones currently under discussion, but it is likely that one or other reflects the predominant model of *Homo sapiens* evolution.

These two hypotheses that describe the evolutionary path from *H. erectus* to anatomically modern humans are known as the **Noah's Ark hypothesis**, the **Garden of Eden hypothesis** or the **Out-of-Africa model** and the **Multiregional Evolution hypothesis**. Other models and slightly different versions that propose the same general pathways with varying degrees of detail are also known. Weiss and Mann (1985, p. 372), for example, list four hypotheses: **Neandertal Phase hypothesis**, **Evolutionary Population model**, **Punctuation model** and **Center-Edge model**. Day (1986, p. 414) groups the basic explanations into three hypotheses, **Neandertal-Phase hypothesis**, **Preneandertal hypothesis**, and **Presapiens hypothesis**. Day, however, added that these three are not distinct and may well be oversimplifications of some previous ideas.

NEANDERTAL-PHASE HYPOTHESIS

One widely supported and long-standing model for the origin of anatomically modern humans is the straight-line model that has become known as the Neandertal-Phase hypothesis. Hrdlicka and his followers accepted as Neandertals all of the hominids from the appearance of *H. sapiens* during the middle Pleistocene to the end of the early Würm glaciation when they were succeeded by Cro-Magnons. To Hrdlicka and his followers, the Neandertals were a relatively poorly known type that seemed to fit between the ancestral *H. erectus* and modern humans. As such, they represented an evolutionary stage, phase or grade through which all hominids passed as they evolved in a straight line from the ancestral *H. erectus* to *H. sapiens*. These Neandertals were seen as a population with variable characters distributed throughout the Old World. Some of the variations in the population can be attributed to their phenotypic responses to pressures of local conditions. As this stock slowly changed or evolved, it advanced directly from the ancestral *H. erectus* into Neandertals and finally into modern humans. This path resulted from the improved structures and features that arose with the simultaneous development of culture.

According to the proponents, supporting evidence for the Neandertal-Phase hypothesis is found in the numerous fossils from sites in the Middle East—Tabun and Skhül at Mount Carmel, Jebel Qafzeh, and Wadi Amud, all in Israel, and Shanidar in Iraq. The fossils from Tabun, Shanidar and Wadi Amud are more anatomically similar to Classic Neandertals, many of which are found in this region. The fossils found at Skhül and Jebel Qafzeh, however, show aspects that are found in modern humans. Howell emphasized this and said that the Jebel Qafzeh fossils were, ". . . in all respects like a proto-Cro-Magnon." Because the fossils from Skhül and Jebel Qafzeh are more modern looking, they are interpreted as being closer evolutionarily to modern humans than the less modern Classic Neandertal fossils from the other Middle Eastern localities. This large collection of Neandertals from the Middle East, therefore, represent two stages or phases of modern human evolution.

The discovery of the badly crushed skeleton and skull at Saint-Césaire, France in 1981, however, created difficulties for this hypothesis. This skull (Fig. 12.3) has the low vault and projecting face of a Classic Neandertal. These remains, however, were found with Upper Paleolithic tools from the Perigordian culture. Perigordian industry tools are believed to be associated only with modern humans, Cro-Magnons, and are not older than about 32,000 years. If the dates and reconstructions are correct, the Saint-Césaire specimen would indicate that at least some Classic Neandertals were still living in Europe after the more modern appearing people had evolved and were inhabiting Skhül and Jebel Qafzeh in the Middle East. Accordingly, the Saint-Césaire specimen suggests that some Neandertals

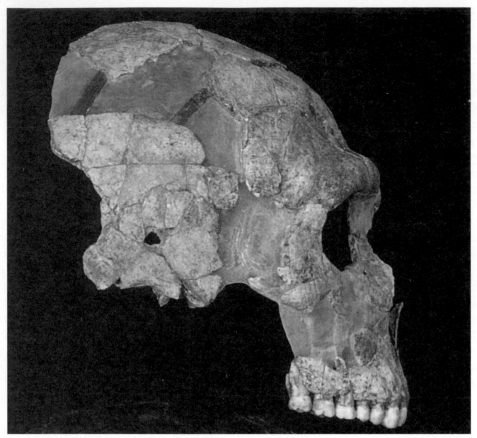

Figure 12.3 Saint-Césaire Neandertal skull.

retained their conservative aspects and remained as the Classic type even as others were evolving and becoming more modern. This argument, therefore, weakens the Neandertal-Phase hypothesis that proposes a straight-line model.

PUNCTUATION MODEL

The Punctuation model proposed that evolution from archaic sapiens to modern humans may have occurred very rapidly with a sudden spurt or punctuation, probably somewhere in the Middle East or eastern Europe. Trinkhaus and Howells restricted the term Neandertal to the Classic Neandertals and pointed out (1979, p. 123) that these fossils have an "anatomical pattern, or combination of skeletal features" that can distinguish them from

. . . modern human populations and from the
patterns of the European Upper Paleolithic

and the Near Eastern late Mousterian. The Neanderthals can also be consistently differentiated from the modern human beings who lived at the same time in Africa and eastern Asia.

As such, these humans represent a distinct group of archaic sapiens that are identified as Neandertals. These people evolved from the early sapiens populations that included the Steinheim and Swanscombe individuals (see Fig. 10.10) who lived in western Europe about 250,000–300,000 years ago. Sometime during the Upper Pleistocene, probably about 100,000–130,000 years ago during the last interglacial, these archaic sapiens evolved into the Neandertals as seen in several fossils, including the Saccopastore skulls (see Fig. 10.9) that exhibit clear signs of Neandertalism. In the span of about 50,000 years, these evolving people reached their full status as Neandertals as shown by the fossils from Krapina in Yugoslavia, a collection of fully robust Neandertals who lived during the Würm Glacial Stage.

Several experts have suggested that climate was influential in the development of some of the Neandertal robustness, especially the massive brow ridges of the skull. This brow ridge would have effectively placed the internal sinus and nasal cavities farther away from the brain, which is temperature sensitive. These larger cavities could have warmed inhaled cold air before it reached the lungs. We cannot verify that climate actually played any part in this robust adaptation, and Trinkhaus and Howells argue that the cold of the Würm glaciation could not have initiated the robust phenotype of the Neandertals because these aspects had already begun to develop during a warm interglacial stage. However, even during the interglacials, the seasons alternated between warm summers and cold winters, and it may be that these cold seasons were, nonetheless, an important factor.

The robust nature of the Neandertals appears to be directly related to and much more emphasized than the robust physique of their ancestor, *H. erectus*. Trinkhaus and Howells maintain that the Neandertal robustness represents an apparently successful adaptation that appeared fully about 100,000 years ago and survived until to about 30,000–35,000 years ago. Evidence for the latest Neandertals in Europe places them about 35,250 years ago in France and at 34,000 years ago in Yugoslavia. The principal question, however, is what initiated such robust adaptations in these people at this time.

About 100,000 years ago an anatomically modern type of human appeared in the Middle East at Skhūl and Qafzeh. This was an abrupt appearance that occurred in the short span of about five thousand years in contrast to the general Neandertal evolution which took perhaps 50,000 years. The earliest evidence in Europe of more modern people is represented by the Aurignacian culture and dates from about 33,000 years ago. Evidence in Australia places modern humans there at 32,000–40,000 years ago and evidence from sub-Saharan Africa is even earlier than that. Obviously, modern humans had already begun to make their appearance in several places prior to the time when the Neandertals were disappearing. Therefore, modern humans could not have arisen from a single large stock of Neandertals.

CENTER-EDGE MODEL

The Center-Edge hypothesis is essentially a multiregional model. It proposes that the evolution and relationships between archaic and modern humans can be explained by the flow of genes and resulting morphological variation from different segments of the population. In this model, the *H. erectus* population expanded its geographic range by migrating into various environments in East and South Africa, along both sides of the Mediterranean, through the Middle East and across southern Asia into China and Southeast Asia. Some of these environments were rich and hospitable while others were relatively poor. The populations in these different environments were in close contact and had direct access to the varied gene pools. The species, thus, remained fully operational with individuals able to freely move back and forth and exchange cultural developments and genes and, with the genomes, natural variations.

Evolutionary changes are more likely to occur in areas where greater morphological variation exists. Such greater morphological variability is more likely to develop at the center of rich environments than at the edges, which tend to be relatively poorer environments. This is simply because the richer environments are more hospitable and exert less stress on the inhabitants than the edge environments. Under less-stressful conditions in the center, more of the less-suitable variants are able to survive and reproduce and the gene pool of morphological variants can increase. The poorer environments at edges of the richer environments, however, experience just the opposite effects. These **edge environments**, being relatively less-suitable for their survival, would tend to have fewer successful variants because of more-stressful circumstances. These greater pressures make survival more difficult for many of the ordinary variants, and they would tend to die off more quickly than in better environments. Such poorer areas would, therefore, experience fewer suitable evolutionary changes because the pool of morphological variants would be smaller. These edges would be more likely to receive new genetic variations from the centers, the sites of richer environments. As a consequence, the populations in these edges would tend to consist of both old and new variants.

The Middle East could have been one such center of richness where archaic sapiens encountered less stress.

These people developed a variety of genetic morphological variations probably some time after 100,000 years ago, including some modernizing genes that caused the inhabitant Neandertals to develop more modern aspects. These individuals then carried their modernizing genes outward from this center into poorer environments whose inhabitants had not yet developed modern aspects. This spread, however, took time and the various modern aspects appeared in poorer areas at successively later times, much like ripples from a center of disturbance in still water. Thus, the Skhül and Qafzeh specimens at the center of such a rich area became relatively modern in appearance very early in time. The Saint-Césaire specimens, though younger in time, are much more primitive because they were more distant from the center and had not yet received the modernizing genes at this edge.

NOAH'S ARK OR GARDEN OF EDEN HYPOTHESIS

The Noah's Ark (Howells, 1976) or the Garden of Eden hypothesis, proposes that modern humans arose from a single origin and then dispersed around the world as modern humans. It assumes (Stringer and Andrews, 1988, p. 1263)

> ... that there was a relatively recent common ancestral population for *Homo sapiens* which already displayed most of the anatomical characters shared by living people.

It is generally accepted among researchers that *H. erectus* arose in Africa. This model proposes that these modern humans began to evolve from these earlier people, exhibiting some changes perhaps 120,000 years ago. These evolving modern humans then began to migrate or radiate outward from the source area slowly throughout Africa into the Middle East, Europe and Asia. Some specimens in the Middle East with modern aspects are of the right age to support this model. These modern humans continued their migrations displacing the existing local populations of previous migrants, *H. erectus* or archaic sapiens, and continued to evolve. This is, therefore, a **replacement** model. In time, perhaps less than 100,000 years, they adapted to the different local environments. Because of differences in local conditions, the local populations evolved somewhat differently and established slightly different characters that we recognize as "racial traits."

Stringer and Andrews maintain that the variety of biochemical DNA evidences do not support the multiple-origin model but do support, along with the fossil evidence, the single-origin model. One difficulty is the complete separation of the invading modern population from the resident *H. erectus* or archaic sapiens people. The mtDNA model does not admit any genetic enchange between the invaders and the residents, a difficulty that has not been fully explained.

Recent study of two old fossil skulls in China (Gibbons, 1992) has added another element of confusion into the single-origin model. These skulls show flattened faces on otherwise typical *H. erectus* individuals. Their age, about 350,000 years, contrasts with the view that these flattened faces are relatively modern. If correct, they indicate a trend toward modernization much further back in time than any other specimens, especially those of Africa upon which the single-origin model is based. Such a view offers support for the Multiregional Evolution model by arguing that these skulls had started a modernizing trend far removed from Africa. Such flattening is not considered modern by Stringer, however, and he notes that some specimens from Africa also show the same characteristic.

MULTIREGIONAL EVOLUTION HYPOTHESIS

The Multiregional Evolution hypothesis uses the same basic fossil evidence as the Noah's Ark model but it proposes the origin of modern living human populations through a different mechanism. It argues that *H. erectus* had already spread across Africa and widely into Asia and Europe (Thorne and Wolpoff, 1992, p. 76) by about one million years ago. These early people then experienced different local environmental conditions wherever they lived. Under the differing pressures each local population began to adapt and evolve according to their available genetic material. In time, each of these local populations evolved independently and somewhat differently under their local conditions and in accord with their limited gene pools. Eventually they became the modern humans we find living today in the different parts of the world.

The recent analyses of DNA from cell nuclei and mitochondria do not offer support for this multiple-origin model. With regard to the mitochondrial DNA analyses that seem to support a single-origin hypothesis, Thorne and Wolpoff (1992) hypothesize a multiregional origin of humans and strongly argue (p. 77) that

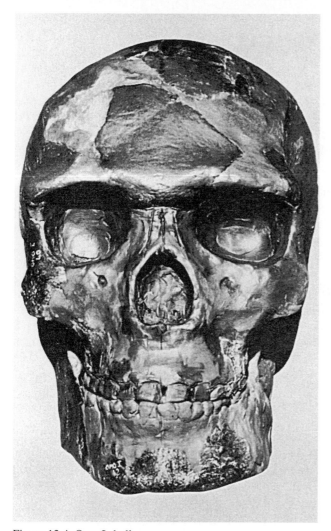

Figure 12.4 Omo I skull.

Mitochondrial DNA is useful for guiding the development of theories, but only fossils provide the basis for refuting one idea or the other. At best, the genetic information explains how modern humans might have originated if the assumptions used in interpreting the genes are correct, but one theory cannot be used to test another.

They are of the view that tests of DNA materials from living species should not be used as a test of fossil relationships; only fossils can supply the necessary data. They, too, are disturbed that a single group of humans "totally replaced all others worldwide."

FOSSILS

Whichever of these differing hypotheses is correct, or another one as yet unformulated, the relatively recent fossil evidence cited by some workers clearly supports the view that Africa is the site of modern human origins. One skull (Omo I) from the **Omo Valley**, Ethiopia (Fig. 12.4) is tentatively dated at about 130,000 years. It is thin boned, with bulging sides and rounded back, quite small brow ridges, a prominent chin and large brain capacity of about 1,400 cc. It is considered by many to be a modern human, although some experts believe it to be an archaic sapiens. Some specimens from South Africa are nearly as old. Bones of essentially modern humans have been found in strata in caves at the mouth of the **Klasies River**. These have been dated at 70,000–100,000 years, although some workers have challenged these dates. Another South African cave, **Border Cave**, has yielded a fragmentary cranium (Fig. 12.5) and other remains believed to be of anatomically modern humans (Rightmire, 1979) that are dated at about 90,000 years ago. With this group of fossils providing important information, Stringer (in Putman, 1988, p. 456) argues for

> ... the origin of modern people in Africa more than 70,000 years ago; then a gradual spread, first to the Middle East, then probably into Europe and to the other end of the world, to the Far East and Australia and America. In my view, all living people have a common ancestor who lived within the past 150,000 years. And it looks like that common ancestor lived in Africa.

Thorne and Wolpoff (1992, p. 82), however, note with emphasis that

> The evidence for a great antiquity of modern-looking people is based primarily on the interpretation of bones from three sites: the Omo site in Ethiopia and the Klasies River and Border Cave sites in South Africa.

They point out that most of the Omo remains are fragmentary and resemble modern humans, but most pieces are undateable because they were found on the surface; that the Border Cave fossils may be recent burials dug into a cave and may be younger than the strata of burial; and that only the Klasies River fossils are well dated. However, these fossils are also fragmentary and, therefore, are not good evidence for modern humans.

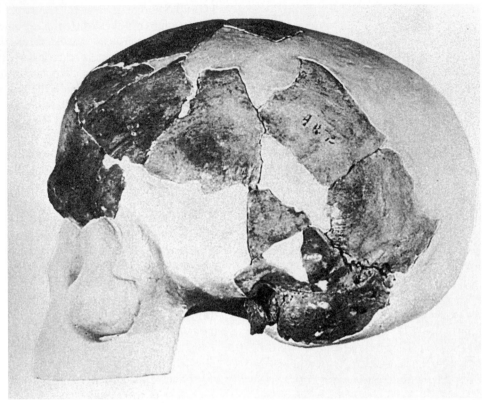

Figure 12.5 Border Cave I cranium.

The fossils found at several Middle Eastern sites are somewhat younger than the South African specimens. Qafzeh Cave in Israel had early modern human remains that are very old; the Qafzeh 9 skull, for example, is dated at about 92,000 years ago. Nearby Kebara Cave held a Neandertal dated at about 60,000 years ago and another Neandertal from Wadi Amud has been dated at about 70,000 years. Together these specimens indicate that early modern humans probably coexisted with Neandertals in the Middle East for as long as 30,000 years. In Europe, however, the only modern specimens were Neandertals until the arrival of Cro-Magnons about 35,000–40,000 years ago. Clearly, the distinctions between Neandertals in Europe and Cro-Magnons is such that there was no evolutionary transition there. We must assume, therefore, that modern humans had evolved elsewhere before they appeared in Europe.

If the dates at these very old sites withstand scientific scrutiny and if these fossils demonstrate a trend toward modern humans, then the single-origin model for modern human origins would be strengthened. It would show that specimens of early humans becoming more modern lived in South Africa before they appeared in the Middle East and long before they lived in Europe.

CLIMATIC INFLUENCES

It is difficult to believe that climate could not have had some effect on the evolution of modern humans, even though some workers argue that anatomical changes were taking place without significant climatic changes. Whatever the model, it must explain how the older, more modern appearing archaic sapiens in Europe apparently became the more Classic Neandertals. The likely explanation is that the archaic sapiens types had been evolving everywhere into the more modern appearing Progressive Neandertals. This stock did not evolve in Europe but spread there only after they had evolved elsewhere. Probably they appeared initially in Africa and spread normally

238

through the Middle East into Asia. Their particular gene pools reflected the local hospitable conditions where they originated and, once started on this evolutionary path, they gradually became more modern.

Europe was inhabited by the archaic sapiens, chronologically early Neandertals who had some relatively modern characters. When the climate of Europe began to change from the relatively moderate third interglacial stage to the much harsher conditions of the Würm Glacial Stage, some or all of these people may have been isolated by the harsh temperature and frigid conditions. These local populations had limited gene pools from which to produce genetic combinations necessary for survival in the harsh conditions. At the same time, the climatic conditions probably reduced the chances of possible contact with neighboring populations and lessened the free exchange of new and possibly innovative genetic materials. These early Neandertals in Europe, accordingly, suffered selective pressures that were much different than those experienced by their ancestors and by their cousins elsewhere.

With their potential for adaptations limited to the available, limited gene pool, the European stock probably could not evolve with the full range of genomes as did the Neandertaloid populations in the more environmentally hospitable parts of the Old World. Accordingly, the European archaic sapiens population evolved phenotypes that were more suitable for the demanding life under glacial conditions. Their rugged outdoor life favored the sturdier bodies with thicker and heavier bones, more powerful muscles to operate this skeleton and the other features that we commonly ascribe to the Classic Neandertals. Such anatomical changes are not likely to have been necessary in a population that depended primarily on a plant-food way of life, but would have been advantageous if their principal activity was hunting with all its strenuous physical demands. These aspects, fortunately, were already present in their ancestral stock, *H. erectus*. Accordingly, the European archaic sapiens types may simply have responded to the local circumstances by emphasizing their more conservative aspects; they evolved into the Classic Neandertals. Their evolution was simply the result of natural selection acting on a limited gene pool with limited available genetic traits; it was the enhancement of those conservative traits that worked best under the existing harsh environmental conditions.

In the Middle East and elsewhere, the environmental climatic conditions may not have degraded significantly during the cooling glacial interval and the region may not have been as inhospitable as Europe and much of Asia. The Middle East environments, therefore, did not exert the same pressures for adaptations on its archaic sapiens inhabitants as the European environments did, and this Middle East population likely continued to evolve normally with full access to genes from Africa and Asia. The unrestricted archaic sapiens population and gene pool continued to evolve and develop variants, and these gradually became more modern in character. In time they became the types we know as the early Progressive Neandertals.

The discovery of Classic Neandertal characters in some of these non-European Progressive populations created a problem when they were first noted. However, these characters may well have been the result of interbreeding between the native Progressive types and Classic Neandertals if and when the latter migrated into the Middle East from Europe. In this way, the Classic Neandertals could have introduced their more primitive genetic characteristics into the more modern Progressive native gene pool. Some of their descendants likely exhibited anatomical characters of both the Progressive and Classic Neandertal types, and these descendants were found as fossils. It is also possible that these Progressive-Classic types were the result of normal variation of the Progressive population in which recessive genes reappeared from time to time.

Whatever the mechanism for evolution of the Progressive Neandertals, if indeed they should be called Neandertals, the facts remain that individual human specimens with more modern features had begun appearing some 120,000 years ago in Africa and nearly 100,000 years ago in the Middle East. Their transition into *H. sapiens sapiens* cannot be argued without doubt because too few fossils have been found to demonstrate an unequivocal pathway. A major question about the spread of these modern humans is why it took so long for them to move into Europe and Asia. The answer cannot be that they were not intelligent enough or that living conditions were too good at home. It may be that the harsher climatic conditions found in Europe and Asia were not appealing enough to attract them. Once into Europe, however, these people spread so rapidly that they quickly became the population that we know so well from the many discoveries.

Several hypotheses have been proposed to explain the evolution of Cro-Magnon people from an earlier species. Though they may describe slightly different pathways or offer different factors that created the impetus for change, they all require an understanding of the ancestral stocks that provide the foundations upon which modern humans have built their civilizations. *Homo erectus* is obviously the ancestral species, though the exact pathway is still being debated. Did the evolutionary trail pass from *H. erectus* only once or did it happen several times in different parts of the world? This evolution took place during the last few hundred thousand years, when the great ice ages of Pleistocene time dominated the world's climates. Such a major factor must be considered in any evolutionary hypothesis.

If *H. erectus* is the ancestral species, how did the evolutionary change to modern humans take place? And where did it occur? Africa seems to be the favored site of this change. Several places have been identified as the locale where very old specimens of *H. erectus* have been found. That these people spread into Europe and Asia is accepted; there are no barriers to prevent such a migration. Once in place in Europe and Asia, these immigrants from Africa were exposed to different environmental conditions and adapted as well as they could. Undoubtedly new phenotypes appeared, because that is the way of organic evolution. This evolution was not radical in physical terms; their form did not change greatly, but their intellectual capacity and potential increased. They became more intelligent and culturally diverse. Hunting, for example, became an essential means of obtaining food. At the same time, they also learned how to transmit much more information more efficiently. Their beneficiaries, the Cro-Magnons and modern humans, became the most efficient recorders and transmitters of information in history. We constantly improve on the various methods of doing this.

Local populations undoubtedly budded and produced new groups of migrants that spread the new items and methods that they were developing. In Europe as elsewhere, the *H. erectus* populations evolved into the archaic sapiens. For one reason or another, these European archaic sapiens did not give rise to the anatomically modern people. Instead they evolved into the Classic Neandertals, perhaps in response to the harsh climatic conditions that prevailed.

Elsewhere, the archaic sapiens were probably already beginning to evolve into more modern appearing types that we know as the Progressive Neandertals. Only after these types had appeared did the Classic Neandertals begin to appear with them. Several hypotheses address the distribution patterns of these Neandertals and the Cro-Magnons. Each has its adherents and each its critics.

The history of the Cro-Magnons from these ancient times to the present day has filled many books; it is the history of our own portion the human race. It has been recorded, reexamined and reviewed many times by anthropologists, archeologists, historians, paleontologists, philosophers, theologians and others; it is also beyond the scope of this work. In this history many more important developments were yet to take place. The bow and arrow were still to be invented, and the dog will become a valuable companion. Agriculture had not yet become a practice, wild animals were not yet domesticated and kept for their meats, milk, furs and strength. Machines will not be produced for many more years. The human population will not for many millenia give up its nomadic lifestyle and settle into villages and towns. Writing is very far ahead, as are mathematics, seafaring, books and so much more. Electricity, light, airplanes and space exploration were beyond the imagination of these Cro-Magnons. Yet these Cro-Magnon people were far superior to their hominid ancestors. And their ancestors were far superior to their early primates ancestors.

We are fortunate enough to be able to view so much of the history of humans, from our primate ancestors through the Cro-Magnons. Though the story is vague in some places and missing in others, it seems clear that the major pieces are in place and the story outline is basically settled. Though many of the details will never by discovered because of the limitations of fossilization and preservation, many other details will be found by careful searchers. Fortunately, like most of science, the story is not yet finished and perhaps never will be. The history of our species is one in which the most recent chapters merely record most of the events of yesterday and today.

Each of the hypotheses explaining the origin of modern humans is based upon the same set of information, fossil and biochemical. Each accepts the geographic distribution of modern human fossils and their predecessors, and each explains how modern humans could have arisen from the earlier forms. However, they employ radically different ideas with contrasting emphases in arriving at the same result: modern humans. Each hypothesis has its adherents and opponents, and each new bit of evidence, a fossil find, a new biochemical test, is examined critically to see if it supports one model or the other. But which viewpoint is correct? At this point in time there is no clear hypothesis acceptable to all. Only future discoveries will clarify this viewpoint.

We should be well aware that the evolutionary trail from the early humans to archaic sapiens to ourselves cannot be determined with scientific exactness that may be demanded by some. The anatomical changes that took place in this relatively recent evolutionary sequence are gradual. Today, we have inherited the great advantage of accumulated knowledge that has been passed from generation to generation for tens of thousands of years. We do not have to reinvent the wheel or the printing press in each generation, though we can improve upon them. The early Cro-Magnon people had this advantage, but they did not have it long enough to capitalize fully on their inventions and discoveries. They had no writing and were unable to record this information with any ease; they had to rely on memory and oral tradition. We, however, have been blessed by having this knowledge for a much longer time and have learned to record it on rock walls, clay tablets, on paper, in books and now digitally using computers. We can transmit it much more efficiently than our ancestors, even around the world instantaneously. As a result each successive generation of humans has had significantly more to learn. We are the beneficiaries of this great accumulation of knowledge and have been able to capitalize on it. The Cro-Magnons lived during a time of hunting and were very good at it. We live in a time of explosion of knowledge; let us hope that we, too, will handle it well.

STUDY QUESTIONS

1. What evidence supports the view that modern humans probably began to evolve in Africa?

2. Describe two evolutionary paths that could have led from *Homo erectus* to modern people.

3. Explain why hunting was so important in the species *Homo*.

4. How did budding affect early human populations?

5. Describe the effects that the glacial stages and the warmer interglacials had on the human populations of Africa and Europe.

6. What cultural advantages could the use of fire have had on early humans?

7. Account for the robust body builds of Classic Neandertals and the more delicate bodies of Cro-Magnons.

8. What is the Eve hypothesis?

9. What advantages did Cro-Magnons have that earlier peoples probably lacked?

10. In what respects did the Cro-Magnons differ from modern humans?

11. Explain the sudden appearance of the modern humans in western Europe.

12. Describe one hypothesis for an evolutionary pathway leading to modern humans.

13. Why is the Middle East sometimes referred to as the crossroads of the world?

References

Alexander, John P. 1992. Alas, poor *Notharctus*. *Natural History* 101, no. 8:54–59.

Andrews, P. 1986. Molecular evidence for catarrhine evolution. In *Major Topics in Primate and Human Evolution*. Wood, B., L. Martin and P. Andrews, eds., Cambridge: Cambridge University Press.

Ardrey, Robert. 1961. *African genesis*. New York: Atheneum.

———. 1966. *Territorial imperative*. New York: Atheneum.

———. 1970. *The social contract*. New York: Atheneum.

———. 1976. *The hunting hypothesis*. New York: Atheneum.

Aronson, Elliot. 1976. *The social animal*. San Francisco: W. H. Freeman and Co.

Avise, John C. 1991. Matriarchal liberation. *Nature* 352 (18 July):192.

Ayala, F. J. 1978. The mechanisms of evolution. *Scientific American* 239, no. 3:56–69.

Azimov, Isaac. 1963. *The human brain*. Boston: Houghton Mifflin Co.

Baker, J. R. 1974. *Race*. New York: Oxford University Press.

Barlow, Nora, ed. 1958. *The autobiography of Charles Darwin, 1809–1882*. London: Collins.

Begun, David R. 1992. Miocene fossil hominids and the chimp-human clade. *Science* 257 (25 September):1929–1933.

Bernstein, I. S. and T. P. Gordon. 1974. The function of aggression in primate societies. *American Scientist* 62 (May–June):304–311.

Bordes, F. 1961. Mousterian cultures in France. *Science* 134:803–810.

———. 1968. *The old stone age*. New York: McGraw-Hill Book Co.

Boule, M. 1911. L'homme fossile de la Chapelle-aux-Saints. *Annales de Paléontologie*. 6:111–172.

———. 1912. L'homme fossile de la Chapelle-aux-Saints. *Annales de Paléontologie*. 7:3–192.

———. 1913. L'homme fossile de la Chapelle-aux-Saints. *Annales de Paléontologie*. 8:1–67.

Boule, M. and H. V. Vallois. 1957. *Fossil men*. New York: Dryden Press.

Brace, C. Loring. 1967. *The stages of human evolution*. Englewood Cliffs, NJ: Prentice-Hall.

Brain, C. K. 1981. *The hunters or the hunted? An introduction to African cave taphonomy*. Chicago: University of Chicago Press.

Bräuer, Günter. 1984. A craniological approach to the origin of anatomically modern *Homo sapiens* in Africa and implications for the appearance of modern Europeans. In *The origins of modern humans*. Smith, F. H. and Frank Spencer, eds. New York: Alan R. Liss, Inc. pp. 327–410.

Britten, Roy J. 1986. Rates of DNA sequence evolution differ between taxonomic groups. *Science* 231 (21 March):1393–1398.

Broom, Robert and G. W. H. Schepers. 1946. *The South African fossil ape-men: The Australopithecinae*. Pretoria: Transvaal Museum.

Broom, Robert. 1950. *Finding the Missing Link*. London: A. C. Watts & Co.

Brown, Frank, et al. 1985. Early *Homo erectus* skeleton from west Lake Turkana, Kenya. *Nature* 316 (29 August):788–792.

Buettner-Janusch, John. 1966. *Origins of Man*. New York: John Wiley & Sons.

Burenhult, Gören. 1993. *The first humans*. New York: HarperCollins Publishers.

Burney, D. A. and R. D. E. MacPhee. 1988. Mysterious island. *Natural History* 97, no. 7:47–54.

Buss, L. W. 1987. *The evolution of individuality*. Princeton: Princeton University Press.

Campbell, Bernard G. 1974. *Human evolution*, 2d ed. Chicago: Aldine Publishing Co.

———. 1988. *Humankind emerging,* 5th. ed. Boston: Little, Brown and Co.

———. 1992. *Humankind emerging,* 6th. ed. New York: HarperCollins Publishers.

Cann, Rebecca L., M. Stoneking and A. C. Wilson. 1987. Mitochondrial DNA and human evolution. *Nature* 325 (1 January):31–36.

———. 1987. Disputed African origin of human populations: Reply. *Nature* 329 (10 September):111–112.

Carpenter, C. R. 1964. *Naturalistic behavior of nonhuman primates*. University Park, PA: Pennsylvania State University Press.

Cartmill, Matt. 1974. Rethinking primate origins. *Science*. 184 (24 April):436–442.

Cavalli-Sforza, L. L., et al. 1988. Reconstruction of human evolution: bringing together genetic, archeological, and linguistic data. *Proceedings of the National Academy of Science* 85:6002–6006.

Cheney, Dorothy and Robert Seyfarth. 1990. In the minds of monkeys. *Natural History* 99, no. 9:38–46.

Ciochon, R. L. 1985. Fossil ancestors of Burma. *Natural History* 94, no. 10:26–36.

———. 1991. The ape that was. *Natural History* 100, no. 11:54–63.

Ciochon, R. L. and J. G. Fleagle, eds. 1987. *Primate evolution and human origins*. New York: Aldine de Gruyter.

Ciochon, R. L., John Olsen, and Jamie James. 1990. *Other origins*. New York: Bantam Books.

Clark, J. Desmond. 1976. The African origins of man the toolmaker. In *Human origins*. G. L. Isaac and E. R. McCown, eds. Menlo Park, CA: W. A. Benjamin, Inc. pp. 1–53.

Clottes, Jean and Jean Courtin. 1993. Neptune's Ice Age gallery. *Natural History* 102, no. 4:64–70.

Colbert, E. H. 1980. *Evolution of the vertebrates*, 3d ed. New York: John Wiley & Sons.

Cole, Sonia. 1975. *Leakey's luck*. New York: Harcourt Brace Jovanovich.

Constable, George. 1973. *The Neanderthals*. New York: Time-Life Books.

Coon, C. S. 1962. *The origin of races*. New York: Alfred A. Knopf, Inc.

———. 1971. *The hunting peoples*. Boston: Little, Brown and Co.

Coppens, Yves. 1976. The great East African adventure. *CNRS Research*:2–12.

Coppens, Yves, et al., eds. 1976. *Earliest man and environments in the Lake Rudolf basin*. Chicago: University of Chicago Press.

Cornelius, R. M. 1987. *William Jennings Bryan, The Scopes Trial, and Inherit the Wind*. Dayton, TN: William Jennings Bryan College.

Cowen, Richard. 1990. *History of life*. Boston: Blackwell Scientific Publications.

Culotta, Elizabeth. 1992. A new take on anthropoid origins. *Science* 256 (12 June):1516–1517.

Darlington, C. D. 1969. *The Evolution of Man and Society*. New York: Simon and Schuster.

Darlu, Pierre and Pascal Tassy. 1987. Disputed African origin of human populations. *Nature* 329 (10 September):111.

Dart, Raymond A. 1925. *Australopithecus africanus*: the man-ape of South Africa. *Nature* 115 (7 February):195–199.

———. 1926. Taungs and its significance. *Natural History* 26:315–327.

———. 1956. Cultural status of the South African man-apes. *Smithsonian Report*, no. 4240:317–338.

Dart, Raymond A. and Dennis Craig. 1959. *Adventures with the missing link*. New York: Harper & Brothers.

Darwin, Charles. 1859. *The origin of species by means of natural selection or the preservation of favored races in the struggle for life*. London: John Murray.

———. 1871. *The descent of man and selection in relation to sex*. London: John Murray.

———. 1872. *The origin of species by means of natural selection or the preservation of favored races in the struggle for life*, 6th ed. and *The descent of man and selection in relation to sex*. New York: Modern Library Edition, Random House.

Dawkins, Richard. 1976. *The selfish gene*. New York: Oxford University Press.

———. 1986. *The Blind Watchmaker*. New York: W. W. Norton & Co.

Dawson, C. and W. A. Smith. 1913. On the discovery of a Palaeolithic human skull and mandible in flint bearing gravel overlying the Wealden (Hastings Bed) at Piltdown, Fletching (Sussex). *Quarterly Journal Geological Society of London* 69:117–144.

Day, Michael H. 1985. Pliocene hominids. In *Ancestors: The Hard Evidence*. Eric Delson, ed. New York: Alan R. Liss, Inc. pp. 91–93.

———. 1986, *Guide to fossil man*, 4th ed. Chicago: University of Chicago Press.

Day, William. 1984. *Genesis on planet Earth*, 2d ed. New Haven: Yale University Press.

Dean, David and Eric Delson. 1992. Second gorilla or third chimp? *Nature* 359 (22 October):676–677.

Delson, Eric, ed. 1985. *Ancestors: the hard evidence*. New York: Alan R. Liss, Inc.

––––––. 1987. Evolution and palaeobiology of robust *Australopithecus*. *Nature* 327:654–655.

de Lumley, Henry. 1969. A Paleolithic camp at Nice. *Scientific American* 220:42–50

Dennell, Robin. 1986. Needles and spear-throwers. *Natural History* 95, no. 10:70–78.

De Valois, R. L. and G. H. Jacobs. 1968. Primate color vision. *Science* 162 (1 November):533–540.

Diamond, Jared. 1993. What are men good for? *Natural History* 102, no. 5:24–29.

Diamond, J. M. and J. I. Rotter. 1987. Observing the founder effect in human evolution. *Nature* 329:105–106.

Dobzhansky, Theodosius. 1955. *Evolution, genetics, and man*. New York: John Wiley & Sons.

Dobzhansky, Theodosius, et al. 1977. *Evolution*. San Francisco: W. H. Freeman and Co.

Dolhinow, Phyllis and Vincent Sarich, eds. 1971 *Background for man*. Boston: Little, Brown and Co..

Dorozynski, Alexander. 1993. Possible Neandertal ancestor found. *Science* 262 (12 November):991.

Dubois, Eugene. 1895. *Pithecanthropus erectus, eine menschenähnliche. Übergansform aus Java*. Batavia: Landesdruckerei.

––––––. 1899. *Pithecanthropus erectus*-a form from the ancestral stock of mankind. *Annual Report for 1898*. Washington, D.C.: Smithsonian Institution., pp. 445–459.

Dunbar, C. O. 1960. *Historical geology*, 2d ed. New York: John Wiley & Sons.

Eaton, G. G. 1976. The social order of Japanese macaques. *Scientific American* 235, no. 4:97–106.

Eckhardt, Robert B. 1972. Population genetics and human origins. *Scientific American* 226, no. 1:94–103.

Eddy, Maitland A. 1972. *The missing link*. New York: Time-Life Books

Eimerl, Sarel and I. DeVore. 1965. *The primates*. New York: Time-Life Books.

Eldredge, Nils and S. J. Gould. 1972. Punctuated equilibria: An alternative to phyletic gradualism. In *Models in paleobiology*. T. J. M. Schopf, ed. San Francisco: Freeman, Cooper and Co. pp. 82–115.

Fabre, J. H. 1901. *Insect life*. London: Macmillan Publishing Co.

––––––. 1918. *The wonders of instinct*. London: T. Fisher Unwin Ltd.

Fagan, Brian M. 1974. *Men of the Earth*. Boston: Little, Brown and Co.

Falk, Dean. 1984. The petrified brain. *Natural History* 93, no. 9:36–39.

Fossey, Dian. 1970. Making friends with mountain gorillas. *National Geographic* 137, no. 1:48–68.

––––––. 1971. More years with mountain gorillas. *National Geographic* 140, no. 4:574–586.

––––––. 1981. The imperilled mountain gorilla. *National Geographic* 159, no. 4:501–523.

––––––. 1983. *Gorillas in the mist*. Boston: Houghton Mifflin Co.

Fox, Lionel and Robin Tiger. 1971. *The imperial animal*. New York: Holt, Rinehart and Winston, Inc.

Galdikas, Biruté M. 1980. Indonesia's orangutans, living with the great orange apes. *National Geographic* 153, no. 4:830–853.

Galdikas-Brindamour, Biruté M. 1975. Orangutans, Indonesia's "People of the Forest." *National Geographic*. 148, no. 4:444–474.

Gee, Henry. 1993. Why we still love Lucy. *Nature* 366 (18 November):207

Ghiglieri, Michael. 1987. War among the chimps. *Discover* (November):67–76.

Gibbons, Ann. 1992, An about-face for modern human origins. *Science* 256 (12 June):1521.

––––––. 1992. Hungarian fossils stir debate on ape and human origins. *Science* 257 (25 September):1864–1865.

Godfrey, L. R. and J. R. Cole. 1986. Blunder in their footsteps. *Natural History* 95, no. 8:4–12.

Goldman, N. and N. H. Barton. 1992. Genetics and geography. *Nature* 357 (11 June):440–441.

Goodall, Jane. 1964. Tool-using and aimed throwing in a community of free-living chimpanzees. *Nature* 201:1264.

––––––. 1971. *In the shadow of man*. Boston: Houghton Mifflin Co.

––––––. 1976. Continuities between chimpanzee and human behavior. In *Human origins*, Isaac, G. L. and E. R. McCown, eds. Menlo Park, CA: W. A. Benjamin, Inc. pp. 80–95.

————. 1977. Infant killing and cannabalism in free-living chimpanzees. *Folia Primatologica* 28:259–282.

————. 1979. Life and death at Gombe. *National Geographic* 155, no. 5:592–622.

————. 1990. *Through a window: My thirty years with the chimpanzees of Gombe*. Boston: Houghton Mifflin Co.

Goodman, Morris. 1963. Man's place in the phylogeny of the primates as reflected in serum proteins. In *Classification and human evolution*. Washburn, S. L., ed. Chicago: Aldine Press. pp. 204–234.

Gould, S. J. 1977. *Ever since Darwin*. New York: W. W. Norton & Co.

————. 1980. The Piltdown controversy. *Natural History* 89, no. 8:8–28.

————. 1983. *Hen's teeth and horse's toes*. New York: W. W. Norton & Co.

Gould, S. J. and Niles Eldredge. 1993. Punctuated equilibrium comes of age. *Nature* 366 (18 November):223–227.

Gowlett, J. A. J. 1984. *Ascent to civilization*. New York: Alfred Knopf, Inc.

Griffin, D. R. 1984. Animal thinking. *American Scientist* 72 (September–October):456–464.

Haeckel, Ernst H. 1868. *The history of Creation*. Translated by E. R. Lankaster. New York: D. Appleton.

Harding, R. S. O. and S. C. Strum. 1976. The predatory baboons of Kekopey. *Natural History* 85:46–53.

Hay, Richard L. 1967. Hominid-bearing deposits of Olduvai Gorge. In *Time and stratigraphy in the evolution of man*. Washington, D.C.: National Academy of Sciences, National Research Council: Publ. 1469:30–42.

————. 1971. Geologic background of Beds I and II. In *Olduvai Gorge, Vol. III, Excavations in Beds I and II. 1960–1963*. Cambridge: The University Press. pp. 9–18.

————. 1976. Environmental setting of hominid activities in Bed I, Olduvai Gorge. In *Human origins*. Isaac, G. L. and E. R. McCown, eds. Menlo Park, CA: W. A. Benjamin, Inc. pp. 209–225.

————. 1976. *Geology of the Olduvai Gorge*. Berkeley, CA: University of California Press.

Hedges, S. Blair, et al. 1991. Human origins and analysis of mitochondrial DNA sequences. *Science* 255 (7 February):737–739.

Hill, W. C. O. 1953. *Comparative Anatomy and Taxonomy: I—Strepsirhini*. Edinburgh: Edinburgh University Press.

Holloway, Ralph L. 1974. The casts of fossil hominid brains. *Scientific American* 231, no. 1:106–115.

Howell, F. C. 1965. *Early Man*. New York: Time, Inc.

————. 1976. Overview of the Pliocene and earlier Pleistocene of the lower Omo Basin, southern Ethiopia. In *Human origins*. Isaac, G. L. and E. R. McCown, eds. Menlo Park, CA.: W. A. Benjamin, Inc. pp. 227–268.

————. 1984. Introduction. In *The origins of modern humans*. Smith, F. H. and Frank Spencer, eds. New York: Alan R. Liss, Inc. pp. xiii–xxii.

———— and Glynn Isaac. 1976. Introduction. In Earliest man and environments in the Lake Rudolf Basin. Coppens, Yves, et al., eds. Chicago:University of Chicago Press. pp. 471-475.

Howells, William. 1959. *Mankind in the making*. Garden City, NY: Doubleday & Co.

————. 1966. Homo erectus. *Scientific American* 215, no. 5:46-53.

Hrdlicka, Ales. 1930. The skeletal remains of early man. *Smithsonian Miscellaneous Collection* 83.

Huxley, Thomas. 1863. *Evidence as to man's place in nature*. New York: D. Appleton & Co.

Irvine, William. 1955. *Apes, angels, and victorians*. New York: McGraw-Hill Book Co.

Isaac, G. L. 1976. The activities of early African hominids: a review of archaeological evidence from the time span two and a half to one million years ago. In *Human origins*. Isaac, G. L. and E. R. McCown, eds. Menlo Park, CA: W. A. Benjamin, Inc. pp. 483–514.

————. 1977. *Olorgesailie*. Chicago: University of Chicago Press.

————. 1978. The food-sharing behavior of protohuman hominids. *Scientific American* 238, no. 4:90–108.

Isaac, G. L. and E. R. McCown, eds. 1976. *Human origins*. Menlo Park, CA: W. A. Benjamin, Inc.

James, S. R. 1989. Hominid use of fire in the Lower and Middle Pleistocene. *Current Anthropology* 30:1–26.

Jerison, H. J. 1976. Paleoneurology and the evolution of mind. *Scientific American* 234, no. 1:90–101.

Johanson, Donald C. 1993. A skull to chew on. *Natural History* 102, no. 5: 52-53.

Johanson, Donald C. and M. A. Eddy. 1981. *Lucy: The beginnings of humankind*. New York: Simon and Schuster.

Johanson, Donald C. and James Shreeve. 1989. *Lucy's child*. New York: William Morrow and Co.

Johanson, Donald C. and T. D. White. 1979. A systematic assessment of early African hominids. *Science* 203 (26 January):321–330.

Johanson, Donald C., et al. 1987. New partial skeleton of *Homo habilis* from Olduvai Gorge, Tanzania. *Nature* 327:205–209.

Jolly, Alison. 1985. The evolution of primate behavior. *American Scientist* 73 (May–June):230–239.

Jolly, C. J. 1970. The seed eaters: a new model of hominid differentiation based on a baboon analogy. *Man* 5:5–26.

Kawai, Masao. 1965. Newly acquired pre-cultural behavior of the natural troop of Japanese monkeys on Koshima Islet. *Primates* 6:1–30.

Kettlewell, H. B. D. 1955. Selection experiments on industrial melanism in the Lepidoptera. *Heredity* 9, pt. 3:323–342.

———. 1956. A resumé of investigations on the evolution of melanism in the Lepidoptera. *Proceedings of the Royal Society of London*, Ser. B. Vol. 145, no. 920:297–303.

King, William. 1864. The reputed fossil man of the Neanderthal. *Quarterly Journal of Science* 1:88–97.

———. 1864. On the Neanderthal skull, or reasons for believing it to belong to the Clydian Period and to a species different from that represented by man. Report of the British Association for the Advancement of Science (1863). *Notices and Abstracts*. pp. 81–82.

Klein, Richard G. 1973. Geological antiquity of Rhodesian man. *Nature* 244 (3 August):311–312.

Korn, Noel, ed. 1973. *Human evolution*, 3d ed. New York: Holt, Rinehart and Winston, Inc.

Kortlandt, Adrian. 1962. Chimpanzees in the wild. *Scientific American* 206, no. 5:128–138.

Koyne, D. E., et al. 1972. Evolution of primate DNA: A summary. In *Perspectives on Human Evolution*. S. L. Washburn and Phyllis Dolhinow, eds. New York: Holt, Rinehart and Winston, Inc., New York. pp. 166–168.

Kurtén, Björn. 1984. *Not from the apes*. Irvington, NY: Columbia University Press.

Lane, N. Gary. 1978, *Life of the past*. Columbus, OH: Charles E. Merrill Co.

Lartet, Edouard and Henry Christy. 1865–75. *Reliquae acquitanicae: Being contributions to the archaeology and palaeontology of Perigord and the adjoining provinces*. London: Williams and Norgate.

Leakey, L. S. B. 1959. A new fossil skull from Olduvai. *Nature* 181:491–493.

Leakey, Louis S. B., et al. 1931. Age of the Oldoway bone beds, Tanganyika. *Nature* 128:724.

Leakey, Louis S. B. and Jack and Stephanie Prost, eds. 1971. *Adam, or ape: A sourcebook of discoveries about early man*. Cambridge, MA: Schenkman Publishing Co.

Leakey, Louis S. B., P. V. Tobias, and J. R. Napier. 1964. A new species of the genus *Homo* from Olduvai Gorge. *Nature*. v. 202, p. 5–7.

Leakey, Mary D., ed. 1971, *Olduvai Gorge, Vol. III*. Cambridge: The University Press.

———. 1976. A summary and discussion of the archaeological evidence from Bed I and Bed II, Olduvai Gorge, Tanzania. In *Human origins*. Isaac, G. L. and E. R. McCown, eds. Menlo Park, CA: W. A. Benjamin, Inc. pp. 431–459.

———. 1979. Footprints frozen in time. *National Geographic* 155, no. 4:446–458.

Leakey, Mary D. and Alan Walker. 1987. *Laetoli: a Pliocene site in northern Tanzania*. Oxford: Clarendon Press.

Leakey, Richard E. 1981. *The making of mankind*. New York: E. P. Dutton.

Leakey, Richard E. and Roger Lewin. 1977. *Origins*. New York: E. P. Dutton.

———. 1978. *People of the lake*. Garden City, NY: Anchor Press/Doubleday.

———. 1992. *Origins reconsidered*. New York: Doubleday.

Leakey, Richard E. and Alan Walker. 1985. *Homo erectus* unearthed. *National Geographic* 168, no. 5:624–629.

Le Gros Clark, W. E. 1959. *The antecedents of man*. Edinburgh: Edinburgh University Press.

———. 1963. *The antecedents of man*. New York: Harper Torchbook, Harper & Row, Publishers.

———. 1964. *The fossil evidence for human evolution*, 2d. ed. Chicago: University of Chicago Press.

Lerner, I. M. and W. J. Libby. 1976. *Heredity, evolution, and society*. 2d ed. San Francisco: W. H. Freeman and Co.

Lewin, Roger. 1984. *Human evolution*. New York: W. H. Freeman and Co.

————. 1985. New fossil upsets human family. *Science* 223:720–721.

————. 1987. The unmasking of mitochondrial eve. *Science* 238:24–26.

————. 1987. *Bones of contention: Controversies in the search for human origins.* New York: Simon and Schuster.

————. 1988. *In the age of mankind: A Smithsonian book of human evolution.* Washington, D. C.: Smithsonian Books.

————. 1993. *Human evolution.* 3d ed. Boston: Blackwell Scientific Publications.

Lewontin, R. C. 1978. Adaptation. *Scientific American* 239, no. 3:212–230.

Lorenz, Konrad. 1966. *On aggression.* New York: Bantam Books.

Lovejoy, C. Owen. 1988. Evolution of human walking. *Scientific American* 259, no. 5:118–125.

Lyell, Charles. 1830–33. *Principles of geology.* London: John Murray.

Margulis, Lynn and Dorion Sagan. 1986. *Microcosmos.* New York: Summit Books.

Marshack, Alexander. 1972. *The roots of civilization.* New York: McGraw-Hill Book Co.

Marshall, J. T., Jr. and E. R. Marshall. 1976. Gibbons and their territorial songs. *Science* 193 (16 July):235–237

Marshall, L. G. 1988. Land mammals and the great American interchange. *American Scientist* 76 (July-August):380–388.

Martin, Paul S. 1975. Sloth droppings. *Natural History* 84, no. 7:74–81.

Martin, Paul S. and R. G. Klein. 1984. *Pleistocene extinctions.* Tucson: University of Arizona Press.

Martin, R. D. 1990. *Primate origins and evolution.* London: Chapman and Hall.

Mayr, Ernst. 1970. *Populations, species, and evolution.* Cambridge, MA: Harvard University Press.

————. 1976. *Evolution and the diversity of life.* Cambridge, MA: Harvard University Press.

————. 1978. Evolution. *Scientific American* 239, no. 3:46–55.

Mayr, Ernst, E. G. Linsley, and R. I. Usinger. 1953. *Methods and principles of systematic zoology.* New York: McGraw-Hill Book Co.

Mazzeo, J. A. 1967. *The design of life.* New York: Pantheon Books

McAlester, A. L. 1977. *The history of life,* 2d ed. Englewood Cliffs, NJ: Prentice-Hall, Inc.

McCown, T. D. and K. A. R. Kennedy. 1972. Pt. 3, 1890–1925. In *Climbing man's family tree. A collection of major writings on human phylogeny, 1699 to 1971.* McCown, T. D. and K. A. R. Kennedy, eds. Englewood Cliffs, NJ: Prentice-Hall, Inc.

————, eds. 1972. Climbing man's family tree. *A collection of major writings on human phylogeny, 1699 to 1971.* Englewood Cliffs, NJ: Prentice-Hall, Inc.

McHenry, H. M. 1986. The first bipeds: a comparison of the *A. afarensis* and *A. africanus* postcranium and implications for the evolution of bipedalism. *Journal of Human Evolution* 15:177–191.

Mellars, Paul, ed. 1990. *The emergence of modern humans.* Ithaca, NY: Cornell University Press.

Mellars, Paul and Chris Stringer, eds. 1989. *The human revolution.* Princeton, NJ: Princeton University Press.

Mendel, Gregor. 1959. Experiments in plant hybridization, 1865. In C*lassic papers in genetics.* Peters, J. A., ed. Englewood Cliffs, NJ: Prentice-Hall, Inc. pp. 1–20.

Merrick, H. V. 1976. Recent archaeological research in the Plio-Pleistocene deposits of the lower Omo, southwestern Ethiopia. In *Human origins.* Isaac, G. L. and E. R. McCown, eds. Menlo Park, CA: W. A. Benjamin, Inc. pp. 461–481.

Moment, G. B. and H. M. Habermann. 1973. *Biology: Full spectrum.* Baltimore: The Williams & Wilkins Co.

Montague, Ashley. 1976. *The nature of human aggression.* New York: Oxford University Press.

Moor-Jankowski, J. and A. S. Wiener. 1965. Primate blood groups and evolution. *Science* 148 (9 April):255–256.

Moore, G. W. and Morris Goodman. 1968. Phylogeny and taxonomy of the catarrhine primates. In *Taxonomy and phylogeny of Old World primates,* B. Chiarelli, ed. Torino: Rosenbery & Sellier.

Moore, Ruth. 1956. *The earth we live on.* New York: Alfred Knopf, Inc.

————. 1962. *Evolution.* New York: Time, Inc.

————. 1963. *Man, time and fossils.* New York: Alfred Knopf, Inc.

Morris, Desmond. 1967. *The naked ape.* New York: Dell Publishing Co.

Morris, H. M. 1977. *The scientific case for Creation.* San Diego, CA: Master Book Publishers.

Napier, John. 1967. The antiquity of human walking. *Scientific American* 216, no. 4:56–66.

Nietschmann, Bernard. 1977. The Bambi factor. *Natural History* 86, no. 7:84–87.

Nitecki, M. H. and D. V. Nitecki, eds. 1987. *The evolution of human hunting*. New York: Plenum.

Nordenskiöld, Erik. 1932. *The history of biology*. New York: Alfred Knopf, Inc.

Owen-Smith, N. 1987. Pleistocene extinctions: the pivotal role of megaherbivores. *Paleobiology* 13:351–362.

Patterson, Francine. 1978. Conversations with a gorilla. *National Geographic* 154, no. 4:438–466.

Pilbeam, David. 1972. *The ascent of man*. New York: Macmillan Publishing Co.

———. 1984. Bone of contention. *Natural History* 93, no. 6:2–4.

———. 1984. The descent of hominoids and hominids. *Scientific American* 250, no. 3: 84–96.

———. 1985. Patterns of hominoid evolultion. In *Ancestors: The hard evidence*. Eric Delson, ed. New York: Alan R. Liss, Inc. pp. 51–59.

Pilbeam, David and S. J. Gould. 1974. Size and scaling in human evolution. *Science* 186:892–901.

Poirier, F. E. 1973. *Fossil man*. St. Louis: C. V. Mosby Co.

———. 1977. *Fossil evidence*, 2d ed. St. Louis: C. V. Mosby Co.

Pope, Geoffrey G. 1993. *Natural History* 102, no. 5:55–59.

Pope Pius XII. 1952. Modern science and the existence of God. In *The Church and modern science*. New York: America Press. pp. 31–48.

Premack, David. 1976. Language and intelligence in ape and man. *American Scientist* 64, no. 6:674–683.

Prideaux, Tom. 1973. *Cro-Magnon man*. New York: Time-Life Books.

Putman, John J. 1988. The search for modern humans. *National Geographic* 174, no. 4:438–477.

Racle, F. A. 1979. *Introduction to Evolution*. Englewood Cliffs, NJ: Prentice-Hall, Inc.

Radinsky, L. B. 1967. The oldest primate endocast. *American Journal of Physical Anthropology* 27:385–388.

Reader, John. 1981. *Missing links*. Boston: Little, Brown and Co.

Reynolds, Vernon. 1963. An outline of the behavior and social organization of forest-living chimpanzees. *Folia Primatologica* 1:95–102.

———. 1965. *Budongo: An African forest and its chimpanzees*. Garden City, NY: Natural History Press.

Reynolds, Vernon and R. Reynolds. 1965. Chimpanzees of the Budongo forest. In *Primate behavior*. I. Devore, ed. New York: Holt, Rinehart and Winston, Inc..

Richards, Graham. 1987. *Human evolution*. New York: Routledge and Kegan Paul, Inc.

Ridley, Mark. 1993. *Evolution*. Boston: Blackwell Scientific Publications.

Rigaud, Jean-Philippe. 1988. Art treasures from the Ice Age-Lascaux Cave. *National Geographic* 174, no. 4:482–499.

Rightmire, G. Philip. 1979. Implications of Border Cave skeletal remains for later Pleistocene human evolution. *Current Anthropology* 20:23–35.

———. 1984. *Homo sapiens* in sub-Saharan Africa. In *The origins of modern humans*. Smith, F. H. and Frank Spencer, eds. New York: Alan R. Liss, Inc. pp. 295–325.

——— 1989. Middle Stone Age humans from eastern and southern Africa. In *The human revolution*, Paul Mellars and Chris Stringer, eds. Princeton, NJ: Princeton University Press. pp. 109–122

Robinson, J. T. and R. J. Mason. 1957. Occurrence of stone artefacts with *Australopithecus* at Sterkfontein. *Nature* 180:521–524.

Rudwick, M. J. S. 1976. *The meaning of fossils*. New York: Science History Publications.

Rukand, Wu and Lin Shenglong. 1983. Peking Man. *Scientific American* 248, no. 6:86–94.

Ruvolo, Maryellen, et al. 1991. Resolution of the African hominoid trichotomy by use of a mitochondrial gene sequence. *Proceedings of the National Academy of Science* 88:1570–1574.

Saitou, Naruya and Keiichi Omoto. 1987. Time and place of human origins from mtDNA data. *Nature* 327 (28 May):288.

Sandmel, Samuel. 1978. *The Hebrew scriptures*. New York: Oxford University Press.

Sarich, V. M. 1967. Rates of albumin evolution in primates. *Proceedings of the National Academy of Science* 58:142–148.

———. 1968. Immunological time scale for hominid evolution. *Science* 158:1200–1203.

———. 1968. The origin of the hominids: an immunological approach. In *Perspectives on human evolution*. S. L. Washburn and P. C. Jay, eds. New York: Holt, Rinehart and Winston, Inc.. pp. 94–121.

———. 1971. A molecular approach to the question of human origins. In *Background for man*. Dolhinow, P. and Vincent M. Sarich, eds. Boston: Little, Brown and Co.

Sarich, V. M. and A. C. Wilson. 1966. Quantitative immunochemistry and the evolution of the primate albumins. *Science* 154:1563–1566.

Sauer, N. J. and T. W. Phenice. 1977. *Hominid fossils,* 2d ed. Dubuque, IA: Wm. C. Brown Co.

Savage-Rumbaugh, E. S., et al. 1980. Do apes use language? *American Scientist* 68 (January–February):49–61.

Schaller, George B. 1961. The orangutan in Sawarak. *Zoologica* 46:73–82.

———. 1963. *The mountain gorilla*. Chicago: University of Chicago Press

———. 1964. *The year of the gorilla*. Chicago: University of Chicago Press.

———. 1972. *The Serengeti lion*. Chicago: University of Chicago Press.

Schaller, George B. and Gordon Lowther. 1969. The relevance of carnivore behavior to the study of early hominids. *Southwestern Journal of Anthropology* 25, no. 4:307–341.

Scheller, R. H. and Richard Axel. 1984. How genes control an innate behavior. *Scientific American* 250, no. 3:54–62.

Science. 1992. Extinct hominid did not veg out, 256 (29 May):1281.

Scientific American. 1987. Family ties, 257, no. 1:16–18.

Self, Herbert, ed. 1971. *The making of human aggression*. New York: St. Martin's Press.

Shapiro, H. L. 1974. *Peking man*. New York: Simon and Schuster.

Sigmon, B. 1971. Bipedal behavior and the emergence of erect posture in man. *American Journal of Physical Anthropology* 34:55.

Simons, Elwyn L. 1964. The earliest relatives of man. *Scientific American* 221, no. 1:50–62.

———. 1965. New fossil apes from Egypt and the initial differentiation of Hominoidea. *Nature* 205:135–139.

———. 1967. The earliest apes. *Scientific American* 217, no. 6:28–35.

———. 1972. *Primate evolution*. New York: Macmillan Publishing Co.

———. 1984. Dawn ape of the Fayum. *Natural History* 93, no. 5:18–20.

———. 1985. Origins and characteristics of the first hominoids. In *Ancestors: The hard evidence*. Eric Delson, ed. New York: Alan R. Liss, Inc. pp. 37–41.

———. 1986. New faces of *Aegyptopithecus* from the Oligocene of Egypt. *Journal of human evolution* 16:273–289.

———. 1989. Human origins. *Science* 245:1343–1350.

———. 1993. Egypt's simian spring. *Natural History* 102, no. 4:58–59.

Simons, Elwyn L. and P. C. Ettel. 1970. *Gigantopithecus*. *Scientific American* 222, no. 1:77–85.

Simons, Elwyn L. and R. F. Kay. 1980. "Dawn Ape" provides clue to social life. *Geotimes* (May):18.

Simpson, George G. 1966. The biological nature of man. *Science* 152 (22 April):472–478.

Simpson, George G., et al. 1960. *Quantitative zoology*, 2d ed. New York: Harcourt, Brace & World, Inc.

Sinclair, A. R. E., et al. 1986. Migration and hominid bipedalism. *Nature* 324:307–308.

Sjøvold, Torstein. 1993. Frost and found. *Natural History* 102, no. 4:60–63.

Skehan, J. W., S. J. 1983. Theological basis for a Judeo-Christian position on Creationism. *Journal of Geological Education*. 31:307–314.

———. 1986. *Modern science and the book of Genesis*. Washington, D.C.: National Science Teachers Association.

Skinner, B. F. 1966. The phylogeny and ontogeny of behavior. *Science* 153:1205–1213.

Smith, F. H. 1984. Fossil hominids from the Upper Pleistocene of central Europe and the origin of modern Europeans. In *The origins of modern humans*. Smith, F. H. and Frank Spencer, eds. New York: Alan R. Liss, Inc. pp. 137–209.

Smith, F. H. and Frank Spencer, eds. 1984. *The origins of modern humans*s. New York: Alan R. Liss, Inc.

Smith, J. M. 1978 The evolution of behavior. *Scientific American* 239, no. 3:176–192.

Solecki, Ralph, S. 1971. *Shanidar: The first flower people*. New York: Alfred Knopf, Inc.

Spencer, Frank. 1984. The Neandertals and their evolutionary significance: A brief historical survey. In *The origins of modern humans*. Smith, F. H. and Frank Spencer, eds. New York: Alan R. Liss, Inc. pp. 1–49.

———. 1990. *Piltdown: A scientific forgery*. Natural History Museum. London: Oxford University Press.

Stanley, Steven M. 1979. *Macroevolution, pattern and process*. San Francisco: W. H. Freeman and Co.

———. 1981. *The new evolutionary timetable*. New York: Basic Books, Inc.

Stebbins, G. L. 1969. *The basis of progressive evolution*. Chapel Hill, NC: University of North Carolina Press.

———. 1982. *Darwin to DNA, molecules to humanity*. New York: W. H. Freeman & Co.

Stern, Philip Van Doren. 1969. *Prehistoric Europe*. New York: W. W. Norton & Co., Inc.

Stone, Irving. 1980. *The origin*. Garden City, NY: Doubleday & Co.

Stringer, C. B. 1974. Population relationships of later Pleistocene hominids: A multivariate study of available crania. *Journal of Archeological Science* 1:317–342.

———. 1990. The emergence of modern humans. *Scientific American* 263, no. 6:98–104.

Stringer, C. B., ed. 1981. *Aspects of human evolution*. London: Taylor and Francis.

———. 1984. Fate of the Neanderthal. Natural History. 93, no. 12:6–12.

———. 1985. Middle Pleistocene hominid variability and the origin of Late Pleistocene humans. In *Ancestors: The hard evidence*. Delson, Eric, ed. New York: Alan R. Liss, Inc. pp. 289–295.

Stringer, C. B. and P. Andrews. 1988. Genetic and fossil evidence for the origin of modern humans. *Science* 239:1263–1268.

Stringer, C. B., et al. 1984. The origin of anatomically modern humans in western Europe. In *The origins of modern humans*. Smith, F. H. and Frank Spencer, eds. New York: Alan R. Liss, Inc. pp. 51–135.

Strum, Shirley C. 1975. Life of a baboon troop. *National Geographic* 147, no. 5:672–691.

———. 1987. The "Gang" moves to a strange new land. *National Geographic* 172, no. 5:676–690.

Sutcliffe, A. J. 1969. Adaptations of spotted hyenas to living in the British Isles. *Mammal Society Bulletin* 31:1–4.

———. 1970. Spotted hyena: Crusher, gnawer, digester, and collector of bones. *Nature* 227, no. 5263:1110–1113.

———. 1985. *On the track of Ice Age mammals*. London: British Museum (Natural History).

Tattersall, Ian. 1993. *The human odyssey*. New York: Prentice-Hall General Reference.

Tax, Sol, ed. 1960. *Evolution after Darwin*. Chicago: University of Chicago Press.

Teleki, Geza. 1973. The omnivorous chimpanzee. *Scientific American* 228, no. 1:33–42.

Templeton, Alan R. 1991. Human origins and analysis of mitochondrial DNA sequences. *Science* 255 (7 February):737.

Thorne, Alan G. and Milford H. Wolpoff. 1992. The multiregional evolution of humans. *Scientific American* 264, no. 4:76–83.

Tiger, Lionel and Robin Fox. 1971. *The imperial animal*. New York: Holt, Rinehart and Winston, Inc.

Tobias, P. V., et al., eds. 1987. *Hominid evolution: past, present and future*. New York: Alan R. Liss, Inc.

Trinkhaus, Erik. 1984. Western Asia. In *The origins of modern humans*. Smith, F. H. and Frank Spencer, eds. New York: Alan R. Liss, Inc. pp. 25–293.

Trinkhaus, Erik, ed. 1989. *The emergence of modern humans*. Cambridge: Cambridge University Press.

Trinkhaus, Eric and W. W. Howells. 1979. The Neanderthals. *Scientific American* 241, no. 6:118–133.

Tullar, Richard M. 1977. *The human species*. New York: McGraw-Hill Book Co.

Van Allen, Leigh and R. E. Sloan. 1965. The earliest primates. *Science* 150 (5 November):743–745.

Vandermeersch, Bernard. 1985. The origin of the Neandertals. In *Ancestors: The hard evidence*. Eric Delson, ed. New York: Alan R. Liss, Inc. pp. 306–309.

van Lawick-Goodall, Jane. 1963. My life among wild chimpanzees. *National Geographic* 124, no. 2:272–308.

———. 1965. New discoveries among Africa's chimpanzees. *National Geographic* 128, no. 6:802–831.

———. 1967. *My friends the wild chimpanzees*. Washington, D. C.: National Geographic Society.

Vigilant, Linda, et al. 1991. African populations and the evolution of human mitochondrial DNA. *Science* 253 (27 September):1503–1507.

Vollert, Cyril, S. J. 1951. Evolution of the human body. In *The Church and modern science*. New York: America Press. pp. 3–30.

Vrba, E. S. 1985. Ecological and adaptive changes associated with early hominid evolution. In *Ancestors: The hard evidence*. Delson, Eric, ed. New York: Alan R. Liss, Inc. pp. 63–71.

———. 1993. The pulse that produced us. *Natural History* 102, no. 5:47–51.

Waechter, John. 1976. *Man before history*. Oxford: Elsevier Phaidon.

Wainscoat, J. S. 1987. Human evolution: out of the Garden of Eden. Nature 325:13–14.

Wainscoat, J. S., et al. 1986. Evolutionary relationships of human populations from an analysis of nuclear DNA polymorphisms. *Nature* 319 (6 February):491–493.

———. 1989. Geographic distribution of Alpha- and Beta-Globin gene cluster polymorphisms. In *The human revolution*, Paul Mellars and Chris Stringer, eds. Princeton, NJ:Princeton University Press. pp. 31–38.

Washburn, S. L. 1961. *Social life of early man*. Chicago: Aldine Publishing Co.

———. 1978. The evolution of man. *Scientific American* 239, no. 3:194–208.

———, ed. 1963. *Classification and human evolution*. Chicago: Aldine Press.

Washburn, S. L. and Irven Devore. 1961. The social life of baboons. *Scientific American* 204, no. 6:62–71.

Washburn, S. L. and Phyllis Dolhinow, eds. 1972. *Perspectives on human evolution 2*. New York: Holt, Rinehart & Winston, Inc.

Washburn, S. L., and P. C. Jay. 1968. *Perspectives on human evolution 1*. New York: Holt, Rinehart & Winston, Inc..

Washburn, S. L. and Ruth Moore. 1974. *Ape into man*. Boston: Little, Brown and Co.

Weaver, K. F. 1985. The search for our ancestors. *National Geographic* 168, no. 5:560–624.

Weiss, Mark L. and A. E. Mann. 1985. *Human biology and behavior*, 4th ed. Boston: Little, Brown and Co.

Wendt, Herbert. 1956. *In search of Adam*. Boston: Houghton Mifflin Co.

Wenke, Robert J. 1980. *Patterns in prehistory*. New York: Oxford University Press.

White, Edmund and Dale Brown. 1973. *The first men*. New York: Time-Life Books.

White, Randall. 1986. *Dark caves, bright visions: Life in Ice Age Europe*. New York: American Museum of Natural History.

———. 1993. The dawn of adornment. *Natural History* 102, no. 5: 61–67.

Wilson, Allan C. and R. L. Cann. 1992. The recent African genesis of humans. *Scientific American* 266, no. 4:68–73.

Wilson, Allan C. and V. M. Sarich. 1969. A molecular time scale for human evolution. *Proceedings of the National Academy of Science* 63:1088–1093.

Wilson, D. B., ed. 1983. *Did the devil make Darwin do it? Modern perspectives on the Creation-Evolution controversy*. Ames, IA: Iowa State University Press.

Wilson, E. O. 1975. *Sociobiology*. Cambridge, MA: Harvard University Press.

———. 1978. *On human nature*. Cambridge, MA: Harvard University Press.

———. 1992. *The diversity of life*. New York: W. W. Norton & Co.

Wilson, J. A. 1966. A new primate from the earliest Oligocene, West Texas: preliminary report. *Folia Primatologica* 4:227–248.

Wolf, Josef. 1978. *The dawn of man*. New York: Harry N. Abrams, Inc.

Wolpoff, Milford H. 1985. Evolution in *Homo erectus*: the question of stasis. *Paleobiology* 10:389–406.

———. 1988. Modern human origins. *Science* 241 (12 August):772.

———. 1989. Multiregional evolution: the fossil alternative to Eden. In *The human revolution*, Paul Mellars and Chris Stringer, eds. Princeton, NJ: Princeton University Press. pp. 62–108.

———. 1989. The place of Neandertals in human evolution. In *The emergence of modern humans*, Erik Trinkhaus, ed. . Cambridge: Cambridge University Press. pp. 97–141.

Wolpoff, Milford H., et al. 1984. Modern *Homo sapiens* origins: a general theory of hominid evolution involving the fossil evidence from east Asia. In *The origins of modern humans*. Smith, F. H. and Frank Spencer, eds. New York: Alan R. Liss, Inc. pp. 411–483.

Wood, B., et al., eds. 1986. *Major topics in primate and human evolution.* Cambridge: Cambridge University Press.

Young, J. Z. 1971. *An introduction to the study of man.* New York: Oxford University Press.

Zapfe, Helmuth, 1939, Lebensspuren der eiszeitlichen Hölenhyäne. *Palaeobiologica.* 17:111.

———. Helmuth, 1940, Fossil traces of bone-crushing predatory animals. *Research and Progress* (November).

CREDITS

10.2: The Natural History Museum, London; **10.3:** © John Reader/SPL/ Photo Researchers, Inc.; **10.4:** © Wm. C. Brown Communications, Inc./Bob Coyle, photographer; **10.5:** From *Guide to Fossil Man* by M. H. Day © University of Chicago Press; **10.6:** © Dr. E. R. Degginger; **10.7:** Peabody Museum, Harvard University/Photo by Hillel Burger; **10.8:** The Natural History Museum, London; **10.9:** Photo by C. Tarka, © and Courtesy of Museo di Antropologia "G. Sergi" (University di Roma) and Dr. Eric Delson (AMNH); **10.10:** Photo by C. Tarka, © and courtesy of Staatliches Museum für Naturkunde (Stuttgart) and Dr. Eric Delson (AMNH); **10.11:** From *Guide to Fossil Man* by M. H. Day © University of Chicago Press

Chapter 11

11.1: © Lawrence Migdale/Photo Researchers, Inc.; **11.2:** The Natural History Museum, London; **11.3:** © Wm. C. Brown Communications, Inc./Bob Coyle, photographer; **11.4:** © Dr. E. R. Degginger; **11.5:** © Cabisco/Visuals Unlimited; **11.6:** The Natural History Museum, London; **11.8:** Science VU/ Visuals Unlimited; **11.9:** © Unicorn Stock Photos; **11.10, 11.11:** Science VU/ Visuals Unlimited; **11.12:** © Cabisco/ Visuals Unlimited; **11.13a:** © Alexander Marshack; **11.14:** © Alexander Marshack

Chapter 12

12.3: Photo by C. Tarka, © and courtesy of Prof. B. Vandermeersch (Université de Bordeaux) and Dr. Eric Delson (AMNH); **12.4, 12.5:** From *Guide to Fossil Man* photo by M. H. Day © University of Chicago Press

ILLUSTRATORS

Raymond Smith: 2.4, 3.5, 4.1, 4.2, 4.3, 4.4, 4.5, 4.6, 4.7, 4.8, 4.9, 4.11, 5.1, 5.2, 5.3, 5.4, 5.5, 5.6, 7.9, 7.17, 8.1, 9.11, 9.12, 11.7, 11.13*b*
Diphrent Strokes: 1.4, 2.2, 2.3, 3.1, 4.10, 4.12, 4.13, 4.14, 5.2, 7.12, 9.1, 12.1, 12.2

LINE ART

Chapter 2

2.4: After Rudolf Freund in Moore, 1962, p. 30.

Chapter 4

4.4: After Simons, 1964, p. 60; **4.7:** Source: Gudtz in Lewin, 1984, p. 40; **4.8:** Sources: Lewin, 1984, p. 23; and Le Gros Clark, 1959, p. 222; **4.9:** After Tullar, 1977, p. 18, modified from Storer, T. T., et al., 1972, *General Zoology*, 5th ed., McGraw-Hill Book Co., NY; **4.10:** After Lewin, 1993, p. 171; **4.11:** Source: R. Holloway, 1974, pp. 113–114; **4.12:** Source: After Sarich, 1968, p. 113; **4.13:** After Lewin, 1984, p. 19; **4.14:** After Weiss and Mann, 1985, p. 169; **table 4.4:** After Weiss and Mann, 1985, p. 167.

Chapter 5

5.3: Sources: After Simons, 1972, p. 126; After Hill, 1953, p. 391; and Le Gros Clark, 1963, pp. 137, 148; **5.4:** Source: Simons, 1964, p. 56; **5.5:** After Weiss and Mann, 1985, 4th ed., p. 189; redrawn from Le Gros Clark, *History of the Primates,* University of Chicago Press; **5.6:** Source: Simons, 1964, p. 56.

Chapter 7

7.12: Adapted with permission from *Time and Stratigraphy in the Evolution of Man.* Copyright 1967 by the National Academy of Sciences. Courtesy of the National Academy Press, Washington, D.C.; **7.17:** After Eddy, 1972, p. 43.

Chapter 8

8.1: Source: Howell, 1986, p. 103.

Chapter 11

Table 11.1: Source: After Fagan, 1974, p. 125; **11.13b:** © Alexander Marschack.

Chapter 12

12.1: Modified from M. H. Day, *Guide to Fossil Man,* 4th ed. Copyright © 1986 The University of Chicago Press. Used by permission.

INDEX

Abbevillian 146, 210
Abel 8
Abraham 8
Acheulian 145
    culture 188, 196, 210
    industry 166, 167, 173, 196
    Middle Acheulian 192
    tools 147
Adam and Eve 8
Adapidae 83
Adapis 83
adaptations 53
    arboreal 37
    behavioral patterns 54
adaptive radiation 27
adult footprint 170
advanced hominid 135
Aegyptopithecus 33, 78, 85, 86, 90
    A. zeuxis 69, 85, 91
Aeolopithecus 85
Afar 125, 133
afterlife 198
age dating, absolute 136, 139
Age of Cave Bears 204
Age of Reindeer 204
aggression 32, 34, 44
    chimpanzees 45
    primate societies 45
agriculture 148
albumin 64
    immunological distance units 52
alpine glaciers 177
Altamira 202, 216, 217
Altamura 165
altruism 32, 107
amber 211
Ambrona 169, 228

ambush 172
American Indians 6
Ameslan 71, 72
amino acid sequences 66, 67
amputation 197
Amud Cave 190
Amud I 190
An 7
Anaptomorphidae 83
ancestor, common 69
ancestral apes 78
ancestral primates 78
ancestral stock for orangutans 93
ancient environments 80
Andersson, John 159
Andes Mountains 16
Andrews, P. 232, 236
angiosperms 81
animal
    antler
        carvings 219
        picks 211
    bones
        burned 167, 215
        cut marks 117
        engraved 204, 215
        handles 211
        hides 183, 211
        needles 211
        Zhoukoudian 168
    ivory 210
    sinews 211
    skins or hides 171
antelope 211
    femurs 118
    humerus 117
Anthropoidea 25

anthropoids 25, 27, 30, 91
Anthropopithecus erectus 156
antibodies 64
anti-evolution laws 8
Anu 6
Anunnaki 6
anvil 140, 171, 174
anvil percussion 174
apes 27, 33
    ancestral 88
    ankle 60
    Asian 89
    brain 62
    feet 60
    great 33, 54, 85, 90
    lesser 30, 33
    Miocene 91, 113
    pelvis 58
    skeleton 54
Apidium 84
Arago 195
arboreal 27, 34, 85
    behavior 36, 38, 62
    primates 38, 96, 113
archaic sapiens 165, 194, 196, 205, 229,
        232, 234, 237
archeologists 185
Ardrey, Robert 43, 117, 146
art 204, 205, 207, 213, 215
arthritis 184, 197
artifacts 199
ash 167, 228
atlatl 212
auditory stimuli 28
Aurignacian 204, 205, 235
australopithecines 54, 87, 90, 98
    brains 63, 129, 132

footprints 94
gracile 115, 124, 125, 128–130, 132, 133
group tactics 135
robust 115, 128–130, 132, 139
Australopithecus 85, 93, 98, 114, 158
   afarensis 126, 127, 128, 129, 131, 132
   africanus 87, 89, 112, 114–118, 128, 129, 131–133, 140, 158, 182, 211
   boisei 87, 120, 121, 124, 128, 129, 132, 133, 139, 140, 146
   prometheus 117
   robustus 87, 94, 115, 116, 128, 129, 132, 133, 163, 228
   teeth wear 94
   transvaalensis 114
Avise, John C. 232
awls 145, 174
aye-aye. *See Daubentonia*
Azilian 205
Azimov, Isaac 62

baboons 31, 93, 168
   dominance behavior 47
   hamadryas 47, 70
   olive 45
   skulls 114, 117
   troops 31
Babylonian 6
backbone 58
bags 169, 228
   skin 213
Barlow, G. W. 114
Barton, N. H. 232
baskets 169, 211, 213
Beagle, H.M.S. 17
bears 208
behavior 26, 34, 68, 148
   arboreal 27, 34, 36, 38, 62, 85
   bipedalism 29, 34, 36, 53, 59, 87, 95, 98, 100, 108, 127, 131, 137, 142, 199
   bonding 32, 150
   brachiating 30
   cooperative 150
   defense 44, 138
   dominance behaviors 45, 47
   genetic derived 69
   group 28, 44, 68, 69, 103, 134
   inhibitive 75
   instinct 74, 75
   laws 70

living apes 97
organization 27, 28, 44, 70, 148, 205
patterns 69, 74
postures 31
selfish 136
sex-based 70
sharing 107, 146, 168
social 32, 68, 97, 102
solitary 36, 68, 69
terrestrial 27, 34, 95, 96, 100
territoriality 29, 31, 43
Bering Straits 86
Bernstein, I. S. 44
Bible 6
Bilzingsleben 165, 209
binocular vision 26, 27, 29
biochemical tests 52, 64
biogenic law 153
biology 8
bipedal. *See* behavior
bipolar percussion 174
Black, Davidson 159
blanket 171, 174, 184, 215
blood
   antibodies 64
   clotting factors 52
Bodo skull 187
body language 71
bolas 145
Border Cave 237
Bordes, Francois 196
Boucher de Perthes, Jacques 154, 210
Boule, Marcellin 184
bowl 172
brachiate 26, 33
brain 28, 40, 60, 71, 82, 198
   quality 41
   size 41, 62, 97
Brain, C. K. 119, 167
braincase 111
Bramapithecus 92
breccia 111
Breuil, Abbé Henri 205
Britten, R. J. 231
Brixton Cave 209, 210
Broken Hill 186
Broom, Robert 114, 122
brow 54
Brown, Frank 164
brow ridge (supraorbital crest or torus) 54, 183, 189
Bryan, William Jennings 7
Buckland, William 118, 209
budding 227

Buettner-Janusch, John 30, 80
burials 180, 189, 190, 197, 199, 203, 204
burins 145
Burma 84
burnt clay 147
bush babies. *See Daubentonia*
Bushmen 143
butchered animals 138, 147, 166, 170, 228
   sites 143, 149, 150
butcher's block 171
buttock muscles 59
buttons 219

Cain 8
Callithricidae 31
calvarium 55
Campbell, Bernard 133, 157, 169, 183, 193, 219
campers 171
campfires 147, 170, 192, 215, 226, 228
camps 170, 227
canine teeth 31, 34, 46, 57, 69
Cann, Rebecca 231
cannibalism 197, 198
carnivorous 30
Carpenter, C. R. 33, 34
Catarrhinia 31
cave 184, 188, 197, 199, 204, 213, 215, 228
   Aurignac 204
   bear (*Ursus spelaeus*) 198, 211
   Cave of the Kids 189
   Cave of the Oven 189
   drawings 215
   lions 211
   Mugharet es-Skhül 189
   Mugharet es-Tabun 189
Cebidae 31
Ceboidea 30
Cenozoic Era 25, 27, 81
Center-Edge hypothesis 233, 235
Cercopithecinae 31
Cercopithecoidea 31
cerebellum 61
cerebral cortex 40, 61, 62
cerebrum 61
ceremonies 190, 204, 208, 216
charcoal 167, 170, 192, 228
cheekbone 57
cheek teeth 57
Chellean 145
Chesowanja 147

chewing muscles 97, 130
chimpanzees (*Pan*) 27, 30, 32–34, 48, 54, 73, 89, 90
    bands 48
    bipedalism 101
    cooperation 107
    DNA 66
    genocide 42
    hemoglobin 67
    learning 106
    pygmy (*paniscus*) 34
    societies 47
    troglodytes 34, 48
chin 57, 163, 183, 206
China 227
chisels 145, 174, 211
choppers 145, 166
Choukoutien. *See* Zhoukoudian
Christy, Henry 204, 210
cingulum 163
Ciochon, R. L. 84, 90, 94
clavicle 26, 27, 30
claws 38
cleavers 171, 174
climate 238
clothing 171, 174, 183, 184, 207, 211, 213, 215
    fur 183
Cole, Sonia 120
Colobinae 31
colobus monkeys 31
color vision 39, 40
combat 46
communication 71, 146, 172
    nonlanguage 72
competition 20
conchoidal fractures 140
Constable, George 154, 191, 225
containers 174
continental glaciers 177
Conyers 209
cooking 163, 170, 215, 227, 228
    roasted or barbecued meats 226
cooking vessel, 172
Coon, C. S. 163
cooperative behavior 44, 106, 134, 136, 148, 149, 228
    defense 150
    hunters 45, 150, 172
    social 197
cords 213
core. *See* stone tools
correlation, stratigraphic 124
Creation Account 6, 9

Cretaceous Period 27, 80–82
Cro-Magnon. *See also Homo sapiens sapiens* 111, 181, 190, 194, 202, 229
    artists 215
    brains 207
    cave art 215
    culture 204
    skulls 206
    toolmaking 209
Croatia 198
culture 69, 138, 175
Cuvier, Georges 83, 209
cytochrome oxidase 67

Darlu, Pierre 232
Darrow, Clarence 7
Dart, Raymond 87, 111, 122, 146, 158, 182, 211
Darwin, Charles 14, 25, 110, 180, 181, 203
Darwinian Evolution 23
dating, radiometric 124
Daubentonia 30
Daubentoniidae 30
Dawkins, Richard 19, 23, 107
Dawson, Charles 190
Day, Michael 164, 233
decision making 40
deductive reasoning 10
defense. *See* behavior
Deinotherium 144
De Mortillet, Gabriel 204
dental arcade 132
    apes 130
    humans 57
dental formula 28–31, 84
depth perception 27, 39
Descent of Man 110, 181, 203
Developed Oldowan
    hand axes 147
    industries 173
    tools 140, 145, 174
de Vries, Hugo 154
diastema 57, 131, 148
diurnal 28, 31
Djetis Beds 162
DNA 66
    divergence data 52, 64
    hybridization 66
Dolní Vestonice 215, 217, 221
dominance 31, 32, 44, 46, 47, 49
    hierarchy 34, 46
Dordogne District 181, 202, 203, 215

Dragon Bone Hill 159
dryopithecine apes 25, 57, 78, 85–87, 89, 91, 114
    African 91
    Asian dryopithecine 91, 93
Dryopithecus 33, 85, 89, 91
    africanus 89
    fontani 88, 89, 91
    indicus 89, 91
    laietanus 89
    major 89, 91
    nyanzae 89, 91
    punjabicus 89
    sivalensis 89, 91
Dryopithecus (Proconsul) africanus 91, 93
Dubois, Eugene 111, 154, 155, 162
Dutch East Indies 155
dwelling 143

Earth's age 10
East African Rift 87, 122, 123
East Rudolph. *See* East Turkana
East Turkana 155, 164
edge environments 235
Egypt 6, 30, 33, 79
Ehringsdorf 190, 191, 229
Eldridge, Nils 22
elephant 169, 228
empirical method 10
endocranial casts 63
engravers 192
Enuma Elish 8
environments 87
    conditions 78, 91
    influences 54
Eocene Epoch 30, 78, 82, 84
erect posture 58
Esper, Johann 209
Ethiopia 123, 141, 155
Ettel, P. C. 90
Eutheria 82
Eve hypothesis 232
evolution 2, 14
    Cro-Magnons 221
    evolutionary 82
    hominid 98
    mosaic 199
Evolutionary Population model 233
extinction 15

Fabre, J. H. 74
facial expressions 71
facts 11

Fagan, Brian 209
Fayum 33, 79, 84, 85, 111
feces 171
feedback 138
Feldhofner Grotto 180
femur 155
fertility figures 219
finches 17
fingernails 26, 27, 30, 39
fire 117, 147, 163, 167, 213, 226, 228
fired clay 219
First Family 127
fish 211
fishing 213
5-Y dental cusp pattern 57
flat, faces 97
Flood of Noah 8, 180
flowering plants 81, 210
Fontan 87
Font-de-Gaume 216
food gathering 134
foods, terrestrial 96
food sharing 136, 146
foot, baboon 97
footprints 127, 131
foramen magnum 30, 39, 58, 83, 112, 198
forebrain 61
forehead 55
forest 34, 133
Fossey, Dian 34, 48, 53
fossilization 78
fossils 8
Fox, Robin 108
Frere, John 209
frugivorous 30, 93
Fuhlrott, J. K. 180
furs 213

Gagarino 215
galago 27, 30
Galápagos Islands 17
Galdikas, Biruté 34, 103
Galdikas-Brindamour, Biruté 48, 49, 53, 73
garbage dump 143
Garden of Eden hypothesis 8, 154, 155, 180, 233, 236
gathering 148
Gaudry, A. 88
Gee, Henry 131
generosity 197
Genesis 6, 7, 8, 9, 108, 155, 180, 203
genetics 21

geologic time 19
geology 8
gibbon (*Hylobates*) 30–34, 154, 236
Gibbons, Ann 236
Gibraltar skull 111
Gigantopithecus 33, 90, 92, 93, 159
    bilaspurensis 90
    blacki 90
Gilgamesh Epic 8
glaciers 177
gluteus maximus muscles 59
Goldman, N. 232
Gombe National Park 34, 48, 104
Gombe Stream Chimpanzee Preserve 48
Goodall, Jane 25, 33–35, 48, 53, 73, 101, 103, 104, 134, 135
Goodman, Morris 64, 231
Gordon, T. P. 44
Gorge d'Enfer 204
gorilla (*Gorilla gorilla*) 27, 30, 32–34, 48, 54, 85, 90, 113
Gould, Stephen Jay 22
gracile. *See* australopithecines
graminivorous 93
grasping feet 96
grasping hands 26, 29, 83
grass fires 170
grassland 93
graves 196, 199
    shallow 189
Gray, Asa 18
Gregory, W. K. 83
grooming 29, 72
guernons 31
Günz Glacial Stage 177

habitats 54
Hadar 125, 126, 128, 131, 133, 155
Haeckel, Ernst 153, 155
hair 54
hand and foot adaptations 38
Harding, R. S. 45
harem structure 70
harpoons 213
Hay, R. L. 122, 140
hearth 170
Hebrews 8
Hedges, S. B. 232
Heidelberg jaw 195
Heidelberg man 164
hemoglobin 67, 231
herbivorous 34
Herodotus 4
hides 174

hindbrain 61
Holloway, R. L. 63, 141, 142
hominids 30, 32, 78, 85, 96, 114
    bipedalism 100, 109, 125, 135
    brain 69
    evolution 102
Hominidae 36
hominoids. *See also* apes 30, 32, 85, 86, 91, 96, 113
    ancestral 88
    Asian 89
    Miocene 91, 113
Homo 87
Homo (Pithecanthropus) erectus 53, 111, 112, 139, 153, 156, 158, 162, 179, 187, 189, 192, 199, 208, 211, 212, 224, 225, 233
    diet 167
    erectus 162
    erectus pekinensis 159, 162, 163, 228
    KNM-ER 3733 164
    KNM-ER 3883 164
    meat eating 167
Homo habilis 87, 121, 124, 125, 133, 136, 139, 141, 153, 211, 225
Homo heidelbergensis 158
Homo neanderthalensis 180, 193, 194
Homo sapiens 3, 96, 98, 128, 177, 224, 225
    sapiens neanderthalensis 194, 205, 229
        African 186
        Asian 187
        brain 207
    sapiens 91, 205, 239
hooks 213
Hopwood, A. T. 89
Hortus 198
Howell, F. Clark 123, 124, 150, 169, 194, 223, 235
Howells, W. 162, 194, 219, 234, 235, 236
Hrdlicka, Alex 158, 233
humans (*Homo sapiens*) 27, 32, 33
    anatomically modern 235
    canine teeth 57
    diet 57
    DNA 66
    feet 60
    hemoglobin 67
    instincts 75
    legs 53, 59
    pelvis 58

relationship to chimpanzees 66
skeletons 54
hunters 45, 147, 228
hunting 34, 48, 135, 138, 145, 146, 148,
168, 175, 205, 207, 208, 211,
228
bands 170
behavior 136
blinds 143, 144
cooperative 149, 172
Cro-Magnons techniques 212
fowling 213
H. erectus locality 169
meat 47
procedures 169, 170, 211
solitary 172
stalking 172
hunting-gathering 148, 171
huts 170
Hutton, James 19
Huxley, Thomas 22, 110
hyenas 118
Hylobates. See gibbon
Hylobatidae 33
hypothesis 11

Ice Ages 225
ice sheets 177
immunological analysis 64
immunological distance units 65
incisors 28
Indonesia 155, 162
Indri 29
Indriidae 29
inductive reasoning 10
inheritance 21, 66
inhibitory mechanisms 45
insectivores 27, 28
insectivorous 30
instinct. *See* behavior
intelligence 62, 68, 71, 173
interglacial stages 177
intuitive reasoning 10
Iraq 233
Isaac, G. L. 114, 146, 150, 168
Isernia 166
Israel 188, 190, 233

Java 112, 154, 155, 161, 187, 226, 227
Java man. *See Homo erectus erectus* 53,
112, 124, 155, 182, 191, 227,
228
skullcap 155
jaw 56, 97

Jebel Qafzeh. *See* Kafzeh
Johanson, Donald 125, 126, 128, 131,
132, 155
Jolly, Alison 93, 114
Judeo-Christian 6
Jurassic Period 81

Kafzeh 190
Kakombe Valley 48
Kay, R. F. 69
Kebara Cave 238
Keith, Sir Arthur 113
Kennedy, K. A. R. 157
Kent Caverns 209
Kenya 45
Kenyapithecus wickeri 89, 91
Kiik-Koba 199
kill or butchering sites 146
kill sites 149
King, William 194
Kirdkale Cave 118
Klasies River 237
knapping 140, 145, 174
knee joint 59, 125
KNM–ER 1470 124, 142
KNM–ER 3733 164, 225
KNM–ER 3883 164
KNM–WT 15000 128
knuckle walking 34, 60, 99
Koko 72
Koobi Fora 124, 141, 225
Koran 7
Kortlandt, Adrian 73, 101, 103
Koshima 105
Kostenki 215
Krapina 198, 235
Kromdraai 115
Kurtén, Björn 52

La Chapelle-aux-Saints 184, 198
Laetoli 94, 127, 131
La Ferassie 184
Lake Turkana (L. Rudolph) 87,
123–125, 128, 141, 142, 164,
225
La Madeleine 204, 205
La Mouthe 216
land bridges 232
language 71
langurs 31
La Quinta 184
Laramide Orogeny 81
Lartet, Edouard 86, 88, 204, 210
Lascaux Cave 202, 216, 217

latrine 171
Laugerie Haute 204
law 11
leadership 47, 172
Leakey, Louis 48, 87, 89, 119, 120,
139–141, 147, 168
Leakey, Mary 94, 119, 127, 131,
143–147, 168
Leakey, Richard 106, 124, 135, 141,
142, 150, 155
learning 71, 104
Le Gros Clark, Wilfrid 117, 162
Le Moustier 184, 204
Lemuria 154
lemuroid 80
lemurs 27, 28, 29, 83
dental formula 29
Lepilemur 29
Les Combarelles 216
Les Eyzies 203, 210
Lewin, Roger 106, 135, 147, 150
Lewis, G. E. 89
Limnopithecus 33, 86, 90
lines 213
Linnaeus 26, 194
living floors 143, 146, 149
locomotion 37, 53
Loris 27, 30, 83
Lorisidae 30
Lovejoy, C. Owen 125
lower jaw 57
Lowther, Gordon 135
Lucy (*Australopithecus afarensis*) 126,
131, 164
Lydekker, R. 89
Lyell, Sir Charles 210

macaques (*Macaca*) 31, 85
Japanese 74, 105
MacEnery, John 209
Mackintosh, R. H. 117
Madagascar 29
Maddison, D. R. 232
Magdalenian 205
Maglio, V. J. 114
magnetic reversals 177
major 89
Makapansgat 116, 117, 118
Malaysian Archipelago 161
Malaysia 155
Malthus, Thomas 20
mammals, placental 82
mammoths 208

bone supported shelters 215
    skins 215
    tusk 210
mandible 57
Mann, A. E. 46, 68, 131, 195
Markgraf, Richard 84
marmosets 31
Marshall, E. R. 33
Marshall, J. T., Jr. 33
marsupials 82
Mason, R. J. 119
masseter muscle 57, 58, 130
mating 32
Mauer jaw. *See* Heidelberg jaw
Mayr, Ernst 162
McCown, T. D. 157
meat 167
meat-eating 134
medulla oblongata 61
Megaladapis 37
meiosis 66
memory 28, 40
Mendel, Gregor 21
Merrick, H. V. 146
mesencephalon 61
Metatheria 82
midbrain 61
Mike 104
Mindel Glacial Stage 177, 227
Miocene Epoch 33, 85–87, 89, 113, 154
missing link 112, 154, 155, 156
mitochondria 236
mitochondrial DNA (mtDNA) 67, 232,
    236
mitochondrial Eve hypothesis 232
molars 57, 93
monkeys 30
monotremes 82
Monte Circeo 198
Moore, Ruth 52, 117
Morocco 164
Morris, Desmond 54
mosaic of varied environments 114, 133
Moslem 7
Mount Carmel 188, 193, 233
Mousterian 190, 234
    culture 189
    tools 188, 196, 204
Multiregional Evolution hypothesis 233,
    236
murder 34, 45, 48, 198
    chimpanzees 45, 48
    lions 48
    orangutan 49
mutations 21, 54, 67, 232

Napier, Alan 99, 140
nasal septum 30
natural selection 19, 110
Neandertal man. *See Homo sapiens*
    neanderthalensis 54, 111,
    153, 179, 202–204, 208, 221
    burials 164, 190
    Cave Bear tool culture 204
    ceremonies 190
    Classic 183, 189, 191–193, 197,
    198, 201, 204–206, 225, 229,
    230, 233, 238, 239
    European 182, 190, 191, 229
    early 239
    evolution 235
    Middle East 197
    mixed characters 231
    Progressive 193, 201 205–207, 225,
    230, 239
    robustness 233
    toolmaking 196
    Wadi Amud 238
Neandertal Phase hypothesis 233
Neandertal problem 192, 193
Neander Valley 110, 180
needles 211, 215
neopallium 61
nerve cord 60
nets 212, 213
New World monkeys 25, 27, 30, 69, 86,
    91
Ngandong Beds 161, 162, 187
Nice 170
Noah 8
Noah's Ark or Garden of Eden
    hypothesis 233, 236
nocturnal 28
nomadic 205
nonphysical qualities 68
nonstone tools 212
Nordenskiöld, Erik 153
nose 55
Notharctus 83
numerical counts 219
nuts 138

objectivity 10
occipital condyles 39, 58
occipital lobes 28
Oldowan
    culture 87, 119
    industry 166, 196
    stone tools 120, 138, 139, 142, 145,
    147, 166, 174

Olduvai Gorge 87, 119, 120, 122, 123,
    124, 133, 138, 139, 141, 146,
    162, 163, 225
    Bed I 120, 122, 139, 141
    Bed II 141, 162, 225
    Bed III 163
    Bed IV 139, 140, 163
Old World 86
Old World hominoids 90
Old World monkeys 25, 27, 30, 31, 69,
    84–86, 91
olfactory receptors 55, 56
olfactory senses 41, 56
Oligocene Epoch 30, 31, 78, 79, 84, 91
Oligopithecus 33, 84
Olorgesailie 168, 225
omnivorous 30
Omo district 87, 128, 133, 141, 146,
    147, 237
Omo River Valley 123, 237
Omoto, Keiichi 232
ontogeny 153
opinion 11
opposable fingers 26, 27, 83
opposable great toes 33, 60
opposable grip 38
optic lobes 61
orangutans 32, 33, 36, 49, 73, 85, 92
orangutans (*Pongo pygmaeus*) 30, 32,
    49, 54, 73, 85, 89, 90, 92
orbits 30
Oreopithecidea 32
Oreopithecus 32
Origin of Species 110, 153, 180, 203
ornamentation 211
ornaments 203
osteodontokeratic culture 211
Out-of-Africa model 233

paintings 215
Pakistan 86, 89, 92
Paleocene Epoch 27, 78, 82, 85, 89, 92
Paleoenvironments 80
Paleolithic 234
    Lower 188
    Middle 188, 196, 204
    tools 233
    Upper 204, 205
paleomagnetic timescale 177
paleontology 8
Paleopithecus sivalensis. See
    Dryopithecus sivalensis
Palestine 188
Pan. See chimpanzees

Papio 31
Paranthropus crassidens. See
     Australopithecus robustus
Paranthropus robustus. See
     Australopithecus robustus
Parapithecus 84
patriarch 8
Patterson, Francine 72
Paviland Cave 209
pectoral breasts 27, 30
Pei, W. C. 159
Peking man. *See Homo erectus*
     pekinensis 139, 158, 159,
     226, 228
Peking Union Medical College 159
Pengelly, William 210
Perigordian 205, 233
Petralona, Greece 165
Petralona skull 165, 194
Phenice, T. W. 191
philosophy 10
philtrum 33
phylogeny 153
physical attributes 26
phytoliths 94
pigments 216
Pilbeam, David 89, 90, 91, 92
Pilgrim, G. E. 89
Piltdown man 190
Pithecanthropus. See Homo erectus
Pithecanthropus alalus 154
placental mammals 26
Platyrrhini 31
Pleistocene Epoch 25, 119, 122, 158,
     162, 177, 214, 221, 225, 230,
     232
plesiadapid 30
Plesiadapis 82
Plesianthropus transvaalensis. See
     Australopithecus africanus
Pliocene Epoch 85–87, 114, 119, 120,
     122, 125, 127, 154
Pliopithecus 33, 86, 90
Poirier, Frank 194
pollen 197
Pondaungia 84
Pongidae 33
pongids. *See* apes
Pongo pygmaeus. See orangutans
potassium-argon age dating 87
pots 219
potto 27, 30
predators 136
predictability 11
prehensile tail 31

premolars 93
Preneandertal hypothesis 233
Presapiens hypothesis 233
Prideaux, Tom 211, 217
primates 26, 27
    ancestors 78
    behavior 41
    brain 40
    classification 27
    earliest 25, 27
    Eocene 82
    grasping hands and feet 37
    Miocene 86
    oldest 82
    Oligocene 84, 85
    Paleocene 82
    Pliocene 87
    primitive 82
    terrestrial 38, 96
Proconsul. See Dryopithecus
    africanus. See D. africanus
    major. See D. major
    nyanzae. See D. nyanzae
    sivalensis. See D. sivalensis
prognathous 41, 55, 56, 130, 182, 227
Pronycticebus gaudryi 80
Propliopithecus 33, 85
prosimians 25, 27, 82
protein analyses 52, 64
Protheria 82
protohominid 102, 103
Ptah 6
punctuated equilibria 22
Punctuation model 233, 234
Purgatorius ceratops 27, 82
Putman, J. J. 207

Qafzeh 231, 233, 235, 238
quadrupedal 97, 99, 199

Rabat 227
rabbits 211
races 4
racial traits 236
Radinsky, l. B. 84
radioactive age dating 87
radioactive minerals 139
    potassiun–argon dating 126
Ramapithecus 86, 89, 92
    brevirostris 89
    punjabicus 89
random mutations 64
recessive 21
Reck, Hans 120, 122, 139

red ocher 171
reduced prognathism 130
religion 205
religious practices 215, 219
Reliquiae Aquitanicae 204
repeatable 10, 11
Replacement model 236
reptile, mammal-like 114
Reynolds, R. 101
Reynolds, Vernon 101
Rhodesian man 186
rhombencephalon 61
Rift Valley 128, 225
Rift Zone 91, 153
Rightmire, G. P. 237
Riss Glacial Stage 177, 229
Riss-Würm Interglacial Stage 191, 196,
    229
rites of passage 216
ritual burial site 198
rituals 190, 204, 216
Robinson, J. T. 116, 119
robust. *See* australopithecines
Rockefeller Foundation 159
rock overhangs 204, 213
rock shelters 184, 213, 215
roots 138
rope 213
Rukand, Wu 162, 174
Ruvolo, Maryellen 67

sabre-tooth tigers 211
Saccopastore skull 190, 191, 229, 234
sacks 213, 228
sagittal crest 56–58, 97, 130
Saint-Césaire skull 233, 236
Saldhana Bay 187
Salé 227
Salmons, Josephine 111
Sandmel, Samuel 8
Sangiran 161, 162
Sarich, V. C. 64, 231
Sauer, N. J. 191
savanna environments 31
savannas 43, 93, 96, 133, 135
savanna-woodland environment 34, 48
scavengers 136
scavenging 149
Schaaffhausen, Hermann 111, 180
Schaller, George 34, 35, 48, 53, 135
Schoentensack, O. A. 158
science 10
Scopes, John 7
Scopes evolution trial 7

sculptures 217, 221
seals 211
sea shells 211
seeds 138
selection, natural 108
selective breeding 19, 22, 23
selective pressures 54
semiterrestrial 34
semitic 6
Serengeti Plain 135
Sergi, S. 191
serum antibody reactions 52, 64
Seth 8
sexual dimorphism 31, 85, 131
Shanidar 197, 231, 233
Shanidar I 197
Shanidar IV 197
Shapiro, H. L. 161
shelters 143, 163, 170, 174, 175, 207,
    213
Shenglong, Lin 162, 174
shovels 215
shrines 216
Shungura Formation 147
siamangs (*Symphalangus*) 33
Sidi Abderrahman 227
Sigmond, B. 101
sign language 71, 172
silverbacks 35
simian shelf 57, 163
Simons, Elwyn 32, 69, 79, 83, 84, 85,
    89, 90, 91
simple pebble tools 139
Simpson, G. G. 110
*Sinanthropus pekinensis. See Homo
    erectus pekinensis*
site
    DE/89 168
    DK 143
    FLK 143
    FLK North 144
Sivapithecus 85, 86, 89, 91, 92, 93
    indicus 89
    sivalensis 89
Siwalik Hills 89
Skehan, James W., S.J. 8
Skhül (Cave of the Oven) 189, 233, 235,
    236
skin clothing 213
slings 169
Smilodectes 83
Smith, Sir Grafton 113
snout. *See* prognathous
social structure 31, 148
social systems 42

society 69
sociobiology 42, 45, 68
Solecki, Ralph 197
Solo man 187
Solo River 155, 161
Solutrean 205
South Africa 87, 111
spatulas 219
spears 170, 211, 212
species 14
speech 68, 71
stasis 23
statuettes 215, 216
Steinheim man 190, 191, 229, 234
    skull 192
stereoscopic vision 27, 39, 40
Sterkfontein 115, 117, 147
stimuli, chemical 75
stimuli, visual 75
stone tools 87, 102, 128, 134, 135, 160,
    166, 169, 174, 181, 211
    axe 209
    bifacial 145
    choppers 149
    cores 140, 171, 174, 196
    dressed 209
    flaked 196
    flakes 120, 140, 145, 147, 174, 192,
        211
    flint flakes 166
    hammers 134, 138, 174
    hammer stone 121, 147, 174
    hand axes 145, 154, 166, 168, 174,
        192, 209
    meat-cutting 146
    missiles 212
    pebble 119, 171
    perforators 211
    points 145
    projectile points 171
    rough bifacials 171
    saws 211
    saw-toothed cutters 192
    scrapers 145, 166, 171, 174, 192,
        211
    shaped 192
    spheroids 145, 174
    technology 174
    weapons 168
stratigraphic correlations 139
Straus, L. 98
strength 54
string 213
Stringer, C. B. 230, 232, 236, 237
structural advantage 40

structure 143
Strum, S. C. 45
submission 45
Sugrivapithecus 92
Sumatra 155
Sumer 6
Sumerians 6
supraorbital crest (ridge or torus) 54, 58,
    130
Sutcliffe, A. J. 118
Swanscombe skull 190, 192, 229, 234
Swanscombe tools 192
Swartkrans 116, 163, 167, 228
Systema Natura 26

Tabun (Cave of the Kids) 231, 233
Tanzania 34, 87
tarsiers 30, 83
Tarsiidae 30
tarsiod 80
Tarsius spectrum 30
Tassy, Pascal 232
Taungs 111, 117
Taungs child. *See Australopithecus
    africanus*
T-complex pattern 93, 94
teeth 56
telencephalon 61
Templeton, A. R. 232
temporalis muscle 57, 58, 130
temporal lobe 28
tents 184, 211
Ternefine, Algeria 164, 176, 227
Terra Amata 170, 227
terrestrial. *See* behavior
territorial. *See* behavior
territory 43
Teshik-Tash 199
Tetonius 83
thalamus 61
theology 10
theory 11
Theropithecus 93
Thorne, A. G. 231, 236, 237
thought 62, 68, 108
thread 211, 213
threat display 31, 44, 46, 57, 70
Tiger, Lionel 108
Tinbergen, Niko 75
Tobias, Philip 117, 140
tool cultures 205
tool kits 146, 196, 208
toolmaking 72, 87, 100, 103, 108, 138,
    142, 146, 175, 205

chimpanzee 73, 100
  percussion 174
tools, *see* stone tools 119, 139, 171,
      207, 208
  Acheulian 187
  animal bones 160
  antlers 160
  Cro-Magnon 202
  culture 205
  flake 204
  Mousterian 188
  Oldowan 133, 173
  stone 119, 154
  unworked 138
  weapons 171
  wooden 134, 140, 189, 211
  worked 136, 169
tool use 72, 100, 102, 146, 150
Torralba, Spain 169, 228
tortoises 17
Transvaal 115
trapdoor spider 74
traps 213
trauma 198
tree shrew 26–28
Trinkhaus, Erik 194, 207, 234, 235
Trinl 155, 157, 162
Trinl Beds 155, 162
tubers 138
Tupaia 27

Ubaidian 6
Übergangsform 156
Uniformitarianism 19
upright posture 59, 128
Ursus spelaeus (cave bear) 213
U-shaped dental arcade 57

Ussher, Archbishop James 9
Utnapishtim 8

Van Bemmelmen 161
variability 23
variants 19, 21, 82
vegetarian 30, 34, 48
Venus figurines 219
vertebrae 60
vertebral column 58
vertebrates 61
Vértesszöllös 194, 227
vertical climbing-leaping (VCL) posture
      27, 28, 38, 39
Vézère River 203
Vigilant, Linda 231
violence 189, 197
Virchow, Rudolf 185
visual reception 40
visual signals 28, 71, 75, 172
visual stimuli 28
vocal communication 28, 34, 71, 172
volcanic ash 124
von Schlotheim, Baron 209
von Koenigswald, G. H. R. 90, 118, 159
V-shape dental arcade 57

Wadi Amud 231, 233, 238
Wadjak skulls 155, 157, 158
Waechter, John 189
Wainscoat, J. S. 231
Walker, Alan 128
walking 125
  feet 96
  gait 59
  knee 128, 131
  leg 125, 131

Wallace, Alfred 14, 155
Washburn, S. L. 52
Washoe 106
wasp, psammocharid 74
wasps, ichneumonid 74
weapons 103, 117, 138, 145, 149, 170,
      211
weapon use 104
weaving 213
Weidenreich, Franz 159
Weiss, Mark L. 46, 68, 131, 195
White, T. D. 127, 131
wild flowers 197
Willendorf Venus 219
Wilson, Allan C. 231
Wilson, E. O. 23, 40, 43, 45, 69, 108,
      231
windbreak 143, 170
wolf 211
Wolpoff, M. H. 231, 236, 237
woodlands 113
woolly mammoth 209–211
woolly rhinoceros 209, 211
woven grasses 169
writing 214
written notations 219
Würm Glacial Stage 177, 195, 196, 204,
      207, 229, 233, 235, 239

Young, J. Z. 54

Zapfe, Helmuth 118
Zhoukoudian 139, 159, 162, 168
Zinjanthropus boisei. See
      Australopithecus boisei
zygomatic arch 57, 130
zygotes 66